Cybersecurity for Commercial Vehicles

Cybersecurity for Commercial Vehicles

EDITED AND WRITTEN BY GLORIA D. D'ANNA

With co-authors in order of chapter:

Doug Britton, Lisa Boran, Xin Ye, Dr. André Weimerskirch, Steffen Becker, Bill Hass, Daniel DiMase, Zachary A. Collier, John A. Chandy, Bronn Pav, Kenneth Heffner, Steve Walters, The American Trucking Association, Dr. Jeremy Daily, Glenn Atkinson, Governor Rick Snyder, Karl Heimer, Lee Slezak, Christopher Michelbacher, Michael Ippoliti, Dr. Jennifer Guild, Joe Saunders, Lisa Silverman, National Motor Freight Transportation Association, Simon Hartley, Gerardo Trevino, Marisa Ramon, Daniel Zajac, and Cameron Mott

SAE INTERNATIONAL®

Warrendale, Pennsylvania, USA

400 Commonwealth Drive
Warrendale, PA 15096-0001 USA
E-mail: CustomerService@sae.org
Phone: 877-606-7323 (inside USA and Canada)
          724-776-4970 (outside USA)

Library of Congress Catalog Number 2018940856
SAE Order Number R-464
http://dx.doi.org/10.4271/R-464

R-464 Cybersecurity for Commercial Vehicles

ISBN-Print 978-0-7680-9257-8
ISBN-PDF 978-0-7680-9540-1
ISBN-ePUB 978-0-7680-9259-2
ISBN-PRC 978-0-7680-9260-8
ISBN-HTML 978-0-7680-9258-5

To purchase bulk quantities, please contact: SAE Customer Service

E-mail: CustomerService@sae.org
Phone: 877-606-7323 *(inside USA and Canada)*
          724-776-4970 *(outside USA)*

Visit the SAE International Bookstore at books.sae.org

dedication

This book is dedicated to my son, Joe, the Computer Scientist. Joe said, "You know, Mom, you should write a book about this." He also reminds me that, "Each line of code is an opportunity to make a mistake."

contents

CHAPTER 2

Should We Be Paranoid? by Doug Britton 21

CHAPTER 3

What Cybersecurity Standard Work Is
Applicable to Commercial Vehicles? by
Lisa Boran and Xin Ye   35

CHAPTER 4

Commercial Vehicle vs. Automotive
Cybersecurity: Commonalities and Differences by
André Weimerskirch, Steffen Becker, and Bill Hass   47

CHAPTER 8

Telematics Cybersecurity and Governance by Glenn Atkinson 143

CHAPTER 9

The Promise of Michigan: Secure Mobility by Karl Heimer 159

CHAPTER 10

How the Truck Turned Your Television Off and Stole Your Money: Cybersecurity Threats from Grid-Connected Commercial Vehicles by Lee Slezak and Christopher Michelbacher 177

CHAPTER 14

Heavy Vehicle Cyber Security Bulletin 229

CHAPTER 15

Law, Policy, Cybersecurity, and Data Privacy Issues by Simon Hartley 235

CHAPTER 16

Do You Care What Time It Really Is? A Cybersecurity Look into Our Dependency on GPS by Gerardo Trevino, Marisa Ramon, Daniel Zajac, and Cameron Mott

foreword to the reader

I hope you find this book of interest, and that this book doesn't just sit on your desk, as purely reference material. Cybersecurity for Commercial Vehicles has been a rather intense topic. And, I expect that it will continue to be.

Each chapter is a topic unto itself. Here is a preview of what each chapter will be about as well as a copy of Chapter 5, Malicious Software in Embedded Hardware. I hope you enjoy this sample of the book.

Hacking is a big business, and yes, this can affect you and your business operation.

CHAPTER 1

What Do You Mean by Commercial Vehicles and How Did We Happen on This Path of Cybersecurity?

Gloria D'Anna

1.1 I'm an Engineer and a Strategist

I'm an engineer and a strategist. I like to solve problems. *Big problems.* I was trained at General Motors Institute as an engineer. At the time, the school was owned by General Motors. That's where they trained their engineers and upcoming managers. I then attended the University of Chicago for my Masters degree in Business Administration (MBA), as a General Motors Fellow. I have worked in the automotive and commercial vehicle industry for over 25 years. Typically, I have focused on the trends in the industry-new engines, new transmissions, new plastics, new aftertreatment technology, modeling commercial vehicles or defense vehicles for optimizing fuel economy/hybridization, electrification, and cybersecurity risks. I have since moved into telecommunication, antennas, phone apps, and the Internet of Things (IoT). Technology continues to fascinate me.

I started working on the topic of Cybersecurity for Commercial Vehicles in September 2013. I have since cochaired a technical committee for SAE Commercial Vehicle Engineering

Congress in calendar years 2013, 2014, 2015, 2016, 2017, and now in 2018 as we go to press. I have continually moderated cybersecurity technical panel discussions, not only at SAE Commercial Vehicle Engineering Congress but at WCX: SAE World Congress Experience and SAE AeroTech Congress and Exhibition and at SAE Government/Industry meetings. The goal has been to take "best practices" in cybersecurity across verticals-planes, trains, automobiles, commercial vehicles. This has then lead into the IoT in these verticals, where I continually ask, "Have you thought about cybersecurity for your Internet-of-Things (IoT) devices?"

But, back to the story.

"Hi! This is Marc LeDuc from SAE!"

And so it began. 2013.

The phone rings in my home office.

"Hi! This is Marc LeDuc from SAE."

"Hello, Marc," I reply, "How can I help you?"

"Well, you know something about commercial vehicles."

"Yes…"

"Well," he says in his enthusiastic voice, "I'm putting together something for SAE/IEEE. I want you to be on a cybersecurity panel."

"Hmmm… Who is on the panel?"

"Boeing, The Association of American Railroads, Ford, Bombardier, and you. You'll need to give a presentation on cybersecurity for commercial vehicles, and then we will open it up for questions."

"And, when is this?" I ask.

"September 23rd in Montreal. Oh, and you are doing a video for us too."

I now needed a presentation for cybersecurity of commercial vehicles. At that time, I was engaged in a consulting project on improving school safety-after the massacre of the elementary students in Newtown, Connecticut. We were working on forensics relating to personal computers, phones, and law enforcement and how that plays into school safety. I happened to have good relationship with AccessData, a leader in forensic solutions. I called up one of their friendly hackers, Mary Ellen for a conversation.

Mary Ellen and I created a rather hard-hitting PowerPoint presentation that I presented at the IEEE/SAE CyberSecurity for the Transportation Industry Conference (C4TTI) on September 23, 2013, in Montreal, Quebec, Canada.

It was a rather controversial PowerPoint that I presented. It included 18 wheelers crashing at the entrance of the Ambassador Bridge in Detroit, MI. Taxis that were totally stopped, creating havoc in New York City-with an introduction from my favorite sci-fi film, "The Day the Earth Stood Still."

And more photos of concerns.

Mary Ellen of AccessData then presented this same topic, "Vehicle Cyber Security and Forensics" to an esteemed audience at the New York-New Jersey Electronic Crimes

Task Force. This presentation deck was greenlighted by AccessData for sharing. See Mary Ellen's blog [1].

The IEEE/SAE Cybersecurity for the Transportation Industry (C4TTI) on September 23, 2013, had the panel discussion listed at the website in [2].

1.2 Panel Discussion: Cybersecurity Risks and Policies for Transportation

As consumer demand for connected devices increases, traditional transportation manufacturers will by default become manufacturers of connected devices. As these devices become subject to the same cyber threats with which IT industry have long been accustomed to, transportation sector companies will need to take steps to integrate security considerations into their designs, processes, and policies. This panel will address these considerations and issues.

Moderator: Mike Dudzik, SAE Fellow

Panelists*

- Bob Gary, Willis

- Gloria D'Anna, GlobalBusinessProfessor.com

- André Weimerskirch, ESCRYPT

- Thomas Farmer, Association of American Railroads (AAR)

- Faye Francy, Boeing Commercial Airplanes

- Guy De Langis, Bombardier

- Jim Buczkowski, Ford

FIGURE 1.1

C4TTI in Montréal, Quebec, Canada

Conference Overview
- What are cybersecurity implications for my product or services?
- What cyber risks are real and what are fictitious?
- Are there common cybersecurity issues between my industry and other areas of transportation?

This one-day technical symposium will primarily focus on the critical issues in securing current and future networked vehicles, manned and unmanned, moving on land and in the air. Academic and industry cybersecurity experts in automotive, aerospace, rail and commercial vehicle sectors of transportation will present recent advances, including standards and best practices, and the cutting-edge technologies and research for developing comprehensive solutions. Regulatory agencies will discuss recent relevant developments in cybersecurity policies, directives, regulations, guidelines, and memorandums for transportation.

Conference Venue

C4TTI was co-located next to:

SAE 2013 **AEROTECH** CONGRESS/EXHIBITION

SAE Aero Tech Congress & Exhibition
Palais des congrès de Montréal, Montreal, Canada

on 23 September 2013

Important Dates
- **13 September 2013:** Early registration
- **23 September 2013:** C4TTI Conference

© SAE International

---

* Each panelist will get 5 min to describe their current challenges and developments regarding cybersecurity for their industry. This will be followed by Q&A with the moderator and audience.

So, after the panel presentation, I was able to meet with the various other panelists. There were two large tables with everyone eating lunch. I was running late. Not that I usually run late, but I had a video interview to do with SAE. I was supposed to be at one table, but ended up at the other table. That's where I met Dr. André Weimerskirch. I like to meet the smartest person in the room, particularly when I am solving problems. He was it. Dr. André Weimerskirch had helped found ESCRYPT—a cybersecurity consulting company, which he was just in the midst of selling to Bosch.

So, why is André important? Dr. Weimerskirch is an expert in automotive cybersecurity. He had just spent his career in Europe and Asia working on cybersecurity for automobiles. I wanted to take his knowledge and extend that to commercial vehicle cybersecurity, not that they are exactly the same, but they are similar. Dr. Weimerskirch explains commonalities and differences for commercial vehicles versus automotive vehicles in Chapter 4.

Dr. André Weimerskirch worked with me to cement the technical papers and cybersecurity panels for commercial vehicles at SAE Commercial Vehicle Congress—the annual SAE Congress for the commercial vehicle industry. We have also worked with SAE at World Congress to put together provocative discussions across vehicle sectors—to aid in sharing of knowledge. He has been my technical rock for this topic. He's helped calm my nerves when I kept proclaiming that "we are pushing a rope" on this topic. "It takes time," he said. I moderated the panels to make them controversial and keep them lively.

I also struck up a friendship with Faye Francy, then of Boeing. She has been on our cross-sector technical panels for cybersecurity since our meeting. Faye Francy now heads the Automotive Information Sharing and Analysis Center (Automotive-ISAC), which has now invited commercial vehicles to be part of that ISAC. And, yes, in a friendly way, Faye says, "It's all your fault!"

So, what did I learn from the SAE/IEEE meeting in Montreal? I learned not to present such a controversial hard-hitting contentious presentation. It was a worrisome presentation-and at the time, I did not have a solution to solve the problem. I'm an engineer. I like to solve problems, and this was a rather large problem.

[Aside: Ironically, Mary Ellen and I had talked about choosing an automobile to hack-and we picked the Jeep Grand Cherokee. That vehicle famously became the Charlie Miller and Chris Valasek hack that set the automotive industry on its head with regard to cybersecurity. It launched a massive recall for Fiat Chrysler [3].]

1.3 How Do We Define Commercial Vehicles for This Book?

And, that's how I started getting formally involved with cybersecurity for commercial vehicles. I had worked for Navistar on future engine programs in the early 1990s, and had worked for Ricardo Engineering Services concentrating on the commercial vehicle and defense sectors in the early 2000s. Having worked in a start-up company focusing on commercial buses and medium-duty trucks after Ricardo—I knew my way around the commercial vehicle world. I looked at the cybersecurity for commercial vehicles, and thought, "Where can we make the most progress, the quickest?"

I looked at cybersecurity as "clean data" in a similar fashion to how the industry had to deal with clean emissions. I had an opportunity to have the John Deere and Caterpillar experts on cybersecurity discussion panels. Those two companies were ahead of where the over-the-road guys were in cybersecurity. So, we started to focus on where we had the most homogeneous fleets and determined that was for Classes 7 and 8 vehicles. We didn't forget about the Classes 4, 5, and 6 vehicles. We just put our energy where we could make the quickest improvement.

Basically, if you are running your vehicles neural network on a J1939 CAN Bus, then this book is for you!

1.4 What I Love about the Cybersecurity World

I've had an opportunity to meet with many experts in the cybersecurity world while working on this problem. What I love about the cybersecurity world is the personalities of the people who work this part of a problem. The hackers are bright. They look at the world a bit "differently" as they think out of the box. Then, of course, there is some personal branding:

- The guy with the black baseball cap.
- The guy with the long gray bushy beard.
- The guy who likes to talk about racing his WRX.
- The guy who loves his Ford Raptor.
- The girl with the blonde hair with a pink streak.
- The girl who likes playing with knives.
- The guy from the Israeli intelligence agency, who puts his back up against the wall, and surveys the room.
- The academic who wears shorts and tall white socks.
- The guy who likes to crash trucks.
- The guy with the bow tie and stylish hat.
- The guy or girl who can fit it anywhere and you won't notice them.
- And, the guys or girls from automotive or commercial vehicle OEMs-who don't say much.

I have run into more women in cybersecurity meetings than elsewhere in my engineering career. They are all smart. The cybersecurity people are *very smart*. In addition, they are strangely introverted, until they get to know one another. Then, the stories roll.

When I told these cybersecurity experts that I was engaged to write a book for SAE International on "CyberSecurity for Commercial Vehicles", they listened. But when I told them that I wanted to write a cybersecurity book that read like a nonfiction novel, their attention "perked." They too wanted to write something that would be relevant and interesting and useful. I didn't want to write a book that just sat on a shelf. I wanted to write a book that you read—cover to cover (albeit, not at one sitting), or could bounce around and read the various chapters if you had a bit of attention deficit hyperactivity disorder.

This would have been easier to just take a bunch of SAE Technical Papers and put them in a book with a forward. Regretfully, in this space, the SAE Technical Papers have been quite limited since we began this in the fall of 2013. However, more papers are being written along with research that is being funded on commercial vehicle cybersecurity.

And so here it is the nonfiction novel on "Cybersecurity for Commercial Vehicles."

1.5 So, Who Should Read This Book?

If you are in the Classes 7 and 8 world of trucking, you should read this. And then if you are into delivery vans, you should read this too.

The goal of this book is to provide a background on cybersecurity for the commercial vehicle industry that will be used, so that a mechanic or a fleet owner or a CEO can pick up this book and say, "Aha! I see what all the fuss is about with regard to cybersecurity." And then encourage those in the industry—not just the commercial truck original equipment manufacturers (OEMs) who work on the neural network architecture of the trucks, but the back-office people of the fleets and the telematics suppliers and the used truck fleets—to worry that, yes, their cargo could be at risk to a cyberattack. And could it happen? "Yes." Should we protect against it? "Yes, we should."

1.6 And Why You? Why Gloria?

First, I was asked.

I'm a strategist. I'm also an engineer that likes to solve problems, *big problems*. If you have ever taken the Meyers-Briggs Type Indicator (MBTI) personality test, for those people who like to categorize things, I'm an INTJ (introversion, intuition, thinking, judgment). What does that mean? My personality is one that sees the entire picture and helps put the puzzle pieces together. And that is what I hope you find here in the next pages.

I remember outlining a good portion of this strategy for improving cybersecurity of commercial vehicles on Dr. André Weimerkskirch's whiteboard at the University of Michigan Transportation Research Institute (UMTRI).

1.7 The Contributing Writers

1.7.1 Chapter 2: Should We Be Paranoid?—by Doug Britton

We start out with Doug Britton writing on "Should We Be Paranoid". Prior to joining RunSafe, Doug was the founder and CEO of Kaprica Security I adore Doug Britton. He has a fabulous mind. I met Doug through a friend.

Doug Britton, in his own words, is "A helpless cybersecurity romantic at heart." His professional mission is to transform the relationship between innovation and adoption so that newfound risks and vulnerabilities can be cauterized quickly, reliably, and inexpensively...A trained computer scientist, he started his career in the National Center for Supercomputing Applications as the University of Illinois, before serving as a Russian Linguist and Interrogator in the U.S. Army. Doug's more formal biography is at the end of his chapter.

Doug Britton was on our Cybersecurity Panel for SAE World Congress in 2016. There is Doug on the left hand side of this picture below. Doug was flanked by Dr. Dan Massey from the Department of Homeland Security for this panel, along with Kevin Harnett from the Department of Transportation VOLPE. When I asked Doug, what keeps him up at night, he replied, "Patient, understated attackers. They're not in it for the headlines. You'll never know they were there, until perhaps 10 years later, if then. Will you become good enough to see them? Because, they are already here." Chilling.

FIGURE 1.2   Doug Britton of Kaprica Security makes a point during the Cybersecurity panel at the SAE 2016 World Congress. Other panelists shown include Gloria D'Anna of General Telecom Systems (at left background), Dan Massey, and DoT's Kevin Harnett. (Paul Weissler)

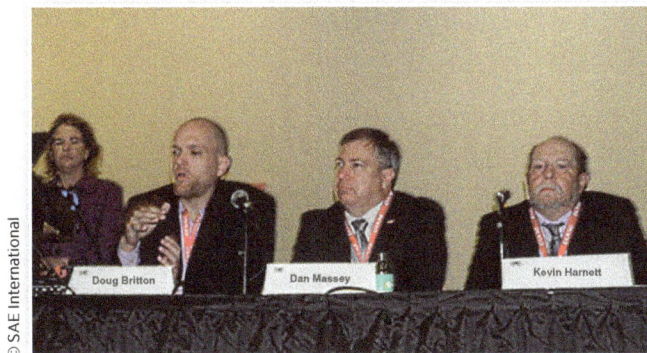

1.7.2 Chapter 3: What Cybersecurity Standard Work Is Applicable to Commercial Vehicles?—by Lisa Boran and Xin Ye

Ford's Lisa Boran and Xin Ye take you through the SAE Engineering Standard J3061. We don't actually print out the standard in this book. But you can order it through SAE.

The PowerPoint below shows you the motivation for creating J3061. Interconnectivity of today's and future vehicles make them potential targets for attack. Noting that commercial vehicles have had telematics much earlier than automobiles, we still need to look at this standard, and understand it.

SAE provides training sessions on this standard. Please note that this standard is continually updated [4].

FIGURE 1.3

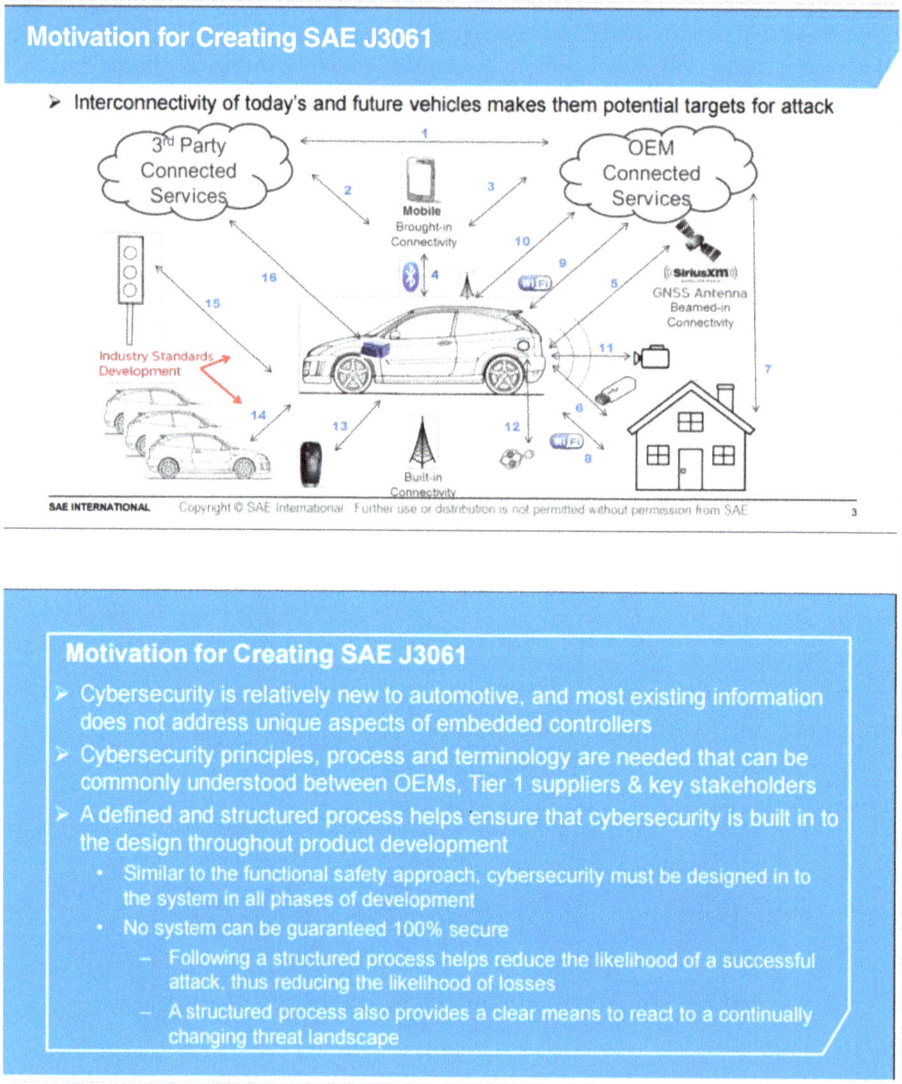
Motivation for Creating SAE J3061
➢ Interconnectivity of today's and future vehicles makes them potential targets for attack
3rd Party Connected Services
OEM Connected Services
Mobile
Brought-in Connectivity
SiriusXM
GNSS Antenna Beamed-in Connectivity
Industry Standards Development
Built-in Connectivity
1
2
3
4
5
6
7
8
9
10
11
12
13
14
15
16
SAE INTERNATIONAL
Copyright © SAE International. Further use or distribution is not permitted without permission from SAE
3
Motivation for Creating SAE J3061
➢ Cybersecurity is relatively new to automotive, and most existing information does not address unique aspects of embedded controllers
➢ Cybersecurity principles, process and terminology are needed that can be commonly understood between OEMs, Tier 1 suppliers & key stakeholders
➢ A defined and structured process helps ensure that cybersecurity is built in to the design throughout product development
• Similar to the functional safety approach, cybersecurity must be designed in to the system in all phases of development
• No system can be guaranteed 100% secure
– Following a structured process helps reduce the likelihood of a successful attack, thus reducing the likelihood of losses
– A structured process also provides a clear means to react to a continually changing threat landscape

© SAE International

1.7.3 Chapter 4: Commercial Vehicles vs. Automotive Cybersecurity: Commonalities and Differences—by André Weimerskirch, Steffen Becker, and Bill Hass

Dr. André Weimerskirch, Steffen Becker, and Bill Hass then take off on what you have learned in the previous chapter by Ford's Lisa Boran and Xin Ye. If there is ONE chapter that you should read in this book, this is it!!! Read it!

André and his cowriters have a vast array of technical and specialized knowledge in creating security architectures ensuring date integrity, privacy, and authenticity for connected car communications, including vehicle-to-vehicle applications.

André was the leading vehicle cybersecurity researcher at the University of Michigan. He continued to serve as the chair of several industry organizations, including the prestigious escar cybersecurity conference. He was the chief executive officer (CEO) of ESCRYPT GmbH and has further experience at Ruhr-University Bochum. He is currently the Vice President of Cybersecurity for E-Systems at Lear Corporation. André received his Bachelor of Science degree in Computer Science from Darmstadt University of Technology, a Master of Business Administration from Worcester Polytechnic Institute, and a doctorate degree in Electrical Engineering and Information Technology from Rurh-University Bochum [5].

1.7.4 Chapter 5: Engineering for Vehicle Cybersecurity—by Daniel DiMase, Zachary A. Collier, John A. Chandy, Bronn Pav, Kenneth Heffner, and Steve Walters

I have interesting friends. The history of this piece stems from work done by a futurist friend of mine. Yes, those people who live 30 years in the future and write scenarios for the government on such topics. I hence ran into this work, done by Daniel DiMase and colleagues. Daniel works at Honeywell and chairs a committee at SAE.

How did I run into Daniel DiMase? I was putting together a cross-sector panel (trains, planes, automobiles, and trucks) on cybersecurity for SAE 2015 Aerotech Congress and Exhibition on September 22, 2015, in Seattle, Washington. See photo of the panel below. Daniel DiMase from Honeywell was supposed to be on the panel, but he couldn't attend, so I then substituted him with SAE's Bruce Mahone. Bruce is SAE's Director of Washington Operations for Aerospace.

Please see photo below from the session [6].

FIGURE 1.4

Technical Session Schedule

Tuesday, September 22

Executive Management Panel Discussion: Cyber-Physical Security
(Session Code: ATC3008)

Room 6E 1:30 p.m.

Recent cyber attack demonstrations, via a vehicle's infotainment systems and aftermarket devices, have shown an impact to safety critical systems. This panel discusses how these cyber attacks apply to the commercial vehicle sector. Potential risks and available protection methodologies will be addressed. Panelists will also talk on how to better define and, evaluate the threat/risk of CyberSecurity as well as the value of standards currently under development will have in their organization.

Moderators - *Gloria D'Anna, General Telecom Systems, Inc.*
Panelists -
John Craig, Boeing
Thomas Farmer, Association of American Railroads
Bruce Mahone, SAE International
Daniel Prince, GE Aviation
Timothy J. Wallach, Federal Bureau of Investigation
André Weimerskirch, University of Michigan

Daniel had sent me this wonderful PowerPoint presentation that he had presented to NASA on this subject. Fascinating subject. The PowerPoint for NASA raised eyebrows as I sent it off to my friends in the industry. Hence, I wanted Daniel to write a chapter for this book. In simplistic terms, how do you deal with malicious code that is embedded in your hardware that you purchase?

This is a fascinating chapter, albeit, quite lengthy. However, expect to see additional work from SAE in this space. And take note of how it affects your supply chains in electronics in both original equipment and aftermarket equipment.

1.7.5 Chapter 6: "When Trucks Stop, America Stops"—Foreword by the American Trucking Association (ATA). ATA's 2006 White Paper

So, why a white paper by the ATA? Who is the ATA? Per the ATA website: "Since 1933, the American Trucking Associations has been the leading advocate for the trucking industry. Through a strong federation of state associations, affiliated conference and individual members, ATA is committed to developing and advocating innovative research-based policies that promote highway safety, security, environmental sustainability and profitability. ATA's professional staff works to educate policymakers and the general public about the essential role trucking plays in the economy, promote responsible policies to improve highway safety and advance the industry's environmental goals" [7].

The ATA wrote a paper on "When Trucks Stop, America Stops", which I find very relevant to the topic of cybersecurity. I have paraphrased from the beginning of this paper,

"Commercial truck traffic is vital to our nation's economic prosperity and plays a significant role in mitigating adverse economic effects during a national or regional emergency. Our economy depends on trucks to deliver ten billion tons of virtually every commodity consumed-over 80 percent of all freight transported annually in the U.S. In the U.S. alone, this accounts for $700.4 billion worth of goods transported by truck. Trucks hauled 65.6% of the shipment of value into Canada and 67% of the shipment of value into Mexico. It becomes apparent that any disruption in truck traffic will lead to rapid economic instability. The unimpeded flow of trucks is critical to the safety and well-being of all Americans. However, it is entirely possible that well-intended public officials may instinctively halt or severely restrict truck traffic in response to an incident of national or regional significance. History has shown us the consequences that result from a major disruption in truck travel. Immediately following the 9/11 terrorist attacks, significant truck delays at the Canadian border crossings shut down several auto manufacturing plants in Michigan because just-in-time parts were not delivered. The economic cost to these companies was enormous. Following Hurricane Katrina, trucks loaded with emergency goods were rerouted, creating lengthy delays in delivering urgently needed supplies to the stricken areas. Although in the face of an elevated threat level, a terrorist attack, or a pandemic, halting truck traffic may appear to be the best defense, it actually puts citizens at risk.

Officials at every level of government must recognize that a decision to halt or severely curb truck traffic following a national or regional emergency will produce unintended health and economic consequences not only for the community they seek to protect, but for the entire nation" [8].

I think the key takeaway from this is that "over 80 percent of our freight here in the USA travels by truck". If you compromise the trucks, you compromise our economy. I'm not sure if you have personally been affected by the lack of groceries in the grocery store, due to events, but I have. I've seen grocery deliveries that have been compromised by weather in Buffalo due to a snowstorm, or due to the Rodney King riots in Los Angeles. We had truck delays here in the auto industry due to 9/11 attacks, as noted in the quote above.

So, even though this white paper was written in 2006, I still find it quite relevant, and so should you.

A big thank you for the ATA for allowing us to reprint this white paper for this book here on commercial vehicle cybersecurity.

1.7.6 Chapter 7: On the Digital Forensics of Heavy Truck Electronic Control Modules—by James Johnson, Jeremy Daily, and Andrew Kongs

I first met Dr. Jeremy Daily at SAE Commercial Vehicle Engineering Congress in 2015. He's very laid back, until he tells you that he likes to crash things. Then, he gets very animated. Specifically, he likes to crash trucks. He is an expert on a commercial vehicles' black box. He presented a technical paper on "A Digital Forensics Perspective of the Cyber Security of Commercial Vehicles." So, Jeremy's chapter is about the commercial vehicle black box [9].

And since meeting Jeremy in 2015, he has been part of our continuous discussions, panel discussions, and leading thinker on cybersecurity for commercial vehicles. I have continued to advocate to get additional trucks for Jeremy to experiment with.

1.7.6.1 Comments on How We Are All Connected: We have been rather lucky and fortunate to work with Dr. Dan Massey at the U.S. Department of Homeland Security. Dan has been on all of our cybersecurity panels at SAE Commercial Vehicle Engineering Congress held in Chicago, and the cross-sector panels that we have put together for the SAE World Congress in Detroit. I asked the DHS, prior to the Distributed Denial-of-Service (DDoS) attack in October 2016 to put together a chapter on how we are all connected. Regretfully for this book, but not for our nation, DHS has been rather busy, and this chapter did not come to fruition. Dan Massey has since left the DHS. I will miss him in that role.

Many of us experience, or heard of the DDoS attack that took place, starting at 7 am EST on the East Coast on October 21, 2016. Per Wired magazine, that morning's attack was aimed at Dyn, an internet infrastructure company headquartered in New Hampshire. That first bout was resolved after about two hours; a second attack began just before noon. Dyn reported a third wave of attacks a little after 4 pm EST. In all cases, traffic to Dyn's Internet directory servers throughout the United States— primarily on the East Coast, but later on the opposite end of the country as well—was stopped by a flood of malicious requests from tens of millions of IP addresses disrupting the system. Late in the day, Dyn described the events as a "very sophisticated and complex attack…" The situation is a definite reminder of the fragility of the web and the power of the forces that aim to disrupt it [10].

1.7.6.2   **IoT: The Internet of Things:** As you may know, SAE has now created an IoT Program Committee to bring this topic to the forefront of SAE and create additional alliances for SAE around the world. I am currently vice chair of this SAE Program Committee, and will be the chair in 2 years.

In April 2017, we had the first IoT Expert Panel at SAE WCX.

FIGURE 1.5   Our Technical Expert Panel on IoT and Cloud Technology at SAE WCX 17. L to R: Michael Keefe, Weather Telematics; Binoy Damodaran, IBM; Gloria D'Anna, General Telecom Systems; Eyal Amir, Parknav; Joe Barkai, Chair of SAE IoT Program

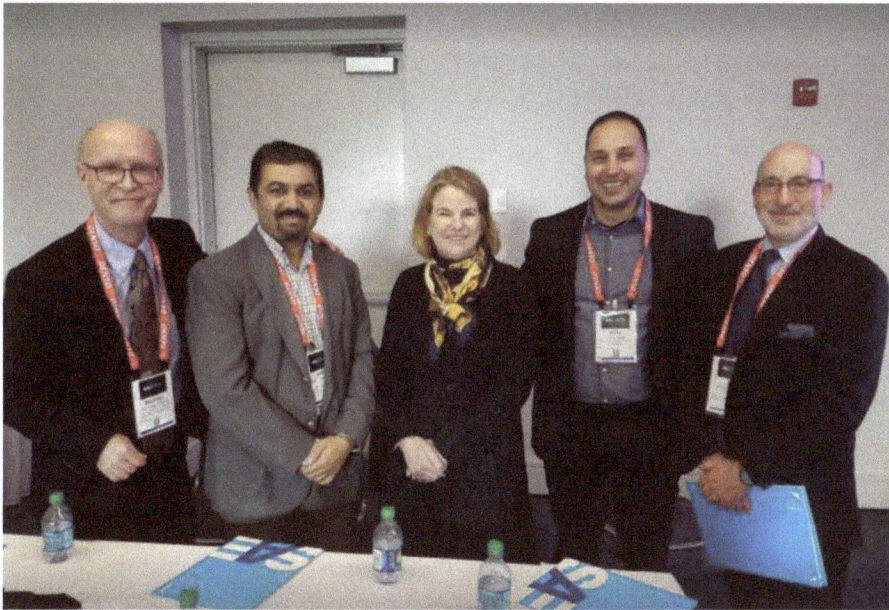

© SAE International

Expect to see more items on the IoT for commercial vehicles in this space. I will continue to advocate for end-to-end cybersecurity for any application.

1.7.7 Chapter 8: Telematics Cybersecurity and Governance—by Glenn Atkinson

In just about every technical meeting that I have been in on cybersecurity for commercial vehicles, Glenn Atkinson from Geotab has been there. Glenn has also been working in conjunction with other telematics providers through industry collaborating in helping secure telematics for commercial vehicles.

This is a very important chapter, as the majority of commercial vehicle fleets are connected by telematics. Yes, your telematics systems need to be thoroughly cyber tested. Have yours?

1.7.8 Chapter 9: The Promise of Michigan: Secure Mobility—by Karl Heimer

This is an extensive chapter on what the State of Michigan is doing with regard to cybersecurity. This chapter starts out with a forward by Governor Rick Snyder. This is rather fascinating as you start to see how the state government can be important in the cyber space. Also, it helps that the U.S. Army Tank Automotive Research Development and Engineering Center (TARDEC) is in the State of Michigan.

If you were at the 2017 SAE Commercial Vehicle Engineering Congress in Chicago, Karl Heimer did an excellent update of the Cyber Truck Challenge. We hope to see more involvement from the commercial vehicle industry in 2018.

1.7.9 Chapter 10: How the Truck Turned Your Television Off and Stole Your Money: Cybersecurity Threats from Grid-Connected Commercial Vehicles—by Lee Slezak and Christopher Michelbacher

This chapter outlines some of the work done by the Department of Energy. I've had the opportunity to review papers by the DOE during their Annual Merit Review and Peer Evaluation Meeting for the past seven years. This year, in 2017, I approached Lee Slezak and Christopher Michelbacher about writing a chapter. I had seen previous work by DOE's Argonne National Laboratory on electric vehicles, and in this year's DOE AMR was another intriguing paper. I asked Lee and Christopher during a dinner meeting if they would like to write a chapter. And here it is.

1.7.10 Chapter 11: CALSTART's Cyber Mission: HTUF REDUX—by Michael Ippoliti

Who is CALSTART? CALSTART originated the Hybrid Truck Users Forum (HTUF) with the assistance of the U.S. Army National Automotive Center (NAC) and TARDEC, nearly two decades ago.

I worked with CALSTART when I was at Ricardo Engineering, working on the hybridization of commercial vehicles. They had an excellent HTUF. In these CALSTART meetings I was able to meet with technologist and engineers from Cummins, FedEx, and UPS, to name a few. I also had worked with the NAC and TARDEC on future hybrid vehicles.

I ran into CALSTART again, at a reception in Novi after the TU-Automotive in March, 2016. I thought, "Wow! Their HTUF was fantastic! What a way to get at Class 4 to 6 for cybersecurity." And hence this chapter.

1.7.11 Chapter 12: Characterizing Cyber Systems—by Jennifer Guild

I ran into Dr. Jennifer Guild at a cybersecurity meeting in D.C. I thought to myself, "Cool. Another woman in cybersecurity AND she plays with commercial vehicles (in the U.S. Navy). We struck up a conversation, and she sent me her PhD thesis. I thought, "This is interesting. I should share this."

So, here is a shortened version of Dr. Guild's PhD thesis. If you love algorithms, this is a must read. Just a note that this chapter is the most technical in the book.

1.7.12 Chapter 13: "...No, We Should Be Prepared"—by Joe Saunders and Lisa Silverman

Joe Saunders and Lisa Silverman outline a "What If" scenario for a commercial fleet in this chapter. This is properly labeled Chapter 13. This is when things go wrong.

"All of a sudden, you notice that seven of the trucks on the East Coast have slowed down to 10 miles per hour... Next, you get a panic call..."

1.7.13 Chapter 14: Heavy Vehicle Cyber Security Bulletin

This chapter is a reprint of the National Motor Freight Traffic Association Inc.'s Cybersecurity Best Practices. Please read. This is how you should be prepared, in case things go wrong.

And per NMFTA's website, "NMFTA member motor carriers perform a vital service to our nation's economy by delivering the goods necessary to keep commerce flowing. NMFTA represents over 500 carriers who collectively operate close to 200,000 power units generating approximately $100 billion in freight revenue."

And if you have a chance, please read NMFTA's Cybersecurity overview on their website.

And, in my humble opinion, NMFTA has been amazing with regard to this cyber-security topic for commercial vehicles!

1.7.14 Chapter 15: Law, Policy, Cybersecurity, and Data Privacy Issues—by Simon Hartley

This is also one of my favorite chapters. I met Simon Hartley, again, through a friend. Since our initial meeting, we have had many conversations on cybersecurity and what it means next. Simon has also joined the SAE IoT Program Committee on my request.

I've put this chapter towards the end of the book, as a bit of a wrap-up. It's a breezy read, and you will find it easy to read, and overly informative.

1.7.15 Chapter 16: Do You Care What Time It Really Is? A Cybersecurity Look Into Our Dependency on GPS—by Gerardo Trevino, Marisa Ramon, Daniel Zajac, and Cameron Mott

We always want to know where we are. Just put in the location on your Facebook page. But, in logistics, we want to know where the vehicles are. We want to know when the freight will arrive. What happens if GPS is not right? Southwest Research Institute briefly explains our dependency on GPS.

1.7.16 Chapter 17: Looking Towards the Future—by Gloria D'Anna

This is my final chapter of the book to look ahead as to what else is in store for us for cybersecurity for commercial vehicles. This really is just the beginning. The sexy part of cybersecurity was when Charlie Miller and Chris Valasek hacked the Jeep Grand Cherokee. The cybersecurity world was new for vehicles.

At 2016 SAE World Congress, there was an article written about the Cybersecurity Expert Panel. I always start off the panel with "What keeps you up at night?" [11] I'll let you, the Reader, start off with this article, and let the thoughts just seep into your brain. Then, after a long deep breath, please enjoy the rest of the book!

SAE World Congress, Detroit, April 12, 2016

Cybersecurity Experts Confess Their Greatest Worries

What keeps you up at night? Cybersecurity experts at the SAE 2016 World Congress were asked about their deepest worries during a panel session on the levels and nature of cyber challenges. Their responses:

"Identity theft, compromised commercial vehicles that affect domestic logistics systems, and a horrific event that affects our first responders' ability" to deliver aid, said panel moderator Gloria D'Anna, President of General Telecom Systems.

"Patient, understated attackers" were cited first by Doug Britton, CEO of Kaprica Security. They're not in it for headlines, he said, and "you'll never know they were there" until perhaps 10 years later, if then. Will you become good enough to see them, he asked, because "they're already here."

Such attackers aren't looking for a few social security numbers; they want an opportunity to get them all. Britton added there is a no uniform understanding of attack methods or distribution of secure coding capabilities in the world of cybersecurity.

The need for available mature security solutions that can be immediately deployed was cited by Brian Murray, Global Director, Safety and Security Excellence at ZF TRW. An overarching concern, he said, is getting the level of design focus to provide both the functional requirements of componentry along with reasonable safety as well as security.

Worry about nation-state-sponsored attackers causes insomnia for Dr. André Weimerskirch, a University of Michigan researcher. And he raised the issue of efficiently testing many possible new solutions in erecting defenses.

Dr. Weimerskirch currently has projects in automotive intrusion detection, providing secure CAN/on-board communications, management of the many issues associated with vehicle key fob systems) and V2X (vehicle to infrastructure). He also has a project to develop a resilient reference electronic architecture for cybersecurity systems development.

Vulnerability of the U.S. government's vehicle fleets is a logical worry for Dan Massey, Program Manager in the Dept. of Homeland Security's cybersecurity division. And he also sees this issue: "As cyber physical systems and the Internet of Things add new features to devices on which we all depend, are security design issues being overlooked?"

What about the motivation of cybercriminals to perform zero-day (no warning) attacks? That gives nighttime jitters to the Department of Transportation's Kevin Harnett, Program Manager at the Volpe National Transportation Systems Center. He tosses and turns thinking about how to implement interim countermeasures to provide security for government vehicles, including application of best practices and industry standards—and use of aftermarket devices when appropriate.

A cyberattack on U.S. aviation from a nation-state makes Faye Francy, Executive Director of Aviation-ISAC (Information Sharing and Analysis Center), a poor sleeper.

Author:

Paul Weissler

References

1. http://manhattanmennonite.blogspot.com/.

2. http://sites.ieee.org/c4tti/.

3. https://www.wired.com/2015/07/hackers-remotely-kill-jeep-highway/.

4. http://www.sae.org/events/ces/2016/attend/program/presentations/forest.pdf.

5. http://www.prnewswire.com/news-releases/lear-appoints-dr-andre-weimerskirch-a-renowned-industry-expert-to-lead-cyber-security-for-e-systems-300317192.html.

6. http://www.sae.org/servlets/techSession?EVT_NAME=ATC3008&GROUP_CD=PANEL&SCHED_NUM=240875&tab=sessionDetails&REQUEST_TYPE=SESSION_DETAILS&saetkn=VbZMuBSCwv.

7. http://www.trucking.org/What_We_Do.

8. http://www.trucking.org/ATA%20Docs/What%20We%20Do/Image%20and%20Outreach%20Programs/When%20Trucks%20Stop%20America%20Stops.pdf.

9. https://www.sae.org/calendar/techsess/227711.pdf.

10. https://www.wired.com/2016/10/internet-outage-ddos-dns-dyn/.

11. https://www.sae.org/news/2016/04/cybersecurity-experts-confess-their-greatest-worries.

About the Author

Gloria D'Anna is an engineer, entrepreneur, and multiple patent award holder—an expert in vehicle engineering and cybersecurity. She likes to solve problems, from future light commercial vehicles, to beefing up the cybersecurity of vehicles, always rolling out new tech, reducing inefficiencies, and driving business.

She began her career at GM, winning an MBA Fellowship to the University of Chicago, moving on to Ford, Navistar, Textron, Eaton, and Ricardo. Later, she led sales at three successful startups, addressing challenges from school safety to building connected devices for law enforcement.

Gloria has been working with SAE for the last 6 years, creating and moderating popular and educational cybersecurity technical sessions from Commercial Vehicle markets to the Internet of Things. Her 2018 book, SAE's *Cybersecurity for Commercial Vehicles* is her first book.

She is the recipient of SAE International's 2018 Forest R. McFarland Award for Automobile Electronics Activity.

She is currently the CEO of General Telecom Systems, a private telecommunications company, on the Board of Atmos XR, an advisor to 202 Partners, a lead mentor for Techstars Mobility and a member of the Uptane Advisory Board.

She likes trucks.

CHAPTER 2

Should We Be Paranoid?

Doug Britton

Where there is something of value, there is someone that wants to acquire it at a "below market" price. In aggregate, McAfee's second report on the economic impact of cybercrime, titled "Net Losses: Estimating the Global Cost of Cyber Crime," suggests that the total losses annually are on the order of $500b and growing. With the sheer volume of value in motion, highly sophisticated attackers will use every means at their disposal to acquire the property without paying for it, with minimal risk of getting caught. So, if others believe cyber presents an opportunity to get low-cost access to valuable assets, will it be levered to do so? Yes. But, will hackers start crashing my fleet, etc.?

2.1 Why Is Cyber So Hard to De-risk?

Basically, "cyber" is like magic. In other domains, the risks are understood and the nature of protection against them is straight forward. As an example, we'll consider the risks to drivers, trucks, and cargo posed by traditional firearms. Transportation companies usually have safety policies that consider a number of risk factors for the drivers, equipment, and cargo:

- Time of year-Around the holidays, organized crime tends to increase hijacking activity because loads are often fuller and more valuable. Need for cash flow as well for organized crime?

- Population density along route-Will the driver have well-populated rest stops that can increase safety? Or will the route include numerous remote rest stops?

At the high end of this risk spectrum, companies will employ armored cars and multi-vehicle convoys. In medium-risk scenarios, two drivers will travel together. In low-risk scenarios, a single driver will move with nothing more than a traditional padlock on the payload. All of these analyses are based on shared, implicit assumptions about how the risks from firearms present themselves:

- A human must be present to hold a gun and pull a trigger.
- You can recognize a gun when you see it.
- Bullets travel in gravity-adjusted straight lines.
- Bullets can't go through all matter.

In the parlance of existing firearms security, cyber would be like a multi-hop bullet that can turn corners, pause in mid-air, wait for someone to open a door, jump inside, wait for 6 months, and shut down a system from the other side of Earth, all while convincing you it was never there in the first place. Organized criminal elements employing such a magical bullet would be able to inflict incalculable damage and would have little risk of getting caught.

2.2 A Primer on Hacker Economics and Tactics

Just like every other part of the global economy, hackers are subject to the governing dynamics of macroeconomics-how to satisfy unlimited wants with limited resources. Despite what may feel like bottomless pools of evil, magic powers at the behest of, what one politician described as, a "guy sitting on his bed weighing 400 pounds"; hackers aren't omnipotent. It can be instructive to look at hacker activities through the lens of traditional accounting tools, such as an income statement and a balance sheet. My hypothesis is that by demystifying the structure of hacker activities, we can identify opportunities to increase security. I will aim to present a useful analytical construct for translating hacker activities into rational economic behavior, without overtorquing the analogy.

2.2.1 Income Statement

Table 2.1 summarizes the 2016 income for Leet Trading Company, a hacker enterprise doing business in the mobility space. While we don't have Leet's audited financials with full footnotes, we can certainly learn something about operations. Leet's developed and deployed a mobile app that generated revenue in several ways. The first revenue generator is in the form of unauthorized premium SMS messages. In a premium SMS scam, the phone surreptitiously sends text messages to expensive content providers that charge the handset owner $10-$15/mo. Often, the scammer is paid a percentage of the fees collected by the content provider. The app also uses an Android exploit to download other apps and utilities. These in turn steal credit card information, credentials, etc., grossing another $800k.

Leet paid $25k to a compatriot in the supply chain for the actual exploit that allowed their application to wreak the aforementioned havoc. Returning to the firearms analogy, the exploit would be akin to the bullet in a gun. It is the part of the system that actually does the damage. Leet's Evil Android App would be the gun, since bullets don't work by themselves. They need to be aimed and they need to be told when to fire. After the exploit, Leet paid another portion of the hacker supply chain to use phishing and spoofing techniques to distribute the application.

To think about the lexicon from another perspective, we can think of programs having vulnerabilities and those vulnerabilities having exploits. The vulnerabilities can be thought of as an unintended secret backdoor into a program that likely the program's author doesn't even know about. An exploit is the key that opens the secret door. With the key, the attacker can go through the backdoor, potentially gaining control of the program and maybe even the entire system. A large part of the cyber research community is dedicated to finding these backdoors before attackers do. Several websites, such as NIST's National Vulnerability Database, are dedicated to publishing the latest information on vulnerabilities and associated exploits.[1] Exploits are said to be "relevant" if the "locks haven't been changed" on the secret backdoors (vulnerabilities). If the lock is changed, the vulnerability might still exist, but a different key (exploit, cyber weapon) would be needed to open the door. Many companies have active "patch" processes that publish updates to their software as vulnerabilities and exploits are discovered. The result is hopefully a decrease in the secret backdoors and changing all of the locks. (Translate that to bullets in the Balance/Income statement.)

TABLE 2.1 Notional Income Statement for Leet Trading Company, a Hacker Enterprise

Leet Trading Company		
Consolidated Income Statement—2016		
Assets		**Value**
Revenue		
	Gross sales—Evil Android App	$800,000
	Gross sales—premium SMS	$250,000
Cost of goods sold		
	Distribution	($40,000)
	Android exploit	($25,000)
Expenses		
	Mountain Dew	($4,000)
Other income		
	YouTube advertising	$5,000
Total income (loss)		$986,000

© SAE International

2.2.2 Balance Sheet

Table 2.2 presents the Assets portion of Leet's balance sheet. We'll ignore the Liabilities because insights into the cash management and credit practices of the enterprise do not help the defender. We'll ignore Owner's Equity for a similar reason.

While the industrialization of the attack (described below) does suggest forming liquidity in tradable hacker assets, exact pricing of an asset is neither possible nor necessary for our purposes. The Evil Android App that generates "sales" would have an assignable value related to its earning potential, discounted for the amount of time it will remain relevant, complexity in deployment, market share, etc.

Leet made substantial investments of resources into the development and testing of the Evil Android App. Before the app could produce revenue, Leet had to ensure there was a distribution plan, consistent with the

TABLE 2.2 Asset portion of Leet's balance sheet

Leet Trading Company	
Consolidated Balance Sheet—12/31/2016	
Assets	**Value**
Evil Android App	$2,000,000
Android exploit	$40,00
Cash	$986,000
Total Assets	$3,026,000

© SAE International

---

[1] https://nvd.nist.gov/.

addressable market. For example, if the distribution networks at their disposal would not get the app onto phones where the specific exploit worked, then the operation would not work. Next, once the app is on the distribution network, does it work with enough reliability to get downloaded, install correctly, and start "doing its evil activities?" If it isn't reliable, due to subtle variations in the 15 models of phone it targets, the app has reduced its earning potential. It's worth pointing out that these are the exact same sets of questions any producer of software-based systems undertakes (marketing, distribution, reliability).

The exploit used by the Evil App has discrete value of its own. This is because, while the exploit is relevant, it could be used to create other apps. For example, Leet could decide to branch out and create an application that acts as part of a listening post in a command-and-control network. In that case, the exploit could facilitate that app's deployment.

2.2.3 Economic Analysis

These statements allow us to explore several important questions about Leet's outlook and behavior in 2017 and beyond.

- **What are Leet's main enterprise risks?**

  - **Detection as malware**—At some point, the Evil Android App will be detected as malware by any one of the dozens of app rating systems, such as Appthority. Depending on the market the app is operating in, this could take longer. Analysis suggests that between 5% and 10% of smartphones have any type of malware protection (comment: scary in itself). As a result, Leet might be able to operate freely for 6 months or longer.

    Once the app is detected as malware, Leet may have a variety of options about how to continue operations. With only a small slice of devices protected, it is possible that the malware designation won't noticeably impact the distribution rate. Additionally, Leet could make small changes in the app that render it sufficiently unique when compared with the original app, now letting it pass right by many malware detection systems. As defenders move away from signature-based analysis to behavior-based malware analysis, Leet will have a harder time accomplishing this evasion, but it is still possible.

  - **Expiration of the underlying exploit**—The bigger risk to operations is that the underlying vulnerability leveraged by Leet's app will be fixed by the manufacturer. On the Android platform, Google has committed to monthly security updates, which could neutralize this exploit. The spread of the updates across a user base is highly variable and could take months to get everyone, if ever. For businesses using smartphones, the administrators likely have policies that dictate more regular updates (yeah, right). There is no guarantee however that Leet's vulnerability is ever covered in a security update. Many vulnerabilities go undetected by the vendor community for longer than 2 years.

  - **Payment processing**—Leet has risk exposure at the point of money transfer because these systems are designed to track back to an actual human. Additionally, if the premium SMS number utilized as part of the theft is identified as illegal, those funds could be seized and investigations could ensue. If the revenue base of the app itself is Click-Fraud for example, Leet might be frozen out by the advertising payment processor. While cyber currencies, like

Bitcoin, help with financial anonymity for the attacker, there is risk at the point of conversion from dollars (Euros, etc.) into Bitcoin.

- **How long will this income stream last?** This is impossible to forecast. Like other industries, Leet must grow carefully, so that it doesn't starve its assets. In this case though, uncontrolled, viral growth of Leet's app increases the probability of being discovered by some security mechanism. Being discovered means there is a possibility that both the app and the underlying exploit are neutralized and become ineffective. Since Leet purchased the exploit from someone, it is possible (even likely) that the exploit was sold to multiple people. Once researchers discover the vulnerability that feeds the exploit, all attempts to use that same exploit will be rendered inert. This could happen because one of the other app developers using the exploit is "discovered." In highly managed, security-aware IT systems, network operators can plug a known vulnerability in seconds, hours, or days. In systems with low security awareness, where there aren't even mechanisms to measure potential intrusion behavior, vulnerabilities can persist for many years after discovery. Leet would be inclined to work in industries where their likelihood of detection stays low and their income can last as long as possible.

- **How will Leet continue to fuel growth?** Leet will do what other enterprises do:

  - **Organic growth on its base of existing "customers"**—I'm using the term customer somewhat loosely to mean someone surreptitiously compromised and exploited for money and information. But, in the competition for customers, right now, Leet is in the lead on at least a few. Leet can use this foothold to push more "product." The first Evil Android App can install other apps, which can perform new methods of financial theft and information theft. In other words, Leet will try to increase its ARPU (Average Revenue per User).

  - **New customer acquisition**—Leet will also try to get as many customers as possible before the exploit is rendered inert. This will mean additional botnet time if necessary. Potentially at a meet-up, Leet will find someone that is running a new phishing approach, or a SMiShing approach, and they'll try a strategic partnership.

2.2.4 What about Nation-States?

A valid question about this analytical framework is that it doesn't consider state-sponsored actors who aren't necessarily looking for profit. While a foreign, hostile nation isn't undertaking these activities for an immediate cash flow return, they are looking for a return and they too have finite resources to invest. Should they build a $100m ship? Or a $100m suite of cyber tools? Nation-states are among the most active on the cyber arms markets, acquiring exploits technologies. Said another way, a foreign nation hacks to achieve the broader military, political, and economic objectives of the state. For example, the nation-state attack on the Ukrainian power grid was timed to coincide with other military activities in the country at the same time. This nation-state actor must preserve the value of their assets, ensuring a low probability of detection and probability of attribution. Otherwise, the political and/or military objectives of the attack will be forfeit and the response could be kinetic (bombs) for encroaching on the sovereignty of another country. So, cyber attacks tying into overall military strategic planning-think of it like a Gantt chart.

2.2.5 **Steps in a Successful Cyber Attack**

Against the backdrop of earnings, we can see that this enterprise is much more complex than just "the hack." An attacker needs to figure out which systems to get into, how to get into those systems, what they're looking for on those systems, and how to get the target information out, all without getting caught. Perhaps if the defender can interrupt any one aspect of a successful attack, the defender can diminish the assailant's economic prospects. The Kill Chain methodology, developed by Lockheed Martin, is one part of an overall intelligence-driven defense. The Kill Chain describes the following steps for a successful attack:

- Reconnaissance—Which systems inside of an organization are the target? What are the defenses of those systems? What types of hardware, operating systems, applications? Is our balance sheet aligned to support this target? Do we have the right exploitation assets? Can we get the assets in?

- Weaponization—Once the target systems and defensive techniques (firewalls, etc.) are known, the specific "hacks" are crafted that will leverage and/or circumvent the defenses. Building on our balance sheet analysis from before, the organization would need to see which exploits it has in its possession to compromise the target systems. To fill any gaps, it would have to do a MAKE/BUY analysis.

- Delivery—With the perfect weapons crafted, how do they get on target? As with firearms and other kinetic weapons, there are questions of size, handling procedures, and detection. Certain weaponized packages are small enough (under a megabyte) that the planner has many options for how to get it on target. Some packages are so large that they are best delivered via USB drive. Then the operating question becomes how to get a foreign USB drive inserted into the right computer.

- Exploitation—Now occurs "the hack." At this stage, the attacker pulls the trigger on the exploit code and gains privileged access to target systems. This would also be the point at which compromised credentials are used to gain unauthorized access to a system or facility.

- Installation—To ensure reliable access to the necessary levels of system privilege, attackers may choose to install a backdoor. This backdoor would allow continued access even if the exploit was patched or the victim credentials were changed.

- Command and Control—The attacker must have a plan for how to get instructions into target system, get feedback from the target system, and get data out of the target system.

- Action on Objective—All other steps, up to this point, are "investment." This step is "revenue." Here the attacker is using their newfound access to steal something (financial data, intellectual property), break something (centrifuges a la Stuxnet), or insert something (fake credentials).

2.2.6 **Industrialization of the Attack**

The last decade has seen an amazing market formation in the hacking enterprise space. The exact moniker for this self-organization phenomenon has numerous subtle variations (e.g., industrialization of cybercrime, industrialization of hacking), but the idea of industrialization remains constant. For our purposes, I believe the definition by Victoria Baines, in her August 2013 article "Fighting the Industrialization of Cyber Crime[2]" is most appropriate:

> *"The sophistication of this criminal business model is such that members of these networks are able to focus on specific tasks including producing malicious code or delivery mechanisms for attacks. There are even specialists who are dedicated to the generation of payment card authentication numbers and the recruitment of money mules, individuals who turn the proceeds of cybercrime into hard cash-sometimes without knowing that they are engaging in criminal activity."*

Our friends at Leet are beneficiaries of this industrialization. Trend Micro's Chief Cyber Security Officer, Tom Kellermann noted that prices for cyber weapons have dropped about 90% in the last decade.[3] Prices are being pushed down by more transparent, bid-style pricing and the short-lived nature of the weapons in certain market-segments. In Table 2.1, we can see the Leet didn't develop the exploit it used for its application, instead opting to pay an exploit developer $25k. The market for this buy/sell activity isn't quite as simple as Amazon or eBay, but several platforms are reasonably well documented:

- Silk Road—Many clock cycles before "Cyber" existed, the Silk Road referred to ancient trade routes connecting Europe to China and other parts of the Asian subcontinent. More recently, Silk Road, Silk Road 2.0, and Silk Road 3.0 are the names of a series of Darknet commerce platforms, selling everything from drugs, to weapons, to forged documents, to cyber weapons. Silk Road 3.0 is the currently operating form of the market, with earlier versions incarnations having been taken down by law enforcement agencies. To increase anonymity, these sites only allow payment in Bitcoin and can only be entered using a Tor-enabled browser. Tor is a protocol and network designed for network anonymity and privacy.

- Eastern European "Cyber Arms Bazaar"—The Cyber Arms Bazaar has operated for about 15 years through several Eastern European countries. Research suggests that, unlike the Silk Road franchise, the Cyber Arms Bazaar specializes in high-end cyber weaponry.

Another component in Leet's Kill Chain was distribution. Leet hired a botnet for distribution ($40k), accelerating circulation of the app. Services at this layer of the supply chain vary in cost, effectiveness, etc., but the focus is on getting something distributed in a way that breaks up attribution (makes Leet hard to find) and gets scale. Just as Leet invested in an app to generate fake SMS messages, others in the supply chain focus on building their networks of bots to rent them out on a per hour basis. Daniel Smith at RadWare published a well-researched piece in July 2016, titled "Malware and Botnet Attack Services Found on the Darknet[4]," characterizing many aspects of the players and pricing in this market.

Figure 2.1 shows a page from a Darknet publication, done in the style of the *Wall Street Journal*. This site provides Bitcoin market analysis and projections. It helps users find and objectively analyze the best Darknet market for their purposes.

---

[2] https://unchronicle.un.org/article/fighting-industrialization-cyber-crime.

[3] http://thehill.com/policy/cybersecurity/235726-feds-search-for-ways-to-impede-cyber-bazaar.

[4] https://blog.radware.com/security/2016/07/malware-and-botnet-attack-services-found-on-the-darknet/.

FIGURE 2.1   Screenshot from one Darknet market comparison page.

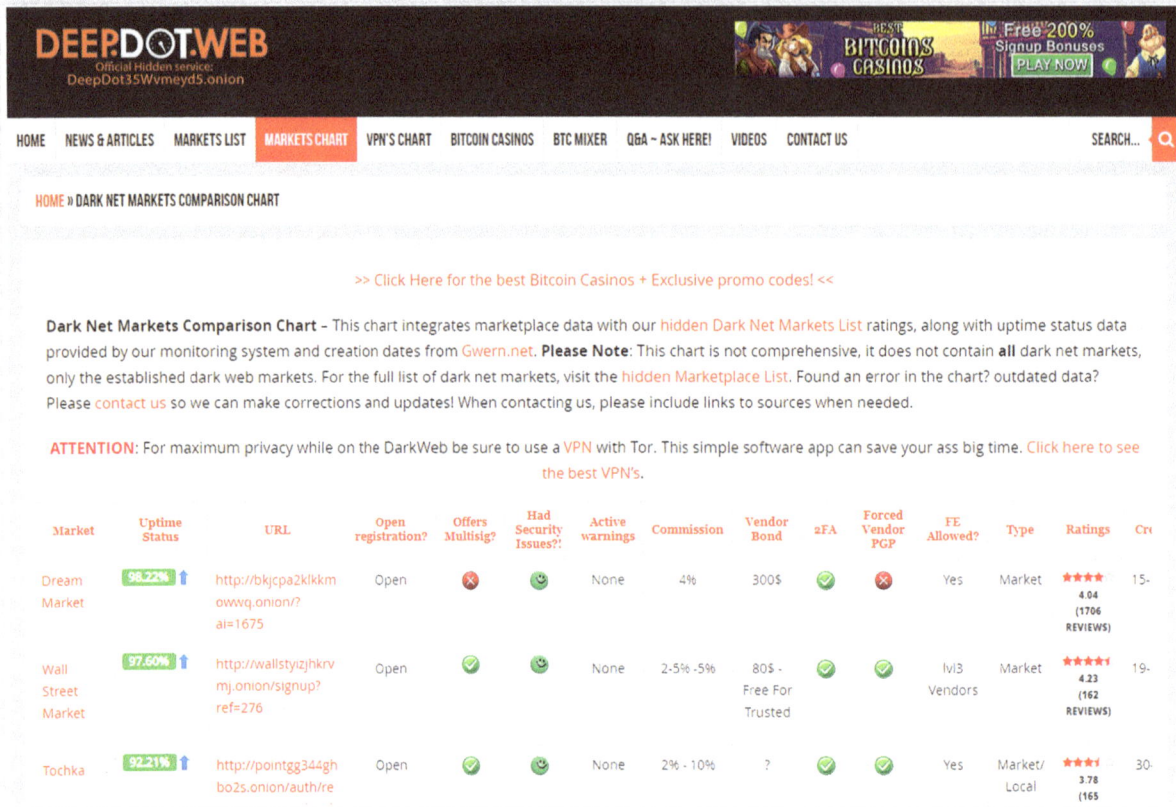

2.3 Hacker Enterprises and Assets Associated with Commercial Trucking

Now we will try to make the economic analysis more focused on the attack surface area associated with commercial trucking and fleets. In response to illicit customer desires to gain access to commercial fleets and cargo, this industrialized attack community will see a market opening and start creating exploits, distribution methods, and other aspects of the aforementioned Kill Chain. The indicators and warnings that this development is underway will be nearly impossible to detect, before a time of the attacker's choosing.

2.3.1 Exploitation Research

- Consumer car vehicle manipulation—Charlie Miller and Chris Valasek are two security researchers that have demonstrated weaponizable vulnerabilities against automobiles. Ranging from remote attacks (no vehicle access) to physical attacks (plug something into the vehicle's Controller Area Network, or CAN bus), these attacks have shown the ability to interfere with vehicle operations, such as steering, acceleration, and braking.[5] The researchers make clear that the software issues that allowed the attacks aren't isolated to one vehicle manufacturer. The vehicle manufacturer's response notes that the attack "require(s) extensive technical knowledge, extended periods of time to write code…." The issue though is that Evil Charlie and Evil Chris Doppelgangers do exist with requisite knowledge, extended periods of time, and will sell those wares on the aforementioned cyber weapons markets. Downstream buyers take the code from the market and adapt it to their own mission (surveillance, hijacking, etc.). While consumer cars are different than commercial trucks, there are many common elements, at an electronic level, under the hood.

- Commercial vehicle manipulation—At the *2016 USENIX Offensive Technologies Workshop*, University of Michigan Transportation Research Institute (UMTRI) demonstrated the ability to disable a brake system, manipulate the instrument panel, and cause unintended acceleration on a tractor trailer, via a laptop plugged into the J1939 bus. The researchers noted that, relative to the CAN bus in noncommercial cars, the J1939 is much more standardized across industrial vehicles. Under many circumstances, standardization is a good thing because it lowers cost and increases reliability. In a cyber security context, standardization means "hack once, attack many." At least some of the demonstrated attacks worked on both a tractor trailer and a school bus. Certainly, one perspective on this research is that it merely demonstrated how CAN J1939 works and does what it was designed to do. Unfortunately, our attackers won't spend much time debating these perspectives, and will instead see opportunity to misuse the system to their advantage.

- Telematics-based attacks—My own company, Kaprica Security, conducted substantial research into the vulnerabilities and exploitability of automotive telematics systems. Telematics systems are essential for modern trucking, because they increase route efficiency, fuel efficiency, and security. But the persistent connection between the vehicle and the Internet means that there is an opportunity for a remote attacker to compromise the telematics system and drive unintended consequences. In our work, we were able to manipulate the brakes, control panel, and several other aspects of the car, from anywhere in the world. This work was quickly scalable to many millions of cars and connected trucks on the road today.

2.3.2 Asset Development

- Telematics exploits—With near certainty, organized criminal elements are buying or stealing versions of the leading telematics systems in use on commercial vehicles. Those devices, once acquired, are being shipped to the smartest vulnerability researchers and exploit developers all over the world to find software weaknesses. Those weaknesses are then being tested for remote exploitability. A remote telematics exploit for a trucking system would be valuable to someone trying to coordinate hijacking, to surveil competitors, or as an entry point into manifest systems, invoicing, etc.

- GPS manipulation—Interfering with the GPS system in a truck would allow an attacker to arbitrarily steer an unknowing victim into whatever trap they wanted. Talk to Eileen on this. Also how easy it is to subvert GPS…

---

[5] http://www.darkreading.com/vulnerabilities---threats/this-time-miller-and-valasek-hack-the-jeep-at-speed/d/d-id/1326468.

- Infotainment exploits—Compromising the infotainment system would allow an attacker to, among other things, record all of the audio in the cab. On most days, the conversations in the cab would not be of particular note. But combine the audio recordings with big data audio analysis capabilities and you now have a powerful weapon. On some days, there might be subtle indiscretions of the driver that would allow for blackmail. On other days, there might be actionable insights into valuable shipments.

- Theft—For many years, hackers have used computer techniques to circumvent the vehicle's security mechanisms, simply in the name of vehicle theft.[6] Even if theft isn't the prime objective, circumvention of the physical security on the vehicle will allow an attacker direct access to the trucking lines' in-cab IT systems. This could be valuable to an attacker to get high-fidelity insights into cargo and cherry-pick the very best loads to steal.

2.3.3 Distribution Development

- Truck Stop Nodes—Truck stops do not exert a counter-cyber defensive posture. If I was trying to create a nationwide network that offered privileged access to the nation's cargo, I would start with a large network of very inexpensive RF surveillance devices deployed to truck stops. Given the lack of awareness, they could surveil for years, without detection. Combining the RF analysis with targeting exploit development, one would have a powerful basis for launching code into thousands of passing trucks from a system costing less than $100.

- In Vehicle Nodes for Air Gap Jumping—A common problem for attackers is how to get close to the really valuable systems. If, instead of running a network of nodes around truck stops, an attacker runs nodes in specific trucks, those nodes could help the attacker get surveillance data on places they couldn't otherwise access. These types of parasites on trucks would work for several reasons. First, the truck would provide plenty of power to keep the node alive. With enough power, the node could support numerous radios and surveillance techniques. Second, the truck is large, making a small computer relatively hard to find.

2.4 Potential Cyber Effects in Transportation

With more shared background, we can now return to the question of "Should we be paranoid?" This section is based solely on my own conjecture. Putting myself in the position of an attacker, with resources and the objective of "utilize the commercial transportation system to achieve my objectives, what would I do?"

- One headline topic when cyber mixes with transportation is that hackers are going to start crashing cars. Will cyber vulnerabilities be used to generate massive traffic jams and crash cars, randomly or in unison? In my opinion, No. The biggest reason I would say this won't happen is that transportation hackers (term of art for hackers that focus on transportation) are currently operating under the best cover of all. Public perception and senior executive decision makers don't believe transportation hackers are either real or, if real, are not a threat. **This lack of focus**

---

[6] http://www.forbes.com/sites/thomasbrewster/2016/03/21/audi-bmw-ford-thief-car-hacking/#191035135941.

gives transportation hackers amazing freedom of movement. A coordinated attack that made clear that cars can get hacked and people can die would quickly shine a very bright light onto that space. The attack vectors used to create the hack would be investigated and shut down.

- Another growing trend is file locking blackmail. For example, will someone disable a piece of the car until you pay? In my estimation, this could be tried, if the attacker patiently placed the bug in a wide swath of cars for a given make and executed all of them at once. In this case, the assets that allowed the attack to occur will be burned and rapidly be closed off. However, in this case, I think some would pay. In a coordinated attack scenario, the attacker will have undermined confidence in just going to the store and buying a replacement. And, by doing it to enough cars, they'll also cause a local disruption in the replacement market for those cars, making it temporarily more expensive to replace. This attack is hard to execute, but the blackmailed funds could be substantial. Additionally, the attack could be run against other vehicles, again and again. Even knowing that this is possible, other car manufacturers wouldn't suddenly be in a better position to prevent the same from happening to them. This approach would also allow an attacker to lock out all or part of a targeted commercial vehicle fleet, taking in blackmail or otherwise damaging a shipper.

- The vectors I see as being most likely are ones that will maximize the useful life of payloads and provide an aura of doubt around the source of the data. Currently, no one is looking "under the hood" to see if an attacker is there. The attackers would be wise to keep it that way and give the authorities no reason to look harder. In that vein, I see a few vectors as likely.

  - Location Surveillance—Running a nationwide surveillance network on truck locations would not be extremely difficult, and, once running, could provide value to the network manager in a variety of ways. They could offer intelligence data to hijackers. They could sell cyber-targeting data to whomever. If someone wanted to get a quick snapshot of the telematics system used by a particular trucking line, armored car carrier, or high-value commercial fleet, this information broker could provide it.

  - Audio Listening Post, Supporting Blackmail, and Policy Insights—An attacker using this vector could likely evade detection for a long time. They could leverage the value of the audio recordings gleaned from the cabin, with low probability that any one victim would make the connection to the source. The blackmail could consist of getting a targeted driver to visit a specific website on his/her in-cab laptop or the blackmailer would release a recording of phone calls with someone other than his/her spouse.

  - Network Intermediary, Distribution Point—The payload can get power.

In conclusion, in my opinion, the question of whether bad actors will use cyber effects to garner an unauthorized outcome in commercial trucking is as inevitable as gravity. Yes, it will happen and very likely is already quietly underway. If you assume that the lack of commercial-truck-hacking headlines mean you're safe, you're wrong. The challenge then becomes adoption of an audacious, alternate perspective about how someone would use your own strengths and systems against you to achieve their outcomes. **Don't dismiss it for its "ridiculousness," but consider it for its possible merits. Build empirical tests that rigorously demonstrate that you are not compromised. Only then, might your paranoia rightfully subside… for 15 min.**

About the Author

A helpless cybersecurity romantic at heart, Doug's professional mission is to transform the relationship between innovation and adoption so that newfound risks and vulnerabilities can be cauterized quickly, reliably, and inexpensively. Having previously built and sold a mobile security and device management company to Samsung, Doug attributes much of his career accomplishment to his actual superhero power of breaking down the barriers of fear that too often lead to cyber-induced paralysis. As RunSafe's CTO, Doug plays an essential role in showcasing how RunSafe's technology changes the economics of cyber defense, giving defenders a fighting chance to build the latest capabilities with confidence in their systems. A trained computer scientist, Doug started his career in the National Center for Supercomputing Applications at the University of Illinois before serving as a Russian Linguist and Interrogator in the US Army.

What Cybersecurity Standard Work Is Applicable to Commercial Vehicles?

Lisa Boran and Xin Ye

3.1 Background

In the past decades, we have seen increasing automation on vehicles. This starts with numerous driving assist technologies, such as lane keeping, adaptive cruise control, parking assist, trailer backing assist, etc. The automotive industry is currently embracing autonomous driving technologies. In the meantime, we are getting used to various connectivity technologies on vehicles. It is common to see vehicles with interfaces such as cellular, WiFi, and Bluetooth. Many vehicles are now able to communicate with other vehicles (known as the V2V technology) and public infrastructure (V2I). There are many benefits of increased automation and connectivity: vehicles are now easier and more fun to drive, deliver better energy efficiency, offer higher customer convenience, and may enhance public safety. Nevertheless, we should also realize that vehicles are now cyber-physical systems and part of the Internet of Things.

Although cybersecurity is a relatively new topic to the automotive industry, it has to be addressed appropriately because vulnerabilities in cybersecurity could impact safety. Fortunately, proactive thinking and design measures can be taken to address cybersecurity issues for the automotive industry.

3.2 **Standards and Information**

Various collaborative efforts have created helpful cybersecurity guidance documents and official ways to share information. The documents and information available are as follows: (This is not a comprehensive list)

- In May 2013, the United Kingdom released PAS 555 Cyber Security Risk Governance and Management. This was created to help an IT (information technology) organization understand and manage its exposure to cybersecurity threats. This has been updated in 2017.

- In January 2016, the first Automotive ISAC (Information Sharing Analysis Center) became fully operational—membership consisting of OEMs and automotive suppliers. This offers the ability to share cybersecurity incident information obtained globally between members for awareness and action.

- In January 2016, SAE (Society of Automotive Engineers) released SAE J3061 Vehicle Cybersecurity Best Practices. This document helps organizations build cybersecurity into their product development life cycle processes, from concept, design, development, production, operation, and maintenance to decommissioning. SAE and ISO have joined forces to further expand J3061 into a cybersecurity standard.

- SAE J3101 Recommended Practice for Hardware-Protected Security for Ground Vehicle applications will be going in for ballot in 2017. It defines the various available cybersecurity protections and hardware-protected security solutions and provides where application may be best suited.

- NIST (National Institute of Standards Technology) has published many information system cybersecurity recommended practices (e.g., cybersecurity and privacy frameworks). Most notable is the NIST Special Publication 800 Series. There are too many released documents to cover here in detail but all the documents have relevant information that can be applied towards any electronic development or organizational structure for inclusion.

- International Standards Organization (ISO) is also providing some cybersecurity guidance direction. They published ISO 27000 Information Security Management Systems, ISO 15408 Common Criteria for Information Security Technology and ISO 26262 for Road Vehicle Functional Safety which has synergies with cybersecurity.

- The Federal Information Processing Standards (FIPS) organization has published many well-known cryptographic requirements and validation documents and requirements for information system technology.

- In July 2016, the Auto ISAC in collaboration with the Alliance and Global Automakers built a framework to create and maintain best practices for motor vehicle cybersecurity. The best practices cover organizational and technical aspects of vehicle cybersecurity, including governance, risk management, security by design, threat detection, incident response, training, and collaboration with appropriate third parties.

- In October 2016, NHTSA (National Highway and Traffic Safety Administration) issued their own set of Cybersecurity Best Practices for Modern Vehicles. It addresses, from a government perspective, cybersecurity issues for all individuals and organizations manufacturing and designing vehicle systems and software. The presented voluntary best practices aim at enhancing vehicle cybersecurity to mitigate cyber threats that could present unreasonable safety risks to the public or compromise sensitive information.

- There are several public vulnerability databases available for reference (Common Weakness Enumeration, Common Vulnerability Enumeration, BugTrac, National Vulnerability Database, etc.) that may apply to your industry and/or products.

- There are many different threat analysis and risk assessment methodologies available for use (Microsoft STRIDE, EVITA, HEAVENS, STPA, OCTAVE, TVRA, DREAD, etc.) that could be applied to new feature/system/module designs.

3.3 SAE/ISO Cybersecurity Standard Development

At the beginning of 2016, we saw the official release of the pioneer work of SAE J3061 *Surface Vehicle Recommended Practices -- Cybersecurity Guidebook for Cyber-Physical Vehicle Systems*. This is the first industry-wide work that provides a cybersecurity process framework and guidance to help organizations to integrate cybersecurity into their business. It includes four major areas of work: (1) defining a complete life cycle process framework that can be tailored and utilized within each organization's development processes to incorporate cybersecurity into cyber-physical vehicle systems from concept through production, operation, service, and decommissioning; (2) providing information about some common existing tools and methods used when designing, verifying and validating cyber-physical vehicle systems; (3) providing basic guiding principles on cybersecurity for vehicle systems; and (4) providing the foundation for further standards development activities in-vehicle cybersecurity.

Cybersecurity guidance should be applicable to all individuals and organizations manufacturing and designing vehicle systems (electrical and electronic (E/E) systems, their interfaces, and communications). The automotive industry has been working diligently on this issue for a number of years, and it is important that all stakeholders consider their role in ensuring a more secure vehicle ecosystem.

The scope of the joint SAE/ISO effort is to expand on the content started within SAE J3061 and develop it to become a Cybersecurity Standard. It will specify requirements for cybersecurity risk management and operation throughout engineering (concept, design, and development), production, operation, maintenance, and decommissioning. It will allow stakeholders to use a common language for communicating and managing cybersecurity risk.

The intent is to provide cybersecurity process requirements around the product life cycle to ensure cybersecurity is built in from the beginning. It will help minimize risks, thereby benefiting stakeholders while fostering needed trust and confidence in the vehicle-owner community, regulatory bodies, and policy makers.

The focus is to harden the electronic architecture against potential attacks to ensure systems take appropriate and safe actions even when an attack is successful. A layered approach to vehicle cybersecurity reduces the probability of an attack's success and mitigates the ramifications of a potential unauthorized access (wording used from NHTSA Cybersecurity Best Practices for Modern Vehicles).

To build a comprehensive and systematic approach to developing layered cybersecurity protections is to follow the NIST Cybersecurity Framework which is structured around five risk management principle functions: "Identify, Protect, Detect, Respond, and Recover."

The design approach should

- Be built around risk-based prioritized identification of threats and protection of safety and cybersecurity critical systems, intellectual property (IP), and personally identifiable information (PII).
- Provide for timely detection and response to potential vehicle cybersecurity incidents in the field.
- Design in methods and measures to facilitate rapid recovery from incidents when they occur.
- Institutionalize methods for accelerated adoption of lessons learned.
- Promote and incorporate published cybersecurity best practices, when applicable.

Expansion in the areas of testing methods and risk management will be the significant emphasis in the standard. Global experts from OEMs, tier 1 suppliers, cybersecurity suppliers, and academic research are participating in the development of this new standard. The standard will be sectioned out into four main categories (risk management, product development, other postproduction and organizational process, and overall introduction and definition) with several pertinent industry informational reports generated for awareness.

The standard will incorporate requirements around risk management, secure design, and operational excellence within an organization. **Figure 3.1** depicts how management of cybersecurity occurs through the product development life cycle.

3.3.1 Secure Design

- Identify threats and associated risks of those threats occurring (severity and likelihood) to module, system, architecture, service, and decommission development.
- Create security goals based off threats identified.
- Perform benchmarking to ensure you meet customer intent and determine industry trends
- Develop security requirements to ensure security goals are met. When developing the security requirements, the normal engineering process should be followed to help ensure the requirements are as complete as possible, taking into account benchmarking, researching public vulnerability databases, following quality robustness initiatives, applying security controls, and meeting regulatory and insurance requirements. **Figure 3.2** shows the cybersecurity analysis steps required in order to develop security requirements that address identified threats and met cybersecurity goals.

FIGURE 3.1   Overall cybersecurity engineering framework, SAE J3061.

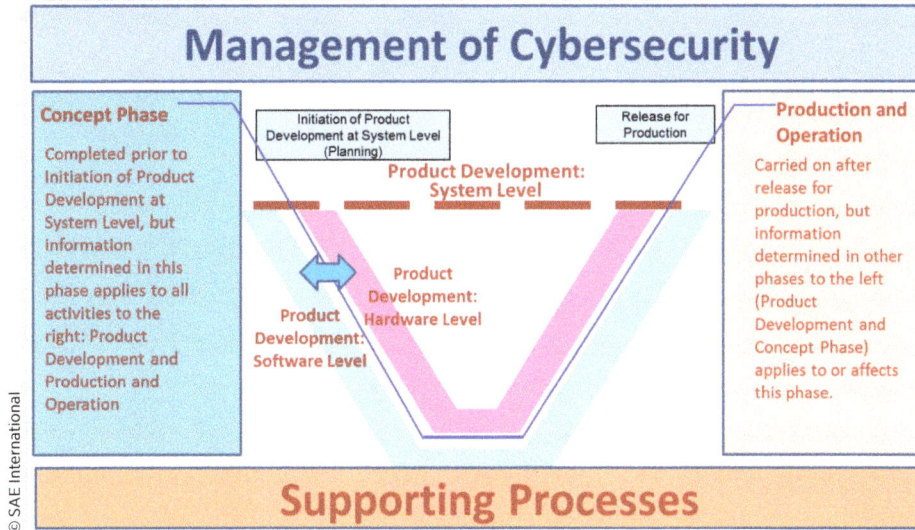

Management of Cybersecurity

Concept Phase

Completed prior to Initiation of Product Development at System Level, but information determined in this phase applies to all activities to the right: Product Development and Production and Operation

Initiation of Product Development at System Level (Planning)

Product Development: System Level

Product Development: Hardware Level

Product Development: Software Level

Product Development: Software Level

Release for Production

Production and Operation

Carried on after release for production, but information determined in other phases to the left (Product Development and Concept Phase) applies to or affects this phase.

Supporting Processes

© SAE International

FIGURE 3.2   Cybersecurity requirement development, SAE J3061.

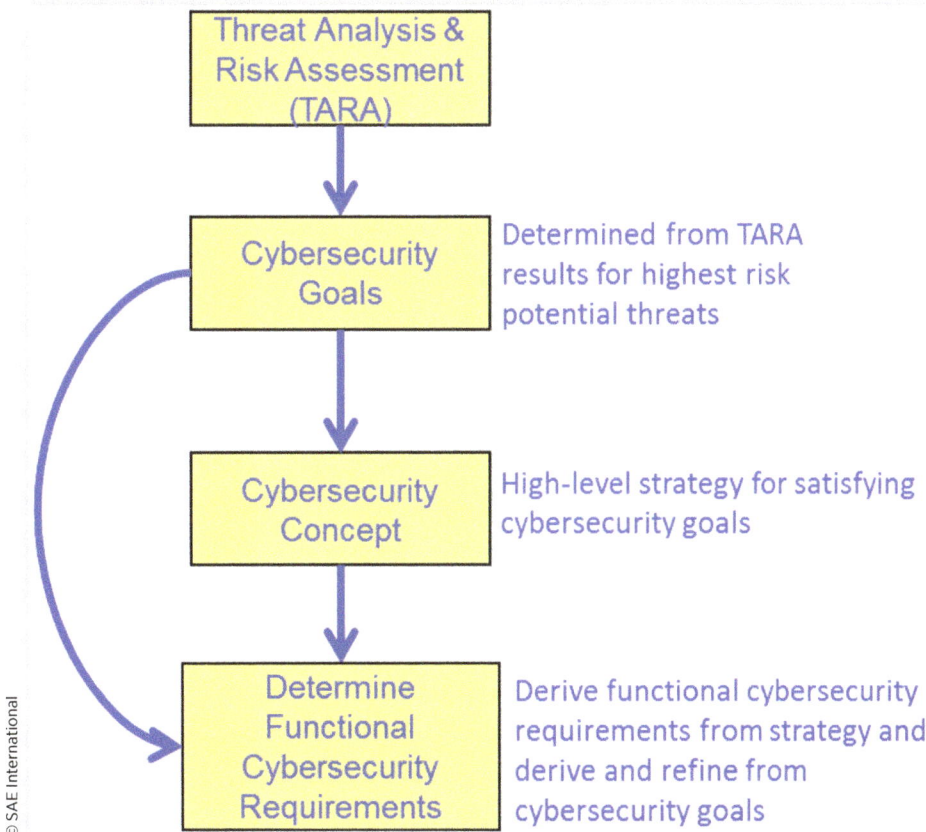

Threat Analysis & Risk Assessment (TARA)

Cybersecurity Goals — Determined from TARA results for highest risk potential threats

Cybersecurity Concept — High-level strategy for satisfying cybersecurity goals

Determine Functional Cybersecurity Requirements — Derive functional cybersecurity requirements from strategy and derive and refine from cybersecurity goals

© SAE International

- Verification and validation testing to ensure the design is doing what was intended.
- Hold Technical Design Reviews—To ask questions to ensure the product went through robust process, meets the cybersecurity requirements, has been validated, and complies with regulations or insurance requirements, if applicable.
- Consider what data should be logged and transported to backend IT database for analysis and evidence tracking.
  - Identify critical systems/modules/architectures that should be penetration tested. This testing will identify any vulnerabilities found in the implementation that may impact safety, security, IP or PII. This may include fuzz testing (random and out-of-the-box inputs) to help find misbehavior.
- Perform security audits both at component and vehicle level.
- Capture requirements and cascade internally to all PD teams, if and when applicable
- Document all security process/action steps taken (threat analysis and risk assessment, verification and validation testing, FMEA, penetration testing, security audits, etc.) for self-certification and due diligence purposes.

Figure 3.3 shows the secure design steps necessary during development.

FIGURE 3.3 Product development of a system, SAE J3061.

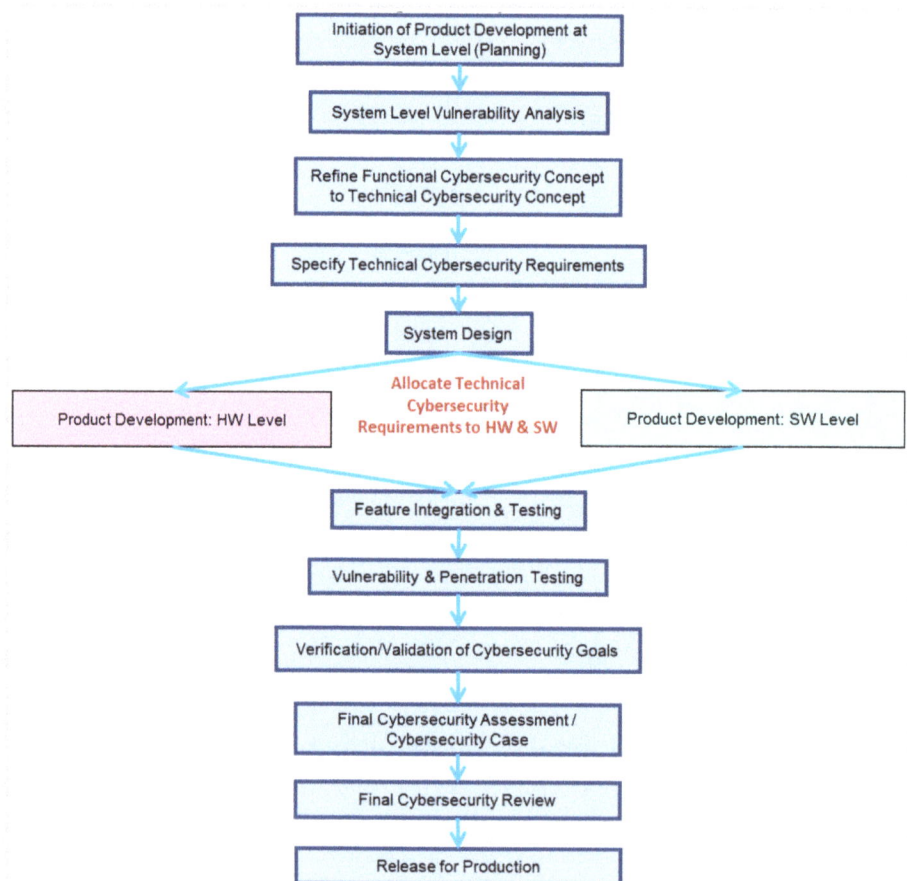

Initiation of Product Development at System Level (Planning)

System Level Vulnerability Analysis

Refine Functional Cybersecurity Concept to Technical Cybersecurity Concept

Specify Technical Cybersecurity Requirements

System Design

Allocate Technical Cybersecurity Requirements to HW & SW

Product Development: HW Level

Product Development: SW Level

Feature Integration & Testing

Vulnerability & Penetration Testing

Verification/Validation of Cybersecurity Goals

Final Cybersecurity Assessment / Cybersecurity Case

Final Cybersecurity Review

Release for Production

© SAE International

3.3.2 Organizational Structure

An organization should have a well-established security structure, process, and strategy in place; see **Figure 3.4** for high-level overview. It is necessary to build security into all aspects of the business to ensure it is applied and designed in upfront. It is necessary to drive a cybersecurity culture into the company and ensure the appropriate teams and processes are in place to run effectively.

- Vulnerability Management

  - Field Monitoring—Provide a process and avenues to obtain information and awareness. (Law enforcement, insurance, Auto ISAC, media, supply chain, warranty, customer hotlines, etc.)

  - Incident response plan—An organization shall have a documented process for responding to incidents, vulnerabilities, and exploits. This process should cover impact assessment, containment, recovery, remediation actions, and the associated testing.

  - Responsible Disclosure Policy—A clear public disclosure communication providing guidance, policy requirements, and avenues to where and how to report vulnerabilities to an organization.

  - Lessons Learned—Capture and create incident reports that are logged. Security requirement development cascaded globally within product development for conscious lessons learned direction adherence.

- Security Governance

  - Formalized governance will support aligned decision making against agreed strategy, consistent adherence to policy, global, and cross-functional visibility.

  - Internal process integration into product development process with specific cybersecurity deliverables at specific Gateway Milestone Reviews.

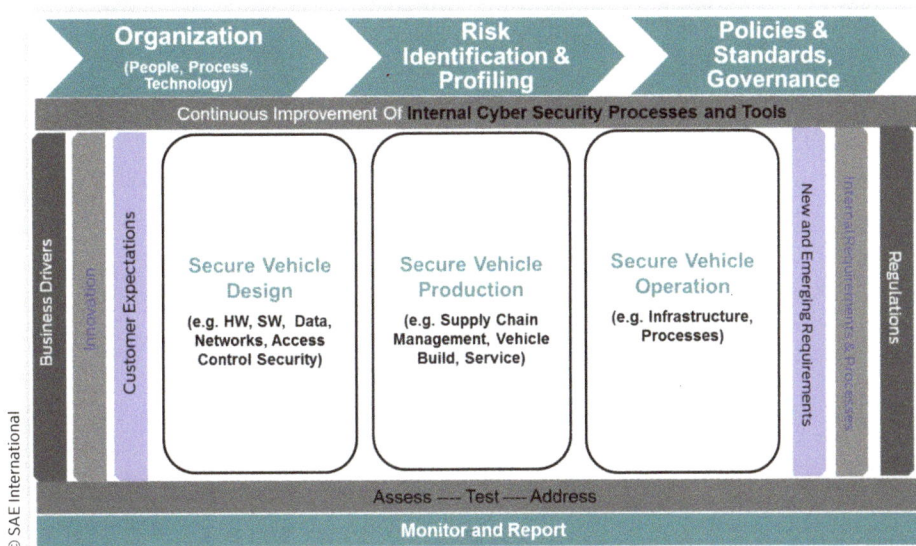

FIGURE 3.4 Overall cybersecurity organizational framework.

- Security Infrastructure (Ensure sufficient headcount)
  - Security team
  - Threat modeling team
  - Red team
  - Regulatory
  - Legal
  - Communications—Public and government affairs
  - Management Structure
  - Evidence tracking and data monitoring team
- Audits
  - External company audit and assessment of security competence around people, products, and process
  - Compliance with security regulation and insurance requirements
- Training
  - Product development engineer training for cybersecurity awareness of threats and risks
  - Product development engineer training to ensure they understand the cybersecurity process deliverables required of them
  - Security team training for skills development, tool usage, and keeping up with cybersecurity trends
- Industry, Government, and Research Consortiums
  - Ensure security team is engaged in these consortiums to stay up to date and influence. design solutions for common industry issues. Effective solutions require a multi-pronged approach and more minds.
- Cross-Functional Communication
  - Ensure there are communication paths between internal functional safety and cybersecurity processes.
  - Ensure cybersecurity plans and processes are communicated globally within a company.
  - Internal communication process identified. Know when escalation should occur and who to do it with.
  - External communication process identified. Know who and why communication is necessary with different parties (law enforcement, media, supply chain, insurance, Auto ISAC, NHTSA, etc.).
- Standards and Best Practices
  - Always be aware of industry best practices and standards that could be applied.

3.4 **Conclusions**

Our vehicles have been and will continue to be integrating all variants of technologies and features to embrace the ever-growing connectivity and mobility. Automotive cybersecurity shall be planned and implemented into every stage of the vehicle life cycle as an enabler to enhanced public safety and security. One should realize that there is no golden key that solves all problems in the field of security. This is also true for the automotive world. In fact, the situation is exacerbated by complexity and long life in cars. As a consequence, an incident response plan is necessary. We as a society should also capture and cascade all lessons learned.

We should adopt a defense-in-depth approach due to the variety of attack surfaces. This includes a foremost mission of identifying threats, a combination of multiple protection mechanisms to reduce vulnerabilities as well as onboarding comprehensive verification and validation of functionality. The layered approach can significantly reduce the attack vectors and makes the remaining hacking more difficult and more expensive.

We advocate applying aforementioned cybersecurity standards and best practices across the product life cycle. We believe an essential factor to the success is to build a cybersecurity mindset within an organization. Only in this way can we improve cybersecurity in automotive design, supply chain (manufacturing/management), production, operation, and decommissioning.

About the Authors

Lisa Boran is a Vehicle Cybersecurity Manager at Ford Motor Company focused on defining and instituting security functions within product development embedded systems and processes (both cybersecurity and physical security). Lisa has both a bachelor and master's degree in Electrical Engineering. Lisa has 28 years at Ford with experience in a wide variety of roles but the majority of her time at Ford has been in the security area. External to Ford, she is on the US ESCAR program committee, advisory board for the SANS Automotive Security Summit, editorial board for the Transportation Cybersecurity and Privacy Journal, SAE Motor Vehicle Council Advisory Group for Transformative Technologies, and she chairs both an SAE (J3061) and joint ISO/SAE Cybersecurity Committee (ISO/SAE 21434) to develop a Vehicle Cybersecurity System Engineering standard. She is also a member of many joint industry research project or information sharing programs with the government, academia, OEMs, and suppliers.

Xin Ye is a Vehicle Cybersecurity Technical Expert at Ford Motor Company where his major role is to provide security architecture and design for in-vehicle systems and network communications. He earned his Ph.D. in Electrical and Computer Engineering at Worcester Polytechnic Institute with a specialization in applied cryptography and embedded system security. External to Ford, he has served as a program committee member for Embedded Systems in CARs (ESCAR 2017 and 2018) and ACM Cyber-Physical System Security (CPSS 2018). He has published multiple research articles in various conference proceedings. He is also one of the authors for the draft authentication service in unified diagnostics service (UDS), which is intended to be included in the next iteration of ISO 14229-1.

CHAPTER 4

Commercial Vehicle vs. Automotive Cybersecurity: Commonalities and Differences

André Weimerskirch, Steffen Becker, and Bill Hass

4.1 Introduction

Automotive cybersecurity is becoming increasingly important, especially in the age of information and connectivity as vehicles' capabilities are becoming more and more connected. While automotive cybersecurity has been around since the 1990s as part of theft protection, odometer manipulation, and chip tuning, it recently gained a new level of attention as cybersecurity breaches have been shown to impact the safety of passenger and commercial vehicles. Over the last 5 years, it has been repeatedly demonstrated that it is possible to hack into passenger vehicles to modify driving behavior or to locate and steal the vehicles. Such attacks have been demonstrated by research teams that hacked built-in vehicle interfaces, and that hacked into OBD2 (on-board diagnostics) dongles with wireless capabilities that are plugged into vehicles. Then in 2015 the first research about hacking commercial vehicles was released to show that commercial vehicles are prone to similar vulnerabilities and can potentially be compromised.

Over the last years, many processes and technologies were developed for automotive cybersecurity. This chapter will provide a background of passenger vehicle cybersecurity and commercial vehicle cybersecurity and describe what the commonalities are, how automotive cybersecurity solutions can be applied for commercial vehicles, and where there are limitations and differences. The chapter at hand will focus on technical solutions only, and we refer to Chapter 3 for an overview of the equally important cybersecurity engineering process. The scope for commercial vehicles in this chapter is comprised of medium-duty and heavy-duty trucks and off-road heavy-duty vehicles, which also includes agricultural and construction vehicles and possibly military vehicles.

4.2 **Background**

In the past 15 years, research on automotive cybersecurity has shown a multitude of vulnerabilities affecting safety-critical features such as steering, braking, and acceleration as well as security features such as unlocking and starting a vehicle. Several sophisticated attacks combine these vulnerabilities with remote exploits leading to a full attack chain, where an attacker can manipulate the driving behavior of a vehicle remotely. Furthermore, in the past 2 years, findings in the area of commercial vehicle cybersecurity indicate the potential for comparable attacks. The most important publications and their results are discussed in this section.

In 2002, a study by the National Highway Traffic Safety Administration (NHTSA) revealed that around 3.5% of all passenger vehicles are affected by odometer manipulations during their lifetime, costing their purchasers an estimated amount of 1 billion US dollars every year in the United States alone [1]. Also, the authors state that fleet vehicles are eminently affected because they typically accumulate a high mileage in a short period of time which makes odometer rollback more believable to a potential buyer. While there are increasing efforts on the regulatory side to prevent odometer fraud, technical solutions to cryptographically secure odometer readings are still not widely implemented. Therefore, many detailed manuals available on the Internet continue to enable criminals to conduct odometer manipulations on a variety of vehicles. This is a prime example of illegal cybersecurity activities due to a financial gain.

Researchers from the University of Washington and the University of California, San Diego were the first to demonstrate safety-critical automotive vulnerabilities in 2010 by sending malicious in-vehicle network messages, so-called Controller Area Network (CAN) messages, to the in-vehicle network of a modern automobile [2]. Assuming direct access to the in-vehicle network (i.e., through the on-board diagnostics (OBD) port), they could control the speedometer and instrument cluster display, and, even more worrisome, they could stop the engine from running and disable or individually control the brakes. Then, in their subsequent work in 2011, the authors conducted an experimental analysis of potential attacks via remote communication interfaces [3]. In that work, they gained full access to the in-vehicle network of a modern automobile utilizing three different remote attack vectors: the CD player via a manipulated mp3 file, the telematics unit via Bluetooth, and the telematics unit via the cellular network. Besides the potential injection of malicious CAN messages, they pointed out location tracking, theft, and wiretapping as other arising threats.

In 2014, private researchers developed several attacks on the in-vehicle network of two modern cars, affecting the dash display, the steering behavior, and braking and acceleration [4]. Contrary to former publications, the authors explained their methods in detail, including source code and information on the hardware, software, and payloads used.

One year later, the same authors published and publicized the first, complete, end-to-end compromise of a passenger vehicle. The results of their work caused the recall of 1.4 million vehicles [5]. A port left open by the cellular network provider allowed access to a service used for inter-process communication in the vehicle's head unit over the Internet. From there, a chip could be reprogrammed to send out arbitrary network messages. In a demonstration for the press, the researchers controlled both the climate control and the radio systems overriding the driver's inputs, turned on the windshield wipers, and eventually cut the transmission causing the vehicle to decelerate on a highway [6].

In 2015, a popular aftermarket insurance dongle, which connects directly to the in-vehicle network, was examined by researchers from the University of California, San Diego [7]. A minor discovery by the researchers revealed a common cryptographic key and administrative username and password among all devices they tested, enabling secure login after a device was discovered. However, the devices they studied were indirectly protected from such remote compromises because the cellular carrier implemented network address translation (NAT) which maps Internet addresses to local network addresses, hence preventing the devices from being addressable outside of the local network. Their major discovery revealed the ability to issue remote updates from an arbitrary server using cellular text messages SMS. With the SMS attack vector, they demonstrated the dongle's concrete susceptibility to remote attacks by demonstrating its discovery on the cellular network, issuing a remote update from their rogue server, and a targeted compromise to send arbitrary in-vehicle network messages that controlled both the wipers and brakes.

A private research group demonstrated vulnerabilities of an automotive vehicle using more advanced networking technologies beyond CAN in 2016 [8]. Utilizing a series of exploitable bugs, they accessed the car's infotainment system via the Wi-Fi network and remotely activated the brakes of a moving vehicle. The OEM was informed and fixed the underlying vulnerabilities before the exploits were disclosed to the public.

Other crucial yet not safety-critical cybersecurity research in the automotive domain includes several key fob and immobilizer hacks. An immobilizer is an anti-theft device that prevents the engine from starting when the corresponding transponder (i.e., key fob) is not in range. In 2010, researchers from ETH Zürich used techniques to emulate or extend the range of the key fob transponder using relay attacks to enter and start vehicles from over 50 m away, non-line-of-sight [9]. In 2013, it was shown that over 100 models from 26 car manufacturers were susceptible to so-called digital lockpicking when researchers published that they completely reverse engineered the most widely deployed immobilizer at the time. The design of the particular immobilizer studied made use of a weak proprietary stream cipher, suffered from weak entropy of their secret keys, and did not have memory protected from being written remotely. Consequently, the researchers could unlock and start the vehicles they studied. Other key fob and immobilizer attacks are summarized in [10]. The authors disclosed different vulnerabilities in keyless entry systems of major vehicle manufacturers affecting millions of vehicles worldwide. The found weaknesses enabled the researchers to gain unauthorized access to the affected cars. These studies depict an obvious potential for car theft if exploited by criminals, and unsurprisingly there have been many reports of car thefts where the attackers enter, start, and drive away with a vehicle that was previously locked by the owner. It was repeatedly shown that tools to undermine key fob security can be put together at low cost, e.g., as reported in [11].

Recently, the first academic papers on commercial vehicle cybersecurity were published. In 2014, researchers from the University of Tulsa checked various electronic control units (ECUs) for their cryptographic methods to establish forensic soundness and integrity of collected incident data, often used to support the reconstruction of accidents [12]. See the University of Tulsa paper in Chapter 7. They concluded that many implementations allowed the modification of incident data, and they proposed the deployment of cryptographic primitives such as hash functions as a solution. They created a test bench for truck ECUs so that they could perform tests and conduct their analysis without needing a large, expensive heavy-duty truck. This idea resulted in their subsequent work in 2016 where they aimed to build a prototype of a remotely accessible test bed for experimentation with security features [13]. The researchers added real ECUs and nodes simulating sensor inputs required by the controllers to accurately emulate a real truck network. To demonstrate the capabilities of their test bed, they conducted two sample experiments: Generating a look-up table (LUT) for a 16-bit challenge-response protocol and testing of an intrusion detection system.

Additionally, in 2016, researchers from the University of Michigan performed an experimental security analysis of a real heavy-duty truck and passenger bus vehicle network [14]. In their work, the authors assumed physical connection to the vehicle bus through the on-board diagnostics connector, and they discovered that similar attacks against automobiles in 2010 were directly applicable to commercial vehicles. A minor discovery let them gain full control over the instrument cluster to let the gauges display arbitrary values. Their major discovery identified a single message in the SAE J1939 CAN standard that affected both the truck and bus. The message could be utilized by a malicious party to deactivate the acceleration pedal, max out the current gear by raising the RPM, prevent the truck from accelerating by lowering the RPM, and turn off the engine brake at speeds under 30 mph. Besides the potential injection of malicious CAN messages, they provided evidence that all vehicles utilizing J1939 would be affected and that the prevalence of third-party telematics units that plug into vehicle networks provides a unique threat to the commercial vehicle industry.

4.3 The Automotive and Commercial Vehicle Environment

The automotive and commercial vehicle environment are similar in many aspects and widely different in other areas. The following paragraphs compare the passenger vehicle and commercial vehicle space.

4.3.1 Supply Chain

In the passenger vehicle space, the OEM is the owner of security and designs the vehicle's security concept. The OEM then specifies requirements for all suppliers, and the OEM is the sole integrator of all suppliers' components. There is only little variation for each vehicle model, and often components are shared between vehicle models.

In the commercial vehicle space, there are a huge number of variations per model. For instance, buyers can typically choose a particular engine, transmission, and even control modules. Furthermore, OEMs often don't build the entire truck, but they provide the chassis to a bodybuilder, and the bodybuilder then adds mechanical and electrical components for specialized functions, e.g., as they are needed for a garbage truck. The role of aftermarket equipment that directly plugs into the communication bus (CAN) is also much larger in the commercial vehicle space. For instance, a majority of trucks are equipped with aftermarket fleet management solutions.

The supply chain model dictates that passenger vehicle architectures have a single security owner, namely, the OEM and that the system can be closed up. Whereas commercial vehicles have several security owners, including the OEM and bodybuilder, hence requiring more open security architectures.

4.3.2 In-Vehicle Network Architecture and Communication

Related to the supply chain models is the in-vehicle network architecture. Both passenger vehicles and commercial vehicles heavily use the CAN and communication gateways to separate networks. Passenger vehicles are about to introduce Ethernet networks, and it's only a matter of time until commercial vehicles will also widely introduce Ethernet.

A difference between passenger vehicle and commercial vehicle network architecture is that commercial vehicles are typically longer on the road and hence older, leading to network architectures in commercial vehicles that do not offer network separation and that appear to be outdated. Another difference is that commercial vehicle network architectures must be open for bodybuilder integration, whereas no such open interfaces need to exist in the passenger vehicle domain. Note that there is often an exterior CAN access, e.g., between tractor and trailer, where an adversary might hook in, whereas such exterior access points are typically not available in the passenger vehicle space.

A majority of vehicles use CAN. While passenger vehicles use standardized CAN on the physical level with standardized packet structure, the semantics of CAN packets is proprietary. This even applies to vehicle models from an OEM that use different semantics. Commercial vehicles use the standardized SAE J1939 protocol that specifies a common CAN semantics. While an adversary needs to reverse engineer CAN messages for a passenger vehicle to cause an impact, the CAN message semantics for commercial vehicles is public and applies to almost all commercial vehicles.

4.3.3 Telematics

Telematics is widespread in the commercial vehicle space, and a survey of a small sample size indicates that more than 90% of all trucks have remote communication systems installed, and a part of those systems directly integrate to the vehicle's electronics [15]. Many of these telematics solutions come as part of aftermarket fleet management solutions. While telematics is also becoming increasingly popular in the automotive space, commercial vehicle fleet operators are typically willing to make larger investments into telematics since it will enable cost savings. Especially aftermarket telematics systems are susceptible to security vulnerabilities, as CERT has demonstrated [16].

4.3.4 Maintenance and Diagnostics

Passenger vehicles are typically serviced by dealerships and workshops that use original carmaker tools. Commercial vehicles are either serviced by dealerships and workshops or they are serviced in-house by trained technicians that utilize OEM tools. The distinction is less between passenger vehicles and commercial vehicles than it is between individually owned vehicles and fleet-owned vehicles in both categories. Fleet-owned vehicles need to be serviced in an open manner by any technician, possibly with original OEM tools, e.g., to swap mechanical and electronic components. The ability of everyone to service vehicles is legally mandated by the Right to Repair Act in the United States which applies in similar form also to most European countries. Note that this makes cybersecurity in the passenger vehicle and commercial vehicle space very different than cybersecurity in other domains since it requires a certain system openness and accessibility.

Diagnostics commands are used to troubleshoot vehicles and to change the configuration of electronic components, or even update the firmware of electronic components. Diagnostic commands were also shown to be vulnerable to hacker attacks, e.g., to turn off an electronics component or enter modes that are only for a standing vehicle but that are dangerous for a vehicle in motion. It is possible to protect diagnostics commands against adversaries by introducing proper plausibility checks in the vehicle (e.g., not allow potentially dangerous test operations while the vehicle is moving), by securing the diagnostics tools properly, or by requiring a secure online connection of the diagnostics tool to the OEM's server.

4.3.5 Emerging Technologies

It is widely believed that many emerging technologies, such as advanced driver assistance systems (ADAS), and automation will first be widely deployed in commercial vehicles. ADAS is being widely deployed in passenger vehicles and NHTSA includes automatic emergency braking (AEB) as a recommended safety technology in its 5-star rating system. These technologies can save cost due to a reduced accident rate or reduced rate of operation, hence commercial fleet operators are willing to invest significant amounts per vehicle in emerging technologies. At the same time it appears that the automotive cybersecurity community had a head start to the commercial vehicle cybersecurity community, hence exposing emerging technologies to commercial vehicle security concepts that are not as mature as security concepts from the automotive domain.

4.4 **Vehicle Threats and the Cyber Attacker**

Commercial vehicle and automotive platforms have been developed to meet the needs of a nearly disjoint set of customers, with the former being highly diverse and specialized, and the latter generally focused on individual transportation. Still, despite the differences in their requirements, the two have shared various technologies as they have evolved leading to many similarities in offensive techniques. This section describes the threat model, attackers, and offensive techniques to consider in the commercial and automotive cybersecurity settings.

4.4.1 **An Evolving Threat Model**

When the CAN was designed in the 1980s, a vehicle's communication network was assumed to be isolated from the rest of the world. The threat model focused on criminals that would break in and hot-wire a vehicle or somehow physically access and tamper with a vehicle's components. So vehicle defenses both in commercial and automotive sectors relied on the hardening of physical attack surfaces to reduce the threat posed by an adversary, and the CAN had no built-in security measures because it was protected by physical defenses.

Throughout the 1990s and 2000s, the threat model assumptions did not change, but the vehicles did. In automotive markets, consumers demanded more electronic features they grew accustomed to from consumer electronics. The automotive OEMs also saw potential to differentiate and provide safer vehicles for their customers through the use of electronics. On the other hand, while commercial vehicle technologies lagged behind their automotive counterparts in connected driver convenience, the cost-saving benefits of integrated fleet management systems accelerated the adoption of GPS and cellular network connected vehicle components.

Then, as vehicles in both areas became more electrified and eventually wireless, the threat model shifted from completely physical to cyber-physical. What was once an isolated, insecure CAN is now connected to the outside world.

Today, the threat model assumptions have drastically changed and expanded. Beginning in the United States in model year 1996, light-duty passenger vehicles required OBD for emissions testing which introduced an easily accessible standard connector located within the cabin of the vehicle. The OBD connector is meant to give access to particular emissions-related ECUs such as the engine, but it is common for OEMs to wire OBD directly to the internal CAN buses with other ECUs because it enables convenient vehicle diagnostics for mechanics at a low cost.

Similarly for commercial vehicles since 2010 [17], highway heavy-duty vehicles require OBD for emissions, and OEMs utilize the standardized port for other functions on the CAN related to diagnostics. Because the OBD port is located within the cabin of the vehicle, it is protected by a layer of physical security which includes door locks and alarms, but a proper threat model does not make this assumption.

On the automotive side, there are new and used car buyers, valets, employees (e.g., taxi or delivery service), mechanics, renters, and ride-sharing services, while on the commercial vehicle side there are passengers, employees (e.g., truck or bus drivers), mechanics, and renters who gain access to the inside of the cabin with varying degrees of access time. It has been demonstrated that even with a short amount of access time, sometimes referred to as a "lunch-time" attack in the IT community and "valet" attack in the automotive community, a malicious component can be plugged into the vehicle's OBD port either temporarily or left there to compromise safety critical functions.

Furthermore, in both vehicle environments, drivers or vehicle owners willingly plug third-party electronics into their OBD port for some benefit at the cost of an increased attack surface or may even bridge networks that OEMs originally intended to be segregated. Many devices exist with different benefits: vehicle monitors from insurance companies; fleet telematics systems; aftermarket backup cameras, GPS monitors, crash detectors, and performance monitors; and generic interface devices that provide wireless input and output to the OBD port from a smartphone or other connected device.

Some notable examples have led to the remote compromise of a vehicle. One such example is an automotive insurance company that provided lower insurance rates for drivers that plugged a wireless dongle into their OBD port, but that device was later shown to enable remote CAN injection by attackers [7]. Another example is a third-party commercial vehicle fleet telematics system with cellular and GPS capabilities that was used in hundreds of commercial vehicles, but it was installed with an open telnet connection accessible on the Internet from across the globe [18]. In these two instances, a key difference with commercial vehicles is that fleet owners don't really have much of a choice when it comes to the use of telematics systems if they want to be competitive and increase safety for their drivers, and similarly drivers need to comply with their company policies.

Finally, an important difference between automotive and commercial vehicle physical attack surfaces is that with commercial vehicles there are external network entry points for things such as trailers, container ship hook-ups [19], or specialized mechanical equipment attachments that need to be considered as potential avenues for attack.

Beyond physical network entry points, more and more ECUs are being integrated into safety critical systems, so there is a need to ensure the modules themselves are secured. Starting at a vehicle's infancy, the confidentiality of cryptographic keys and integrity of ECU software must be maintained from the early stages in the supply chain, to the time that ECUs are installed in a vehicle, and throughout the 10+ year life of a vehicle. Additionally, various wireless components are now added standard in passenger vehicles and often added to commercial vehicles in different industries. Therefore, wireless attacks against the vehicle must be considered from the short range (up to 10 m, e.g., Bluetooth), medium range (up to 1,000 m, e.g., Wi-Fi and vehicle-to-vehicle dedicated short-range communication), and long range (beyond 1,000 m, e.g., cellular and GPS). Furthermore, with today's semiautonomous features already deployed, the attack surface has grown to include vision, radar, and LiDAR sensors.

4.4.2 **The Adversary**

It is common knowledge that the CAN protocol is insecure, and there are many low-cost software and hardware tools available that let anyone communicate on CAN. Furthermore, the commonplace reliance on electronics by governments, corporations, and individuals has spawned an arms race between white hat and black hat cybersecurity professionals resulting in the development and release of techniques that can compromise a broad range of general electronic systems. Such techniques can either be directly applied or adapted in such a way to be used against both commercial and automotive vehicle systems.

We can imagine various potential attackers on the automotive and commercial vehicle infrastructures. On the one hand, there are sophisticated, large-scale attackers such as criminal groups developing for the black market or nation-state sponsored groups developing cyber weapons. These attackers possess almost unlimited resources, are able to engineer and deploy different exploits at scale, and penetrate dealerships, mechanics, or supply lines. On the other hand, there are freelance or rogue attackers with varying levels of privilege and expertise. These attackers have much less resources and a lesser extent of their influence, but could obtain tools from the black market. While the motivation for either type of adversary consists of the threat potential, which comes with damaging a single vehicle, causing a death, paralyzing the freight industry, or damaging critical infrastructure, it is generally understood that the most common motivation is financial.

Whether large scale or small scale, an adversary will get access to a similar vehicle to the one they are targeting for an extended time to learn the function, features, and attack vectors. Additionally, due to the privileged features that diagnostics sessions enable, an adversary will obtain diagnostics tools for feature exploration, reverse engineering, and exploitation. A difference here between commercial and automotive environments is that a typical commercial vehicle is an order of magnitude costlier than an average consumer automobile which raises the barrier of entry of a would-be attacker. Nonetheless, tools developed by a more motivated or well-funded attacker can be sold for much less, and exploits against the J1939 standard found in commercial vehicles would work across vehicles and industries amortizing the cost of developing an attack. See Chapter 2, "Should We Be Paranoid," for more information.

4.4.3 **Offensive Techniques**

In both automotive and commercial vehicle industries, the approach of an attacker depends on their goals, and their goals will fall into one of several categories that make up a complete attack chain: remote exploit, CAN access, diagnostic session initiation, and module reflashing. Each stage of the attack chain can be developed or purchased independently and assembled piecewise as required.

For remote exploits, there are many different types of devices and attack vectors to consider. In many cases, third-party devices run a variation of Linux which has its own set of exploits from the IT domain that could carry over. These devices are much less expensive than a vehicle so many can be purchased for hardware reverse engineering with low overhead. Through hardware reverse engineering, an attacker could determine things such as the computer architecture, types and manufacturers of chips, and find debug pins to gain access to memory. The computer architecture provides a map for an attacker, and the types of chips give valuable information about embedded features and whether there might be a hardware exploit already available. With a memory dump, an adversary could perform static analysis to determine control flows or read cryptographic keys and passwords.

An attacker can also use the administrative tools and regular use case tools marketed to customers such as web portals and smartphone apps to probe for security vulnerabilities. For example, a phishing attack could be crafted with knowledge of the admin tool web portal to attempt to gain privileged access to fleets, or a smartphone app could expose remote vulnerabilities on the back-end services. Additionally, an attacker could use a low-cost software defined radio and other wireless capture devices to sniff and transmit packets for analysis of the remote channels.

The techniques just described for exploiting a wireless interface are borrowed from the traditional cybersecurity domain which has had a long past of vulnerabilities. Recent attacks against Android, iOS, and embedded Internet of Things devices are a good indicator that no wireless interface is 100% secure. If a remote exploit is found that enables arbitrary code execution and the device has a CAN interface, it's likely the attacker will be able to perform arbitrary CAN packet injection remotely. The same principle holds for vehicles whose networks use some underlying protocol other than CAN. We will focus on CAN-based network attacks due to the popularity and prevalence of CAN in both commercial and automotive vehicles.

With arbitrary CAN packet injection, an attacker needs to understand the topology of the network in order to craft a targeted attack. Under the assumption that an attacker has access to their own similar vehicle, network topology is determined through physical observation. Furthermore, a CAN injection attack can be developed apart from a wireless exploit by simply clipping a low-cost CAN tool into the same bus or buses as the remote device that will be exploited. Some vehicle architectures utilize a single CAN bus which makes it easy for an attacker to influence all of the ECUs. Others use several buses, but bridge them together using a wireless gateway which can still enable an attacker to influence all of the ECUs once the wireless gateway is compromised. Yet others use several buses and segregate wireless devices from the rest of the ECUs with a bridge or gateway. Just as in the traditional IT domain, network segregation can greatly limit the potential damage caused by the security failure of a single component.

Once attached to the network, a first step might be to flood the CAN bus with traffic to perform a denial of service. In most cases, vehicles are engineered to handle a loss of CAN communication or a flooded CAN bus, but this type of attack can put a vehicle into a "limp" mode that decreases performance proving to be hazardous if at high speed. A next step would be to record network traffic during various driving conditions and while utilizing features that interface with safety critical or physical security components. An attacker could then try replaying messages that were recorded to see if there is any component susceptible to replay attacks. In [14] this method found a message capable of controlling engine RPM, preventing the driver from using the accelerator, and degrading engine brake performance in commercial vehicles. Another technique, termed fuzzing, uses randomized permutations of data to send over the network. This type of attack can expose corner cases where modules behave erratically or aid in reverse engineering packet structure. In [2], it was discovered through fuzzing that the brakes could be engaged or prevented from being engaged while their car was in motion. More complex CAN interactions can be reverse engineered to influence safety critical systems. For example, through ECU impersonation, a series of commands sent from the engine to the transmission that cause the transmission to shift could be sent from the exploited module instead. Or a handshake between a telematics unit and body controller could be spoofed to issue a door-unlock or start-engine command.

As mentioned earlier, a reasonable, motivated attacker would obtain a diagnostics tool for reverse engineering diagnostic commands. Diagnostics tools are a major area of interest for attackers because they enable special features that affect safety critical components such as brakes, engine, transmission, and body controllers, and also allow ECU parameters to be configured or memory to be read and written to. Typically, diagnostics commands are protected by a layer of security that only allows authorized mechanics to perform the diagnostics functions. An attacker might first explore the diagnostics software user interface and attempt to exercise all the possible diagnostics commands while collecting network traffic. Then, the attacker can try to replay the diagnostics commands to see if they are vulnerable to simple replay attacks. Next, the attacker might perform analysis on the diagnostics software that interfaces with the diagnostics tool to see if cryptographic keys can be easily obtained or if there are any weaknesses in the back-end services.

Then, depending on the attacker's skills, they might try to perform hardware reverse engineering on the target ECU, perform a memory dump using the diagnostics tool, and analyze the firmware for exploits, or perform network protocol analysis on the diagnostics commands to measure the security level protecting the commands. Many authentication protocols used to secure these functionalities have been shown to be insecure due to flawed algorithms, insufficient key lengths, vulnerable implementations, or no security whatsoever. In one case, a single CAN message could be replayed to reset the master password protecting the diagnostics session for a commercial vehicle ECU. In another commercial vehicle instance, only 16 bits were used in a challenge-response protocol, so [13] generated a 16-bit LUT and were able to bypass the security to access diagnostics commands. Similarities appear in the automotive environment. In [2], they found a hard-coded, static challenge and response and instances where the handshake was ignored altogether.

If the security of diagnostics commands can be successfully bypassed, it oftentimes lets an attacker flash code onto an ECU. With this ability, it is still necessary for the attacker to reverse engineer the update process and firmware already running on the ECU. Once that is complete, however, an attacker could write arbitrary software that does several things then erases itself after being executed making forensics very difficult. Additionally, this type of software could be used for spying, as a backdoor for later exploits, or to bridge internal vehicle networks as part of a link in the attack chain.

4.5 Cybersecurity Approaches and Solutions

There are a variety of cybersecurity approaches and solutions available for the commercial vehicle domain. It appears that the majority of solutions come from other domains, in particular the automotive domain. However, some adjustments will be necessary to account for the unique features of the commercial vehicle space.

4.5.1 Legacy Vehicles

It is very hard, if not impossible, to improve the cybersecurity features of commercial vehicles that are already on the road today. A promising approach to stop the most likely attack scenario is to either not connect vehicles with known vulnerabilities to the Internet via a fleet management solution, and to turn off all wireless interfaces, or to include a filter between all external communication interfaces and the vehicle's electronics network.

The following paragraphs describe cybersecurity solutions that need to be planned for and included in the vehicle's electronics architecture.

4.5.2 Network Architectures and Separation

Modern network architectures separate networks by gateways and implement firewalls and filters between network segments. Such networks are typically separated in a way that no network with external interfaces directly connects to a safety-critical network, such as the powertrain network, but that all communication is routed through a gateway with a firewall first. There might be dedicated networks that only connect two nodes, e.g., between an ADAS controller and a camera. Note that even in the case of commercial vehicles, such dedicated private networks do not need to implement the SAE J1939 standard since there is no need to extend such network segments at any time.

The automotive space already moves towards the next architecture of domain controllers, where there are only a few powerful domain controllers, typically connected by Ethernet, and each domain controller connects to basic electronics components. The same paradigm of network separation for security purposes applies for the domain controller architecture.

4.5.3 Secure On-Board Communication

A major security objective is to introduce security of the on-board communication. In almost all cases, authentication is the required security objective, but in a few cases also encryption is interesting. In the following, we focus on authenticated on-board communication. Such authenticated on-board communication makes it much harder for an adversary to advance further. For instance, if an adversary successfully compromised a telematics unit or an OBD dongle, the attacker is only able to authenticate messages to components on a predefined list but, if properly designed, not to authenticate messages sent to any safety-critical component.

The passenger vehicle space uses a variety of on-board communication networks, including CAN, CAN-FD, Ethernet, and FlexRay. For low-bandwidth communication bus like CAN, which has a payload of up to 8 bytes per packet, introducing authentication is difficult. Nonetheless, AutoSAR specified a standard for CAN authentication called secure on-board communication (SecOC) [20], and Volkswagen announced that they will start deploying authenticated CAN [21]. It appears though that different carmakers go different paths and start introducing proprietary solutions for CAN authentication. A communication bus with more bandwidth, such as Ethernet or CAN-FD, allows to use either standardized or straightforward solutions, such as IPsec for Ethernet and a regular Message Authentication Code (MAC) for CAN-FD.

The SAE J1939 communication standard for commercial vehicle is based on CAN; hence authenticated CAN approaches such as the AutoSAR standard could be applied as well. Any security standard for commercial vehicles would need to be standardized though. The hard problem behind the standardization is not necessarily the format of the CAN messages but the underlying key management. While this applies to both the automotive and commercial vehicle world, the key management problem is significantly harder in the commercial vehicle world. For passenger vehicles, the OEM is the sole owner of security and is able to specify and require a particular key management scheme, and the OEM is able to control all suppliers accordingly. In the commercial vehicle space, a proper key management system needs to be both open enough to account for the supply chain model, but also restrictive enough to avoid that illegitimate parties can learn secret keys. This appears to be not as much a technical problem as an organizational issue, and any solution will reduce the openness of the supply chain model since there needs to be a trusted party that controls which components are able to communicate on the SAE J1939 communication bus. As of today, no such solution is available; however, the authors believe that it should be a high priority to introduce authentication in SAE J1939 and mandate the use thereof.

4.5.4 Secure Computing Platform

Electronic components in vehicles require a secure computing platform. A secure computing platform uses mechanisms such as secure boot to protect integrity at start-up, hypervisor or software containers with secure inter-process communication to separate software components, and a hardened kernel as well as memory protection (e.g., address randomization) to protect the kernel. Mechanisms that allow to monitor the integrity of the computing platform are available as well, e.g., by using a white list of binary files that are allowed to be executed. Note that the white list includes integrity information about those binaries, e.g., by hashing binaries before executing them. It appears that security solutions of secure computing platforms are equal for the automotive and commercial vehicle domain.

4.5.5 Anomaly Monitoring

Protecting a vehicle against cybersecurity attacks and extraction of data is a priority. The ability to monitor the vehicle's network and computing platforms is a second priority to understand potential breaches and continuously improve cybersecurity. The output of on-board network anomaly detection systems (ADS) and ECU platform integrity monitors can be merged; used for local anomaly detection, possibly leading to a basic local reaction; and then provided via telematics to a Security Operations Center (SOC), which then runs analytics over all data to find anomaly patterns in the big picture. This information is then provided to the carmakers and suppliers, who then analyze the information and start the incident response process. Note that a real-time reaction is unlikely, but a reaction that eventually updates firmware or configuration in the vehicle within days or weeks is realistic.

Network ADS monitor the on-board communication, such as CAN, CAN-FD, and Ethernet, and compare the message traffic to a learned traffic pattern. Such an ADS is ideally installed in a central gateway to monitor a variety of network segments. It is possible to connect an ADS to a message filter that discards messages and raises an alarm; however, due to the real-time and safety relevance of messages, this scenario is unlikely. A far more realistic scenario is that ADS is only used for monitoring in order to provide input to the SOC. An ECU monitor is basically an extension to a secure computing platform that cannot only stop an attack but also report the attack.

It appears that an anomaly monitoring system is the same for passenger vehicles and commercial vehicles. In fact, since the commercial vehicle space uses standardized CAN messages, it might be easier to realize ADS in the commercial vehicle space.

4.5.6 Security Operations Center

The SOC receives reports from the network ADS and ECU platform integrity monitors of vehicles. The SOC then takes these reports, analyzes them, and finds relationships between reported incidents. Automated algorithms find certain patterns, and then human analysts explore details to understand whether filtered incidents relate to cybersecurity incidents, other anomalies, or false alarms. Once an actual cybersecurity event is confirmed and the mechanisms are understood, the underlying system can be fixed and software can be updated via firmware over-the-air (FOTA) mechanism. Naturally, the more effective an SOC works, the more reports from different vehicles it receives.

It appears that SOC for passenger vehicles will work differently to an SOC for commercial vehicles. Passenger vehicles are mainly closed systems, so the SOC could be under the control of individual carmakers. Some synergies might be possible if carmakers collaborate and combine their SOCs; however, this doesn't seem to be necessary. Commercial vehicles are open systems and it appears quite necessary to combine the expertise of the OEM, suppliers, fleet operators, and aftermarket solution providers in a single SOC to be successful. Of course, with the landscape constantly changing, commercial vehicles could become more closed systems, hence approaching the model of the automotive manufacturers. Note that today no such SOCs are deployed, and more research must be performed and experience has to be gained to better understand the use of SOCs in the ground vehicle transportation sector.

4.5.7 Secure Firmware Over the Air

FOTA is used to update features and fix bugs, some of which might be security relevant. FOTA has the potential to reduce the number of recalls, and hence it is a widely discussed topic. There are numerous examples though when such FOTA mechanisms have been hacked in the PC network world. A good example is the Flame malware that used a combination of cryptographic vulnerabilities to then hijack the Windows update mechanism to distribute malware [22].

Software updates are typically protected by digitally signing the updated firmware such that an ECU will only update with the received firmware if the verification is successful. The code is signed by the OEM and/or supplier, ideally using a secure computer that is offline and that is not easily accessible. A secure communication channel, such as the widely deployed transport layer security (TLS) between cloud and vehicle, can further increase the security level. The Uptane framework for secure FOTA further increases the security and separates roles of the involved entities such that if a single server is compromised, the overall system can still recover [23].

4.6 Gaps and Conclusions

Recent research results demonstrated that cybersecurity is not only relevant for passenger vehicles but also for commercial vehicles. With the rapidly increasing push to deploy automation in commercial vehicles, and the expectation that heavy truck fleets might deploy fully automated vehicles first, cybersecurity issues will become increasingly important.

The passenger vehicle and commercial vehicle domains can learn a lot from each other. In fact, the Auto-ISAC which was established by passenger vehicle manufacturers was just opened up to include commercial vehicle stakeholders [24]. It is expected that many of the cybersecurity engineering processes and technical solutions can be utilized by the commercial vehicle domain. The ideas of threat analysis, risk assessment, secure development, security testing, and incident response apply to both the passenger vehicle and commercial vehicle industry. Also, technical solutions such as secure FOTA, network anomaly detection, and platform security can be applied in a similar manner as technologies are shared among industries.

However, there are also areas that require different solutions due to the differences of the passenger vehicle and commercial vehicle industry. A good example is secure CAN. Car manufacturers are the security owners of their vehicles and can specify proprietary protocols to protect their in-vehicle networks and, more importantly, key management systems. In the commercial vehicle domain on the other hand, strict standards are required to bring together all the various stakeholders in order to overcome the obstacle that commercial vehicles are an open system. In the long term, passenger vehicle and commercial vehicle industries will likely come together in such areas as well since passenger vehicles will, with increased resources such as offered by automotive Ethernet, also switch to standardized technical solutions.

References

1. National Highway Traffic Safety Administration, "Preliminary Report: The Incidence Rate of Odometer Fraud," Washington, DC, April 2002, available at https://crashstats.nhtsa.dot.gov/Api/Public/ViewPublication/809441.

2. Koscher, K., Czeskis, A., Roesner, F., Patel, S. et al., "Experimental Security Analysis of a Modern Automobile," *IEEE Symposium on Security and Privacy*, Berkeley, CA, 2010, available at http://www.autosec.org/pubs/cars-oakland2010.pdf.

3. Checkoway, S., McCoy, D., Kantor, B., Anderson, D. et al., "Comprehensive Experimental Analyses of Automotive Attack Surfaces," *Proceedings of USENIX Security 2011*, San Francisco, CA, 2011, available at http://www.autosec.org/pubs/cars-usenixsec2011.pdf.

4. Valasek, C. and Miller, C., "Adventures in Automotive Networks and Control Units," Technical White Paper, IOActive, 2014, available at https://ioactive.com/pdfs/IOActive_Adventures_in_Automotive_Networks_and_Control_Units.pdf.

5. Miller, C. and Valasek, C., "Remote Exploitation of an Unaltered Passenger Vehicle," 2015, available at http://illmatics.com/Remote%20Car%20Hacking.pdf.

6. "Hackers Remotely Kill a Jeep on the Highway—With Me in It," *Wired Magazine*, 2015, available at https://www.wired.com/2015/07/hackers-remotely-kill-jeep-highway.

7. Foster, I., Prudhomme, A., Koscher, K., and Savage, S., "Fast and Vulnerable: A Story of Telematic Failures," *WOOT'15 Proceedings of the 9th USENIX Conference on Offensive Technologies*, Berkeley, CA, 2015, available at http://www.autosec.org/pubs/woot-foster.pdf.

8. Keen Security Lab, "Car Hacking Research: Remote Attack Tesla Motors," 2016, available at http://keenlab.tencent.com/en/2016/09/19/Keen-Security-Lab-of-Tencent-Car-Hacking-Research-Remote-Attack-to-Tesla-Cars/.

9. Francillon, A., Danev, B., and Capkun, S., "Relay Attacks on Passive Keyless Entry and Start Systems in Modern Cars," *NDSS Symposium 2010*, San Diego, CA, 2010, available at http://www.syssec.ethz.ch/content/dam/ethz/special-interest/infk/inst-infsec/system-security-group-dam/research/spot/332.pdf.

10. Garcia, F.D., Oswald, D., Kasper, T., and Pavlidès, P., "Lock It and Still Lose It—On the (In)Security of Automotive Remote Keyless Entry Systems," *25th {USENIX} Security Symposium*, Austin, TX, 2016, available at https://www.usenix.org/system/files/conference/usenixsecurity16/sec16_paper_garcia.pdf.

11. "Just a Pair of these $11 Radio Gadgets can Steal a Car," *Wired Magazine*, 2017, available at https://www.wired.com/2017/04/just-pair-11-radio-gadgets-can-steal-car.

12. Johnson, J., Daily, J., and Kongs, A., "On the Digital Forensics of Heavy Truck Electronic Control Modules," *SAE Int. J. Commer. Veh.* 7, no. 1 (2014): 72-88, doi:10.4271/2014-01-0495, available at http://papers.sae.org/2014-01-0495/.

13. Daily, J., Gamble, R., Moffitt, S., Raines, C., "Towards a Cyber Assurance Testbed for Heavy Vehicle Electronic Controls," *SAE Int. J. Commer. Veh.* 9, no. 2 (2016): 339-349, doi:10.4271/2016-01-8142, available at http://papers.sae.org/2016-01-8142/.

14. Burakova, Y., Hass, B., Millar, L., and Weimerskirch, A., "Truck Hacking: An Experimental Analysis of the SAE J1939 Standard," *WOOT'16 Proceedings of the 10th USENIX Conference on Offensive Technologies*, Austin, TX, 2016, available at https://www.usenix.org/system/files/conference/woot16/woot16-paper-burakova.pdf.

15. National Motor Freight Traffic Association, "A Survey of Heavy Vehicle Cyber Security," Washington, DC, September 21, 2015.

16. Klinedinst, D. and King, C., "On Board Diagnostics: Risks and Vulnerabilities of the Connected Vehicle," SEI Whitepaper, 2016, available at https://resources.sei.cmu.edu/asset_files/WhitePaper/2016_019_001_453877.pdf.

17. Environmental Protection Agency, Code of Federal Regulations (Annual Edition), Title 40, Volume 18, Chapter I, Subchapter C, Part 86, Subpart A, Sections 86.010-18 and 86.007-17, 2010, available at https://www.gpo.gov/fdsys/pkg/CFR-2010-title40-vol18/pdf/CFR-2010-title40-vol18-part86-subpartA.pdf.

18. Norte, J.C., "Hacking Industrial Vehicles from the Internet," March 6, 2016, available at http://jcarlosnorte.com/security/2016/03/06/hacking-tachographs-from-the-internets.html, accessed April 30, 2016.

19. Ruiz-Garcia, L., Barreiro, P., Rodriguez-Bermejo, J., and Robla, J.I., "Review, Monitoring the Intermodal, Refrigerated Transport of Fruit Using Sensor Networks," *Spanish Journal of Agricultural Research* 5, no. 2 (2007): 142-156, available at http://citeseerx.ist.psu.edu/viewdoc/download?doi=10.1.1.978.6151&rep=rep1&type=pdf.

20. Werner, P., Happel, A., Fritz, R., and Keul, S., "AUTOSAR Security Modules," *escar USA Conference*, Detroit, MI, 2015, available at https://vector.com/portal/medien/solutions_for/Security/AUTOSAR_Security_Modules_Lecture_ESCAR_2015.pdf.

21. Tschache, A., "Vehicle Security from the OEM Perspective: Securing In-Vehicle Communication—Challenges and Workable Solutions," *escar Asia Conference*, Asia, 2016.

22. Goodin, D., "Crypto Breakthrough Shows Flame was Designed by World-Class Scientists," Ars Technica, 6/7/2012, available at https://arstechnica.com/security/2012/06/flame-crypto-breakthrough/.

23. Kuppusamy, T.K., Brown, A., Awwad, S., McCoy, D. et al., "Uptane: Securing Software Updates for Automobiles," *escar Europe Conference*, Europe, 2016.

24. PR Newswire, "Commercial Vehicles to Join Auto-ISAC," 2017, available at http://www.prnewswire.com/news-releases/commercial-vehicles-to-join-auto-isac-300396844.html.

25. Verdult, R., Garcia, F.D., and Ege, B., "Dismantling Megamos Crypto: Wirelessly Lockpicking a Vehicle Immobilizer," *Proceedings of 22nd USENIX Security Symposium*, Washington, DC, 2013, available at https://www.usenix.org/sites/default/files/sec15_supplement.pdf.

About the Authors

Steffen Becker received the B. Sc. and M. Sc. degrees in IT Security from the University of Bochum. During his graduate studies he spent two semesters as a research assistant and exchange student at the Department for Electrical and Computer Engineering, Purdue University, and was a visiting scholar at the transportation cybersecurity group, University of Michigan Transportation Research Institute (UMTRI). Steffen is currently pursuing the Dr.-Ing. degree with the embedded security group at the University of Bochum, under the supervision of Prof. C. Paar. His research interests include automotive cybersecurity, as well as hardware Trojans and obfuscation.

Bill Hass received the M. Sc. Eng. degree in Computer Science and Engineering from the University of Michigan in 2017. During his graduate studies, he was a research assistant in the cybersecurity group at the University of Michigan Transportation Research Institute (UMTRI) where he developed a key management system for an automotive OEM, coauthored a security research paper on the commercial vehicle network standard, SAE J1939, and was involved in automotive intrusion detection system testing. Prior to graduate school, Bill received the B. Sc. Eng. degree in Electrical Engineering from the University of Michigan in 2013 and was a product development engineer at Ford Motor Company for 2 years.

Dr. André Weimerskirch is Vice President of Cyber Security at Lear Corporation. Before that, André established the transportation cybersecurity group at the University of Michigan Transportation Research Institute (UMTRI), and cofounded the embedded systems security company ESCRYPT which was sold to Bosch in 2012. André is active in all areas of automotive and transportation cybersecurity and privacy. He is a main designer of the vehicle-to-vehicle security system, which will likely be the largest security system ever deployed, published numerous articles in the area of automotive and embedded cybersecurity, and is cofounder of the American workshop on embedded security in cars (escar USA). André is involved in various standardization, cooperation, and research efforts. André can be reached at aweimerskirch@live.com.

CHAPTER 5

Engineering for Vehicle Cybersecurity

Daniel DiMase, Zachary A. Collier, John A. Chandy, Bronn Pav, Kenneth Heffner, and Steve Walters

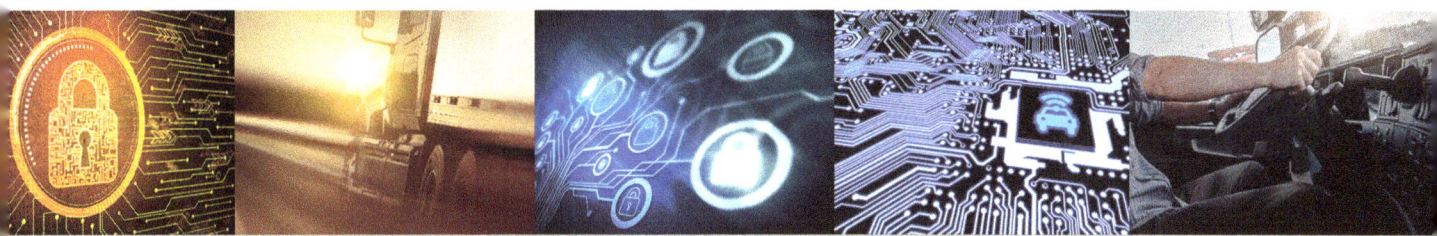

5.1 Introduction

Since inception, automotive vehicles evolved from being purely mechanical conveyance machines to highly integrated cyber-physical systems (CPS) with electronic features designed to improve safety, reliability, and comfort for consumers. Imposed government regulations and the consumer's desire for a wide range of sensing and communications features add so much complexity that computing systems are required to operate and repair or test the vehicle. For example, after 1996, all light-duty vehicles and trucks sold in the United States were required to have an on-board diagnostics (OBD-II) port for emissions testing purposes and to facilitate diagnostics for vehicle maintenance and repair [1].

More recent vehicle designs embed electronic control units (ECUs) responsible for core vehicle functions as well as for convenience and entertainment. In 2009, a typical vehicle had about 50 ECUs. By 2014, that number grew to as much as 100 ECUs [2]. A common bus links different ECUs within the vehicle to transfer necessary data to the OBD-II and enable other value-added features. A central bus system reduces extensive physical wiring that would otherwise be needed to interconnect functions. The most commonly used bus in vehicles is the controller area network (CAN), which was first developed in 1985. Other in-vehicle communications networks used in vehicles include Ethernet, FlexRay, Local Interconnect Network, Media Oriented Systems Transport, and SAE J1850. While enabling many valuable features, the OBD-II introduces unintended safety vulnerability by enabling access to safety-critical systems that can be hacked through physical access to the vehicle's connected OBD-II-ECU communications network.

Vehicles have also evolved with the introduction of telematics systems. Telematics systems blend the vehicle's ECUs and wireless telecommunications technology for both long- and short-range wireless connections. Telematics systems include additional convenience and functionality for vehicles such as hands-free Bluetooth connectivity for cell phones, navigation systems, and wireless access to services (e.g., concierge services, live person driving directions) and features (e.g., automatic crash response, theft response, virtual unlock). The number of vehicle models with Internet access grew from less than 1% in 2011 to over 53% in 2016 [3]. Vehicles connected to the Internet are expected to increase over time, with many consumers expecting models to include the option as a standard offering. Many vehicle manufacturers offer telematics services.

The introduction of vehicle telematics systems and services adds additional CPS security concerns through remote access. Vulnerable telematics systems can be hacked virtually and, in some cases, from anywhere in the world. Researchers have demonstrated that vulnerable telematics systems can be hacked to control a number of vehicle functions, including safety-critical systems such as braking. For example, researchers were able to remotely control the functions of a tractor trailer and a school bus, affecting the gauges on the instrument cluster, engine acceleration and deceleration, and engine braking features [4]. Another group of researchers developed a mobile app to remotely force a passenger vehicle to move forward or backward, display false dashboard readings, control the speed, or cause engine damage [5]. The vulnerabilities and opportunities to introduce attack vectors seem limitless. For instance, a vehicle with a Global Positioning System (GPS) and vulnerable telematics system could be targeted by hackers to kidnap, injure, or even kill passengers by programming their own destinations in the built-in navigation system. Remote starters and keyless entry systems tied to software applications (or Apps) can be hacked by thieves looking to gain entry or steal a vehicle. A security camera that captured one of many robberies in Houston, Texas, showed a thief pulling out a laptop after breaking into a vehicle before starting it up and driving it away [6]. Researchers were able to hack into Jeep Cherokee's telematics unit in 2015 and remotely turn off the engine while it was speeding down the highway in a test environment which disables braking and steering while still coasting at speed [7]. The demonstration led to the first cyber-related recall for vehicles by Fiat Chrysler that affected 1.4 million cars [8].

These capabilities have troubling ramifications. The July 14, 2016, terrorist attack in France, where a man used a 19 ton truck that left 84 people dead and over 200 injured [9], could be small in comparison to a coordinated terrorist attack on a specific vehicle manufacturer's telematics system that could compromise thousands of vehicles at the same time. A coordinated attack on a vulnerable system that disables all of the vehicles' brake systems simultaneously during rush hour on any given day isn't farfetched. A motivated attacker could currently accomplish this terrorist act through vulnerabilities already identified and vetted by researchers. While the likelihood of a terrorist attack on vehicles seems remote, the consequences to national security and to the automaker that was attacked would be substantial. These examples point to a need to apply advancements in human-computer technology wherein the human and vehicle form a partnership where intent and intuition can screen the behavior of the driver's state of mind in real time.

Many manufacturers are considering selling Apps on the telematics system, similar to the Apps available on any smartphone. Malware is defined as malicious software (or alteration of the computer hardware processing software) that disables the intended function of the host system. Malware could be introduced through third-party Apps or third-party hardware add-ons with infected code intended to change the software or firmware of the vehicle for access or control. Apps are likely to have access to personal identifiable information, such as credit cards, bank accounts, personal addresses, and social security numbers. The introduction of Apps to vehicle telematics systems will increase the probability of a successful attack that exploits information leakage from vehicle systems for financial gain.

Ransomware is becoming more prevalent with hackers, with cybercriminals extorting a fee in exchange for relief from an attack. In a vehicle, attacks could range from disabling the engine or locking motorists outside of their vehicles in exchange for a ransom. It is projected that over 250 million cars will be connected to the Internet by 2020, with each car being a ransomware opportunity for cybercriminals [10].

Vehicles are expected to continue to evolve, with vehicle-to-vehicle (V2V) and vehicle-to-infrastructure (V2I) communication and driverless cars. Tesla, Google, and Uber have already introduced self-driving cars, and work is already being done to introduce autonomous trucks [11]. The V2V and V2I technologies rely on data sent between vehicles and other communication devices to improve safety by warning drivers of potential accidents and hazards. While there are numerous advantages to V2V, V2I, and self-driving vehicles, the security risks increase significantly with the addition of complex hardware, software, and firmware with interdependencies and connections to ECUs in the vehicle.

Vehicles in the mid-1990s were not designed to address CPS security, as the concern for hacking was not well known at that time. The CAN first introduced in the mid-1980s is still the dominant bus used in modern vehicles today. The CAN has limitations on bandwidth for data, which would limit potential solutions such as encryption and authentication used in more modern computing systems. Given there are no standard metrics of security and little information is shared among competitors on effective solutions, implementation of features to solve the problem could vary significantly.

Today's vehicles include controls from engine management, crash avoidance systems, infotainment, and self-driving technology that can be hacked to take physical or virtual access of safety-critical operations. Today's modern vehicle could use more than 100 million lines of code to control the electrical/mechanical CPS [12]. For comparison, the F-35 fighter jet has approximately 8 million lines of code, equating to a 12th of the amount of code used in the modern vehicle [13].

Up to this point, the discussion of threats has been focused on remote entry to the vehicle. However, there are threats which may already lurk within the vehicle in the form of compromised hardware. Additional vulnerabilities can be introduced through electronic piece parts or third-party electronic hardware added to the in-vehicle communications network. There are millions of electronic piece parts being fabricated in untrusted foundries. In addition, counterfeit electronic piece parts are prevalent in all sectors of the supply chain from parts extracted off of recycled circuit boards. The cyber threat is further exacerbated by electronic parts with potential hardware Trojans and embedded malware. These counterfeit parts pose a significant risk to the overall functionality, reliability, and security of vehicles. According to a report by VDC Research, it is estimated that 75% of new vehicles will have Internet connectivity by 2020 [14]. This connectivity represents a vulnerability to the over 30 microprocessors embedded within the vehicle. In 2014, only 2% of the microprocessors found inside cars had any hardware security features.

Attack vectors could be introduced through vulnerabilities in electronic parts that could be used to compromise ECU functions or gain access to the in-vehicle communications network. Vehicle ECUs are susceptible to compromising attacks due to electronic parts with embedded malware or hardware Trojans. An attack could be introduced through a malicious component that is permanently attached to the in-vehicle network or embedded within the vehicle's components. A vulnerability could be embedded to the in-vehicle network through a brief period of connectivity that is later disconnected. A point of entry could be presented by counterfeit or malicious components entering the vehicle parts supply chain, such as the purchase and installation of an aftermarket third-party component (e.g., a counterfeit FM radio) [15]. Further unintended vulnerabilities can also be introduced with the integration of complex hardware, software, and firmware supporting the vehicle without a holistic approach to address areas of concern for a CPS. Any third-party component added to vehicle's digital and electronic network could introduce vulnerabilities that could allow a hacker to infiltrate the CPS. An example of this issue was reported by an Israeli cybersecurity firm which demonstrated breaking into the vehicle's internal communication system through a third-party dongle add-on to trigger a signal to disable the fuel pump of a moving car [16]. Dongles are made by several third-party organizations for a variety of reasons, such as to track drivers' performance for insurance purposes or to monitor vehicle diagnostics and alert owners of potential issues requiring repair. The dongle is plugged into the OBD-II port, enabling access to the ECUs attached to the vehicle's communications network. Hackers can exploit vulnerabilities in the dongle to gain unauthorized access to the vehicle.

The motivation to improve security to resolve gaps may not be apparent, as the typical consumer is unaware of the vulnerabilities and potential personal consequences. Costs to address vehicle CPS security vulnerabilities are estimated to increase the cost per vehicle as much as 3–5%. In addition, incorporating technology features into vehicles to address security is estimated to take approximately five years to complete and implement into production. By the time a security feature is introduced, it may be outdated based on the evolution of the cyber hacker. Additionally, adoption of key practices and technologies to mitigate CPS security concerns varies significantly across the auto industry [2]. Industry standard work that includes taxonomy of vulnerabilities and common metrics of security coverage from a systems engineering perspective is needed to level the playing field in industry and build trust and resilience into the supply chain.

The malware challenge is expansive. Unlike systems engineering in critical infrastructures that integrates years of learning to establish scientifically disciplined rules for optimal control systems, cybersecurity requirements are dynamic and lack any rooting in science or empirical models. Therefore, to achieve cyber resilience in automotive systems, the systems engineering process must be adapted to concurrently design a functional yet self-aware, adaptive computing network. Such an approach would use CPS security.

5.2 Introduction to Cyber-Physical Systems Security

A CPS is an electronic system that operates as a single, self-contained device or within an interconnected network providing shared operations. The U.S. National Institute of Standards and Technology (NIST) defines CPS as "smart systems that include engineered interacting networks of physical and computational components" [17]. Similarly, The Networking and Information Technology Research and Development (NITRD) Program's definition is "smart networked systems with embedded sensors, processors and actuators that are designed to sense and interact with the physical world (including the human users), and support real-time, guaranteed performance in safety-critical applications" [18]. Some researchers have explicitly included the human component of these systems, proposing cyber-physical-social or cyber-physical-human systems [19, 20]. CPS affect a tangible output through the use of the command and control electronics embedded within the system devices or network. **Figure 5.1** shows a general schematic of CPS. Within the blue box representing the CPS (or cyber ecosystem), a number of servers, workstations, and other devices are connected and controlled through an internal communications infrastructure. The cyber ecosystem communicates through external devices, networks, and other CPS to produce a system capable of performing operations that yield a tangible output. Under the umbrella of the CPS are industrial control systems and information technology. In this manner, connected vehicles are examples of CPS.

FIGURE 5.1   Schematic of cyber-physical systems.

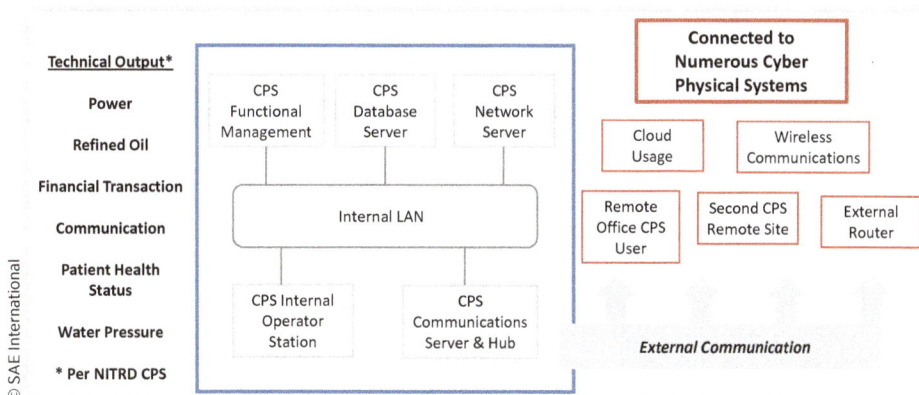

©2019 SAE International

5.3 Systems Engineering Perspective to Cyber-Physical Security

5.3.1 Areas of Concern

The most urgent concern for implementing a systems engineering approach for CPS security is the need to leverage the domain knowledge of skilled personnel highly familiar with the broad range of disciplines used to build and operate the vehicle's systems design. The International Council on Systems Engineering (INCOSE) defines systems engineering as "an interdisciplinary approach and means to enable the realization of successful systems" which "considers both the business and the technical needs of all customers with the goal of providing a quality product that meets the user needs" [21].

Recognizing the many malicious threat factors and considerations across the life cycle of a CPS, DiMase et al. [22] identified ten areas of concern which must be addressed in order to holistically achieve greater security. These ten areas of concern are outlined in the pinwheel diagram shown in **Figure 5.2**. We describe the risks associated with each area of concern with respect to automotive vehicles below.

5.3.1.1 Electronic and Physical Security: The first area of concern we address is electronic and physical security. We define electronic and physical security to mean the defense against insider threats that include physical, technical, and administrative controls including system privileges. It incorporates measures designed to deny unauthorized access to facilities, equipment, and resources and to protect personnel and property from damage or harm (e.g., espionage, theft, or terrorist attacks). It includes protection resulting from measures designed to deny unauthorized individuals information derived from the interception and analysis of noncommunications electromagnetic radiations.

All cars since 1996 contain an OBD-II port that docks into the car's communication bus with access to the electronic control module. The electronic control module (ECM) contains diagnostic information and performance controls for the engine throttle, steering position, speed sensors, and other subsystems such as cruise control, ABS brakes, automatic transmission, anti-theft monitor, and even the ECU which deploys airbags. Since the OBD-II connects to the in-vehicle network which may be a simple 8 byte bus communication, the information and feedback cannot be secured with encryption and authentication without significant system redesign.

Exploits which remain undetected until the day of public deployment are known as "zero-day vulnerabilities." In vehicle systems, these vulnerabilities could be exploited to gain access through a variety of ports that include physical access to the OBD-II port, short-range indirect access through Bluetooth or Wi-Fi, and long-range access through satellite radio, AM/FM radio, or cellular connections. The unfortunate reality is that as more convenient features are added to the vehicle, like keyless entry or remote start, the more there are vulnerabilities that need to be addressed in order to protect the CPS and ultimately its users.

5.3.1.2 Information Assurance and Data Security: CPS are primarily about control of physical systems by computation devices. However, in the course of their

FIGURE 5.2   Systems engineering perspective to CPS security.

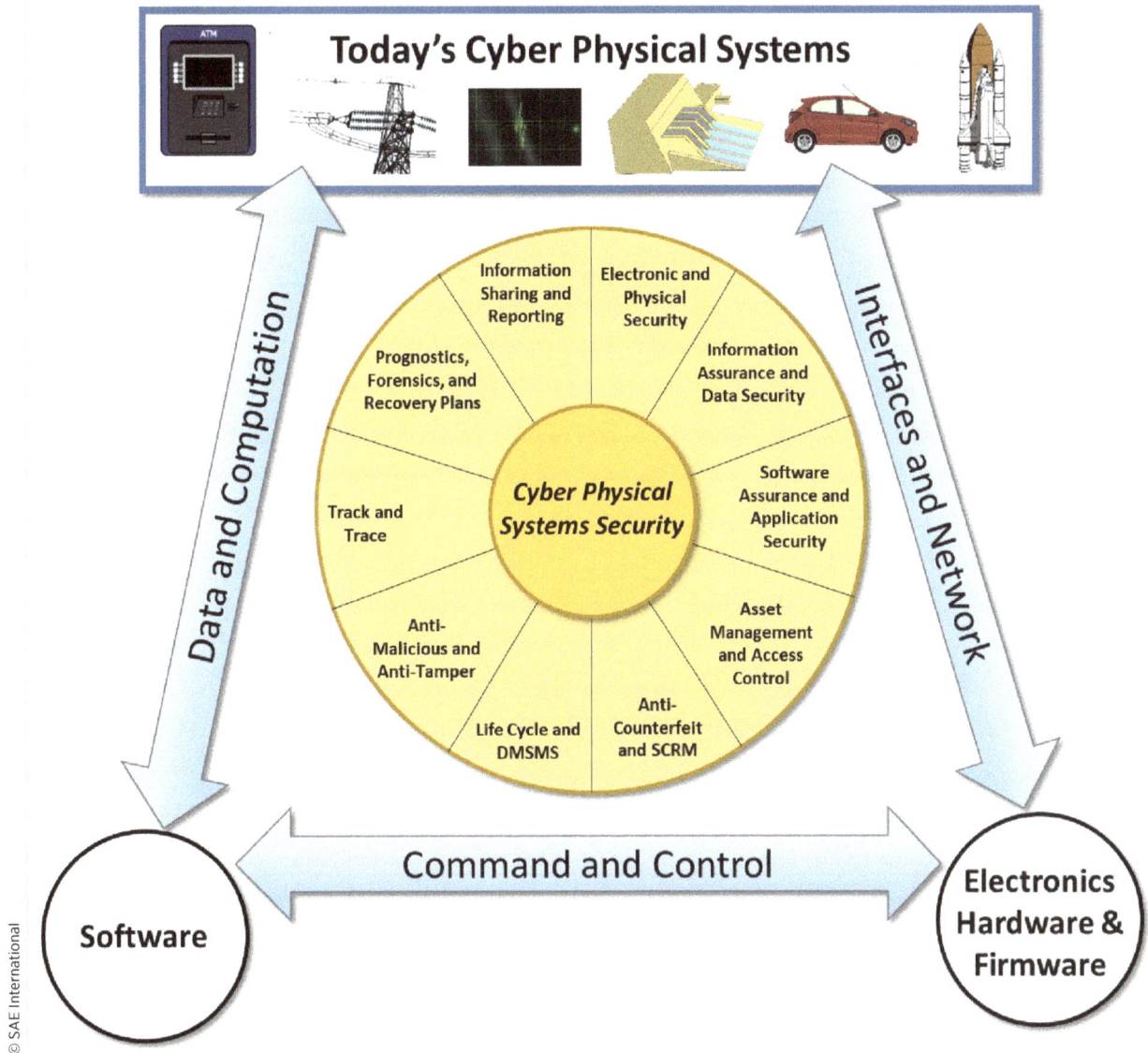

operation, CPS-embedded systems can accumulate a large amount of data from sensors, users, and external communications, and that data is stored in embedded systems memory that may be accessible even after the system is powered off. Assurance of that data is a critical concern. In this context, assurance means that the system ensures the data availability, integrity, authentication, confidentiality, and nonrepudiation. It ensures protection of data from unauthorized (accidental or intentional) modification, destruction, or disclosure. Protection applies to not only data in transit but also at rest. Some of these issues are well understood and can be handled by widely known cryptographic primitives including encryption, which makes data unreadable without the possession of a password/key, and hashing, which enables verification of data integrity.

Consumers often don't realize that the data they store or produce could reveal sensitive information. For example, imagine a scenario where an attacker views GPS information in a car to reveal the owner's home address. A system may reveal calendar information that can tell when the owner is away from home. Since the garage door opener codes are not encrypted and the device is not equipped with biometric authentication, a hacker could use this information to enter the house without needing to break a door. In more sensitive systems, trusting the data is vital to preventing widespread infrastructure collapse. One could insert oneself into a smart grid and fake sensor data that makes it seem like there is a power surge that may cause circuit breakers to trip. Researchers have demonstrated that it is possible to insert fake data into a system to cause automotive ECUs to fail [23]. How do we know which data to trust? There need to be mutual authentication mechanisms between sensors and control.

Many information systems can rely on physical security to protect information. For example, because the server is behind locked doors, we don't necessarily have to worry too much about users rebooting the server or attaching a serial cable to a serial port and getting console access. With CPS, however, the systems are often out in the wild and subject to physical access. As mentioned before, this physical access makes data much easier to steal. Therefore, the same assumptions that we make for standard computers do not apply for CPS. Layers of authentication and access control are needed that make access to information much more difficult. For example, with the OBD or telematics vulnerability discussed earlier, one could add features that require vehicle authentication in a manner that is impossible to duplicate, such as public-key cryptography with tamper-proof key storage or Physical Unclonable Function (PUF)-based challenge-response schemes.

5.3.1.3 **Asset Management and Access Control:** Vehicle cybersecurity requires the continuous management of critical assets in the system. These system assets may be an unintentional point of vulnerability due to their functional role and interaction within the CPS operating environment. This area of concern provides tools and systems for maintaining an inventory of critical assets and monitoring access using verification credentials. This includes management of information relevant to the operation of the asset (e.g., software revision, firmware revision) and the process of granting or denying specific requests: (1) for obtaining and using information and related information processing services and (2) to enter specific physical facilities.

One would need to identify the various ECUs and electronic assemblies which have been added to the vehicle. Each of the subsystems' software and firmware revisions would need to be cataloged and managed to ensure they are up to date, so that zero-day vulnerabilities which have been identified with fixes cannot be exploited once the updates have been applied.

Access controls to various systems also need to be managed. Personal identifiable information that could be accessible through the vehicle's telematics system would need to have appropriate password and access controls. Vehicle GPSs should have appropriate access controls to ensure unauthorized users do not have access to addresses stored in the system, such as your home address. One could imagine providing your vehicle to a valet at the local restaurant, with full access to your home address that includes wireless access to your garage and home through the built-in garage door opener.

5.3.1.4 Life Cycle and Diminishing Manufacturing Sources and Material Shortages (DMSMS): This area provides sustainment processes for assets in a CPS threatened by loss or impending loss of manufacturers of items or suppliers of items, services, or raw materials necessary to sustain availability of the asset. This includes updating the asset to address the latest vulnerabilities and ensuring hardware and software configuration and functionality (e.g., patches in software and updating firmware or hardware to repair or replace broken assets).

Parts vulnerable to counterfeiting and malware can be introduced when they are no longer available from trusted sources of supply due to obsolescence [24]. When one thinks of this problem from a software perspective, a good example to understand the issue is the Windows operating system. Zero-day vulnerabilities are constantly identified and patched through software updates to home computers. When Microsoft sunsets a product, like Windows XP, zero-day vulnerabilities are no longer patched, leaving the system vulnerable to attacks. Any CPS with software and firmware needs to have patches to address zero-day vulnerabilities which have been discovered to avoid exploitation of the system.

5.3.1.5 Anti-Counterfeit and Supply Chain Risk Management: This area maintains systems and processes associated with CPS protection from counterfeit parts and supply chain vulnerabilities. It reduces the likelihood that material is not authentic and that suppliers do not produce or use products that introduce vulnerabilities to the host CPS. Counterfeit parts include components which have been intentionally or maliciously modified from their intended design to enable a disruption in performance or an unauthorized function, which can be introduced anywhere in the supply chain. Supply chain risk management ensures pedigree to the original manufacturer and adequate controls against counterfeiting.

Counterfeit parts introduce concerns with quality, reliability, safety, and security. Tampered counterfeit electronic parts include those with embedded malware and hardware Trojans. These vulnerabilities could be introduced through counterfeit electronic piece parts or assemblies. In the defense sector, a report disclosed a cyber hack of a German Patriot missile system carried out through a tampered computer chip in the system. Experts expressed that the hack could lead to the system failing to intercept incoming missiles or even firing at an unauthorized target [25]. A similar attack could be deployed on an automobile. For example, in a vehicle, a counterfeit radio add-on that contains embedded malware or malicious functionality introduces vulnerabilities to the vehicle.

Supply chain risk management is necessary to ensure quality and trust. Sources should be validated through second- and third-party audits to industry specifications identifying best practices to address concerns. Industry standards to mitigate the risk of counterfeiting include SAE AS5553, AS6171, and AS6174 which were developed by the aerospace sector but could be applied to any sector, such as automotive or medical, to mitigate the risk of counterfeiting. There are a number of SAE standards applicable to different sectors of the supply chain. Additional mitigating strategies in the supply chain include utilizing trusted sources of supply for critical logic-bearing electronic devices.

The Department of Defense (DoD) Trusted Foundry Program [26] was initiated in 2004 to ensure that mission-critical national defense systems have access to leading-edge integrated circuits from secure, domestic sources. It is a joint DoD/NSA (National Security

Agency) program, administered by NSA's Trusted Access Program Office (TAPO). The program includes foundry capability plus full range of microelectronics services from design through prototyping, packaging and assembly, photomask manufacturing, and aggregation, accredited by U.S. Defense Microelectronics Activity (DMEA) to be trusted. Implementation of a similar strategy within the high-risk/vulnerability business sectors including automotive, medical, and others would provide added assurance to the critical supply chain.

5.3.1.6 Software Assurance and Application Security:

As would be expected, CPS suffer from the same type of vulnerabilities that exist with any system containing software, such as a common home or office computer system. These software vulnerabilities can cause critical failures or perform unintended operations. Since software such as word processors and open source software are common in many computer systems, vulnerabilities are generally fixed very quickly and are updated remotely. However, in CPS, these fixes may take a long time to flow down to actual devices or may not be viable since changes to the software may alter or disable functions or features of the CPS. Computers also typically have virus protection software installed that provides defenses to the system that is not typical in CPS. Some systems do not even have mechanisms to automatically upgrade in the field or may not have a means to identify and fix software vulnerabilities whatsoever. For example, after reports of vulnerabilities in Chrysler vehicles [27], Chrysler released an update on a USB stick that had to be manually installed instead of remotely updating, thus leaving many vehicles vulnerable. Any update will need verification of trust to avoid the introduction of malware, install Trojan firmware, or alter critical functionality of the CPS. Signed software solves some of these problems by linking updates to trusted sources, but this approach has challenges in many CPS applications. The manifestation of vulnerabilities should be analyzed against the evaluator's expectations of the code's instructions. In one example, Volkswagen cars had software installed that could change performance of emissions tests when they were being tested. A "defeat device" put the vehicle into a different mode when detected as operating on a stationary test rig by running the engine below normal power and performance to meet regulations [28]. This example demonstrates that vulnerabilities could be introduced from an insider source. Testing only provides a limited level of confidence. Sandbox settings with isolated virtual space for the software enable a test environment to run the application without having to install the software in a traditional way. The applications operate in a virtual environment without affecting the CPS. Formal verification that can prove that the software is "safe" is the ideal. However, most existing techniques haven't really scaled beyond a few thousand lines of code and most CPS are not set up with sandbox settings for testing.

5.3.1.7 Forensics, Prognostics, and Recovery Plans:

This area provides processes and tools for gathering of CPS operations data for use in the examination and analysis of cyber incidents, thereby characterizing the CPS operational cybersecurity baseline. The baseline provides the basis for tools used in forensics (internal to the CPS), prognostics, and recovery plans (including resilience). An adjacency provision for external forensics serves the cooperative effort with other industrial CPS organizations and government agencies responsible for pursuing the root cause of an attack vector external to the CPS operating environment.

Many new technologies are available to monitor the health of your automobile in real time. By connecting a wireless dongle to the OBD-II port, users can monitor various operating parameters via wireless Apps offered directly through manufacturers, or aftermarket products [29].

Forensic data can be obtained offline (after the event) and online (in real time), as well as acquired passively (intercepted) or actively (the party conducting forensic investigation has control over vehicle functions). Forensics can be collected through hardware (ECUs) or software (gateways/telematics) [30]. The fusion of physical and electronic data (video streams, GPS coordinates, visual images, etc.) allows for enhanced forensic capabilities as opposed to having only physical or electronic data separately [31].

5.3.1.8 Track and Trace:
As CPS components (both software and hardware) move through the supply chain, it is essential to understand where the components have been [32]. Hardware mechanisms such as radio-frequency identification (RFID) provide some capability to track devices, but they are not ideal. It is relatively easy to remove a RFID tag and use the tag to authenticate and track a counterfeit part. It is ideal to have a mechanism that is integrated into the device. Botanical DNA applied to devices provides some assurance of source origin. However, to take advantage of its features, it requires cotton swabbing for detection and complicated laboratory analysis that is not practical in production environments. Luminous methods have also been used to provide some assurance of source origin, but are easily spoofed. PUFs provide an untamperable method to authenticate a device. Existing PUFs are silicon based and do not work well for nonelectronic parts or larger electrical or electronic systems. However, new optical tag technologies allow one to store large amounts of information as well as retain a user ID that can be applied for track and trace applications.

5.3.1.9 Anti-Malicious and Anti-Tamper:
Anti-malicious and anti-tamper is a systems engineering process that includes CPS vulnerability to tampered hardware or malware introduction achieved through reverse engineering. It provides tools and processes for the integration and assessment of protective technology features in the CPS electronic systems that mitigate the impact and consequences of reverse engineering attacks that could include an attacker's assessment of vulnerabilities on an otherwise unprotected CPS.

Anti-tamper (AT) integration in a systems design is essential for each subsystem and its relationship in a CPS to self-sufficiently detect and defend against compromise of critical capabilities that is specific to reverse engineering and malicious attacks. It provides a type of self-automated alarm system of an attack with responding designer-designated penalties. AT implementation acts as a castle wall against reverse engineering and malicious attacks by protecting each entry point or crack in the CPS and can include any identified feature such as I/O ports and software backdoors to physical hardware accessibility. A typical AT approach analyzes a CPS from a platform level (entire vehicle) followed by subsystem (Electronic Control Module [ECM] or Transmission Control Module [TCM]), embedded systems (CAN bus), printed board assemblies, and finally integrated circuits or microelectronics. It identifies the vulnerabilities at each point and provides a solution for each with an overall software and hardware AT plan.

Whereas anti-tamper prevents the identification of vulnerabilities and their exploitation, anti-malicious aims to prevent the introduction of vulnerabilities through systems design and layers of security controls. With so many points of potential malicious introduction in a system's life cycle, this anti-malicious goal requires a comprehensive review of trust at every factor related to the product including manufacture processes, toolsets, engineering IT systems, sourcing/procurement, and even some design features of the system. Just as importantly, regular evaluation of current published and suspect system and industry vulnerabilities must be maintained. Some common examples include supply chain trust, incoming part screening, isolated tooling, malware/Trojan detection in piece parts and final products, Operations Security (OPSEC) training, and access controls. As with many insurmountable tasks, approaching it by dividing and conquering is critical while also understanding that the resulting mitigations will inevitably cause significant change upsetting some industry stakeholders. The benefits of reduced liability, financial advantage, and regulatory compliance will eventually support this anti-malicious goal and overcome "the way things have always been done." Utilizing a cyber-physical system approach with full life cycle coverage of identified vulnerabilities is this divide-and-conquer strategy necessary for near-term solutions.

As the probability of malicious tampering of CPS as well as the resulting consequences exponentially increase, the importance of anti-tamper and anti-malicious systems designs is directly critical in addressing the dynamic threats of today's technical revolution.

5.3.1.10 Information Sharing and Reporting: The final area of concern provides tools and shared database resources for reporting and rapid exchange of cyberattack events and the mitigation measures to minimize the breadth of impact of the attack in the CPS network. It addresses the communications plan and information sharing necessary to report a cyber incident and prevent an issue from reoccurring. We cannot recover from attacks until we can first identify them and then develop a recovery plan to mitigate the threat. Without information sharing and reporting, we will never develop systems capable of surviving the evolving cyber hacker. This area is so critical to our understanding as we develop systems resilient to attacks that it is often required through regulation in industries supporting the critical infrastructure. For U.S. DoD, counterfeit part reporting and remediation is required through Defense Federal Acquisition Regulation Supplement (DFARS) regulations. Reports of cyberattacks are also required through regulation for many sectors beyond DoD.

Cyber ranges and penetration testing is imperative to our understanding as we develop systems capable of surviving an attack. Manufacturers of CPS should adopt practices of organizations such as Apple and Microsoft that pay a bounty for identified vulnerabilities, provided the researcher who discovered the problem allows time for the manufacturer to address the gap and correct the problem.

Each of these areas of concern is currently siloed, each with distinct governing bodies and sometimes conflicting policy and guidance documents [22]. This fractured approach results in gaps and vulnerabilities which can be exploited by malicious actors. There is therefore a need to employ a holistic systems engineering perspective across all of the areas of concern. This includes the integration of crosscutting capabilities such as risk assessment and management, decision analysis, employee training and certification, and education and outreach.

Table 5.1 provides notional example scenarios that relate to each of the areas of concern. Associated with each of these scenarios are qualitative estimates of their

TABLE 5.1 Example scenarios and risk assessment for areas of concern

Area of concern	Example scenario	Likelihood	Consequences
Electronic and physical security	• Remotely disable braking system of fleet through zero-day vulnerability in telematics systems • Hacking of keyless entry, remote start of car for theft • Hacking of telematics system for denial of service to engine start for ransomware • A hacker remotely gains access to the electronic control module and displays false dashboard readings	• Low • Medium • Medium/low • In all cases, likelihood increases with the inclusion of additional Bluetooth and Wi-Fi features	• Severe: could result in traffic accidents, injury, and death; DOT, DHS, and potential DOD involvement • Medium: personal property loss • Low: potential monetary loss; inconvenience for loss of service • Severe: could result in traffic accidents, injury, and death
Information assurance and data security	• A hacker gains unauthorized access to stored personal information such as home address, credit card information, etc. • A hacker inserts false data into vehicle system to cause ECUs to fail	• High: assuming the right tools and know-how	• Medium: loss of personal information • Severe: could result in theft, personal injury
Asset management and access control	• Poor software and firmware revision procedures resulting in vulnerabilities exploitable by hackers • Weak password control compromises the personal data of thousands of users	• Low: assuming proper asset management and access controls	• Severe: could result in theft, personal injury • Medium: loss of personal information; high from the perspective of the auto company
Life cycle and diminishing manufacturing sources and material shortages (DMSMS)	• Hackers target certain tainted hardware obtained from the broker market • Hackers exploit unpatched vulnerabilities from unsupported, obsolete software or firmware settings	• Moderate: depends on number of vehicles affected, degree of obsolescence	• Severe: could result in accident, injury, death
Anti-counterfeit and supply chain risk management	• Counterfeit parts fail earlier than intended, causing quality and reliability issues • Counterfeit parts include Trojans, allowing hackers to gain access to the system	• High: depending on supply chain practices (e.g., not using trusted suppliers)	• Moderate: could result in additional costly maintenance and repairs • Severe: could result in accident, injury, death
Software assurance and application security	• Hackers activate dormant malicious code from software that allows remote access, causing malfunctions and safety issues • Hackers are able to guess passwords that do not meet basic hygiene best practices	• Low: assuming good software security measures are in place • High	• Severe: could result in accident, injury, death

TABLE 5.1 *(Continued)* Example scenarios and risk assessment for areas of concern

Area of concern	Example scenario	Likelihood	Consequences
Forensics, prognostics, and recovery plans	• Vehicle does not have capability to address a zero-day vulnerability that is discovered until it is brought in for service • A hacker gaining access to active monitoring could gain control of vehicle functions	• High: published issue currently identified in industry that has already resulted in recalls • Low	• Medium/high: significant financial impact to manufacturer; potential impact to passenger safety if threat is acted on by hacker • Severe: could result in traffic accidents, injury, and death
Track and trace	• Material passes through untrusted part of the supply chain and is altered by an adversary	• High: depending on supply chain practices (e.g., not using trusted suppliers)	• Moderate: could result in additional costly maintenance and repairs
Anti-malicious and anti-tamper	• Adversary is able to reverse engineer critical safety measures and identify vulnerabilities for an entire fleet of vehicles • Based on identified vulnerabilities, adversary is able to introduce new hardware or software that allows unauthorized access	• Moderate: depending on the layers of defense designed into the system	• Severe: could result in traffic accidents, injury, and death
Information sharing and reporting	• Institutional practices are not in place to share reported problems within the entire enterprise and industry • Mechanisms are not in place to detect and remediate threats	• High	• High: could result in costly recalls, lawsuits, and damage to brand

© SAE International

likelihood of occurrence and consequences. The table is not meant to be exhaustive or represent a fully complete risk assessment, but rather to illustrate some of the key considerations when thinking about each of the areas of concern. Note that for some scenarios, the consequences may be felt by multiple stakeholders such as customers (i.e., drivers and passengers) or the manufacturer.

5.3.2 Systems Engineering Modeling

A systems engineering approach to Cyber-Physical Systems Security (CPSS) meshes design requirements at the operational, functional, and architectural levels [22], as shown in **Figure 5.3**. The systems engineering approach facilitates documentation and modeling needed to assess the security at any point in the life cycle of the CPS. The approach incorporates standard work achieved through industry, government, and academic collaboration, developed to mitigate risks utilizing best practice and guidance defined for each area of concern. It can serve as the basis for vulnerability and risk analysis which identifies threats and best practices for mitigation across all ten of the areas of concern and helps to prioritize resources and countermeasures. Additionally, it can be used to perform a gap analysis to identify weaknesses in the system and related supply chain, organizational, and user environment and create a roadmap to achieving the desired level of security. It can also be used to provide an assessment of legacy systems and iterative reassessments, which is

FIGURE 5.3 Systems engineering requirements flowdown.

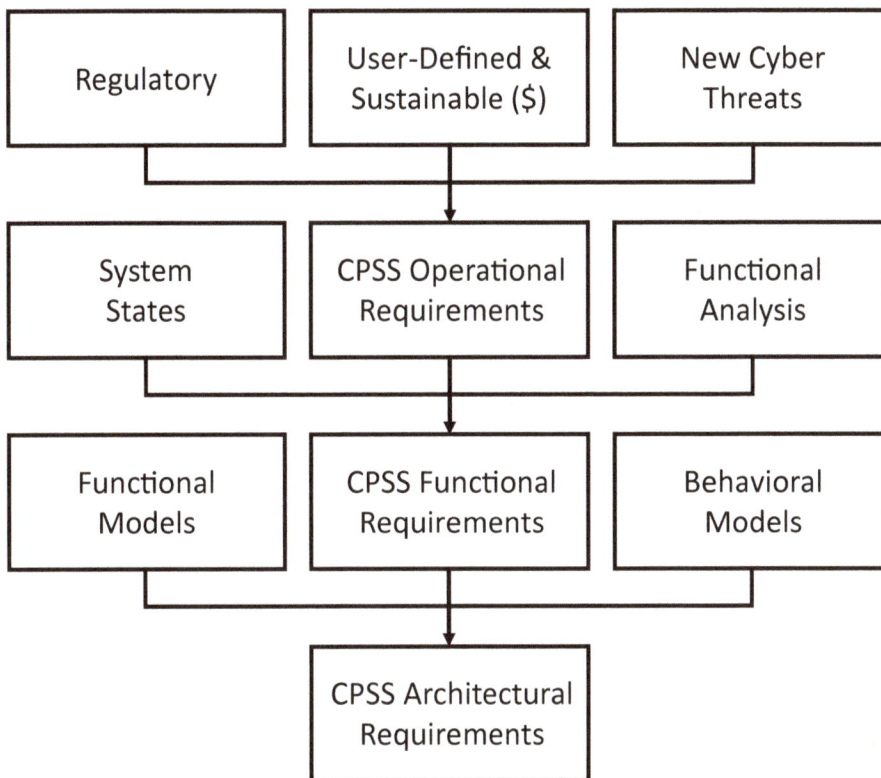

recommended to assess the system's vulnerability to evolving cyber threats. Recommended updates, repairs, or replacements for the CPS can be made to address newly identified threats. Importantly, the analysis requires an investigation of the flow of sensitive data and critical command/control functions affecting the expected CPS' tangible output in the operational environment.

Operational requirements capture customer and organizational needs, regulatory policies, and perceived cyber-related risks. These define the constraints imposed on the system. Customer and organizational needs include programmatic cyber risk requirements and organizational liability risks. They also include the physical and functional needs of the system such as operating environment, threat rejection characteristics, sensitive data security, and encryption needs. Government and industry regulatory requirements must also be addressed to assure compliance. All of these elements must be balanced against potential cyber threats to assess overall operational requirements. Threats are derived from multiple sources such as operating environment, system access points, known technology exploits, system sandboxing, and user interactivity and accessibility with the system.

Functional requirements capture the performance characteristics of the system. This includes system performance requirements which should be addressed for all operational modes. Functional states of the system, both intended and unintended, must also be considered, with unintended modes receiving explicit attention since these often provide the conduit for system attacks. Some tools effective for determination and assessment of hardware and software operating modes are Failure Modes, Effects and Criticality Analysis (FMECA), Fault Tree Analysis (FTA), and Sneak Circuit Analysis. Further assessment of software operating modes and architecture vulnerabilities can be ascertained through vulnerability analysis.

Architectural requirements are the resulting set of requirements comprised of the operational requirements and the functional requirements. This reflects the combination of performance needs and application constraints. Architectural requirements also encompass the physical attributes of the system, blending these characteristics with system performance needs. Physical design attributes should take into account the need for physical anti-tamper measures, design obfuscation to reduce the risk of Trojan insertion, and physical accessibility. Architectural requirements should also include design modeling evaluating the combination of operational constraints and functional needs, bounded by the required physical characteristics.

For example, an operational requirement may be to implement controls that address consumer desires for vehicle security. A functional requirement to satisfy this may be that the safety-critical ECUs in the vehicle, such as those controlling braking or steering, are not accessible by noncritical subsystems without proper authentication. Finally, this functional requirement may be implemented through an architectural requirement to provide physical and virtual partitioning.

The systems engineering model is consistent with Work Domain Analysis [33] in which systems are defined from the top-down in terms of levels of abstraction (system purpose, domain values, domain functions, technical functions, physical resources/ material configuration) and decomposition (system, unit, component, part).

The systems engineering model provides the framework for the development of a system architecture. This structure provides the cornerstone for the development and validation of the security characteristics necessary for a robust CPS. The systems engineering approach therefore facilitates documentation and modeling needed to access the security at any point in the life cycle of the CPS.

The systems engineering model perspective provides guidance on how areas of concern can be identified and addressed through the incorporation of operational, functional, and architectural requirements. This defines the necessary security elements of a vehicle CPS (**Figure 5.4**). The systems engineering perspective provides the framework to mitigate cyber risks utilizing a bottom-up approach, assuring critical aspects are incorporated into the product design to assure minimal cyber risk. The same methodology can also be used to provide an assessment of legacy systems, which is recommended to assess the system's vulnerability to evolving cyber threats. Recommended updates, repairs, or replacements for the CPS can be made to address newly identified threats.

To assess a CPS according to the systems engineering perspective, a six sigma style scorecard tool can be used [22]. The scorecard tool identifies the key system assets and command and control targets at the operational, functional, and architectural levels according to the systems engineering requirements flowdown.

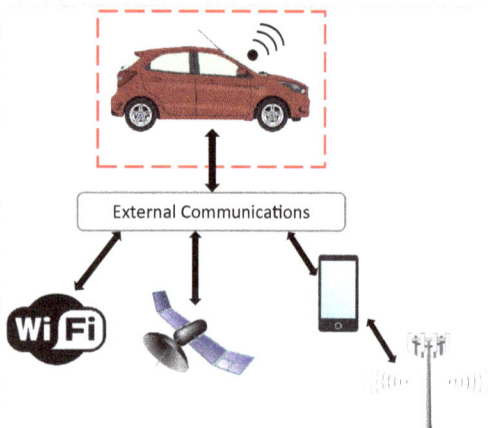

FIGURE 5.4 Example of a vehicle as a CPS.

External Communications

WiFi

© SAE International

FIGURE 5.5 Notional CPSS scorecard assessment.

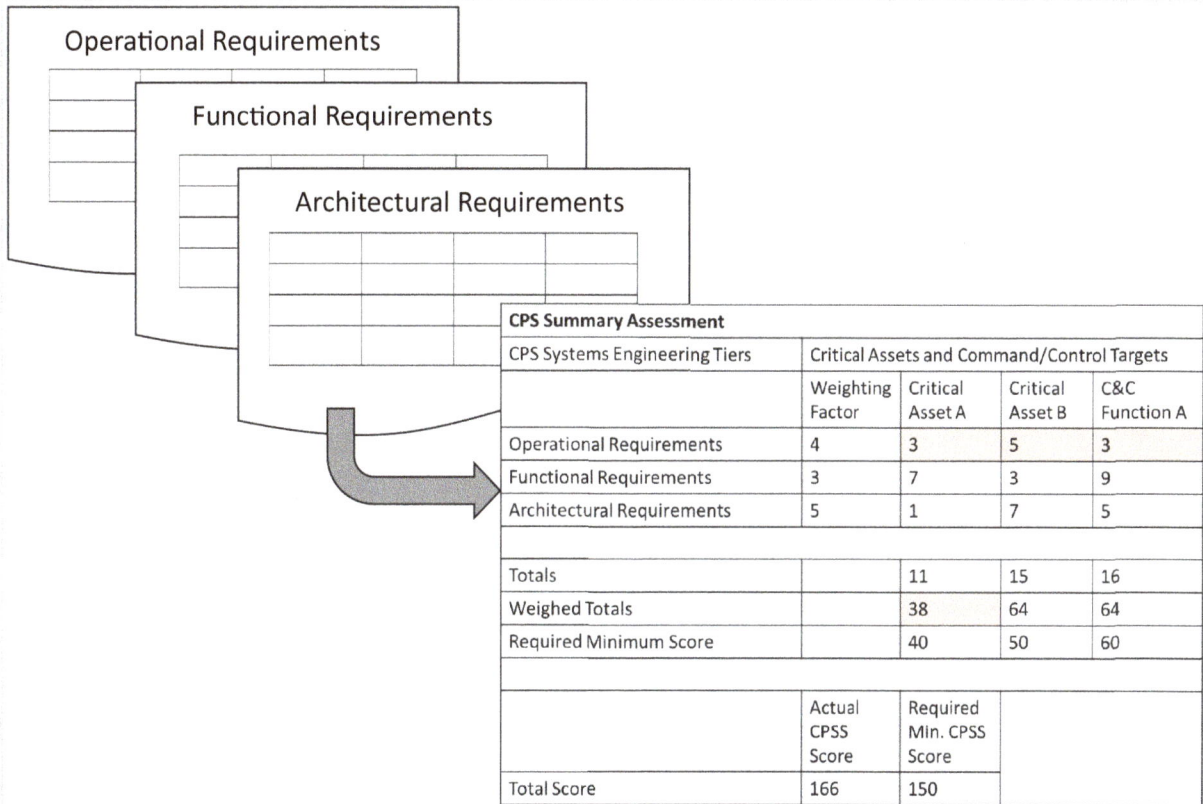

CPS Summary Assessment					
CPS Systems Engineering Tiers		Critical Assets and Command/Control Targets			
		Weighting Factor	Critical Asset A	Critical Asset B	C&C Function A
Operational Requirements		4	3	5	3
Functional Requirements		3	7	3	9
Architectural Requirements		5	1	7	5
Totals			11	15	16
Weighed Totals			38	64	64
Required Minimum Score			40	50	60
		Actual CPSS Score	Required Min. CPSS Score		
Total Score		166	150		

The summary table is used to quickly assess the effectiveness of risk mitigation to the critical assets of the system. In **Figure 5.5**, the red regions indicate a poor performance in the operational level. Drilling down, detailed individual scorecards for the operational, functional, and architectural levels are filled out, which serve as data inputs for the summary scorecard. The operational level scorecard assesses areas of concern in the end user's operating environment. The functional level scorecard assesses areas of concern from an operational and functional perspective, including vulnerability analysis and system performance analysis. The architectural level scorecard evaluates areas of concern that includes evaluation of the piece parts (e.g., microchips). This includes evaluation of the physical layout of the assemblies and components which enable the overall system performance and sustainment of the system throughout the anticipated product life cycle. These multitiered assessments feed into the summary scorecard, which is updated periodically at key design milestones at the operational, functional, and architectural levels. Implementation of such a tool would require guidance on common definitions, criteria, assessment procedures, and reporting requirements to assure analysis uniformity.

One of the critical difficulties in CPS security is the identification and modeling of risks. Traditional risk analysis attempts to quantitatively answer three questions: "What can go wrong?" "How likely is it?" and "What are the consequences?" [34].

While suitable for many engineering applications, risk analysis is uniquely difficult for cybersecurity:

- Adversaries are adaptive-it is difficult to identify threats.

- Once threats are identified, it is unclear how threats, vulnerabilities, and consequences can be reliably measured/quantified.

- High uncertainty makes such quantifications and estimates potentially unreliable.

- Unlike in some other fields of risk analysis, it is not clear what acceptable levels of cyber risk are.

- There is a disconnect between assessment of risks and implementation of mitigation plans.

To help guide test facilities in the determination of appropriate level of risk-based authentication testing, the SAE G-19A Standards Development Committee developed a risk-based method which leverages the limited information available about suspect counterfeit parts and assigns them a risk category which ultimately establishes baseline testing protocols [35]. Given the absence of empirical data on counterfeit frequency and severity, a semi-quantitative risk scoring method is proposed, in which risk levels are defined in well-specified ranges, giving users a shared understanding and common language to discuss risk management decisions. Further methodological innovations have been introduced [36] in the form of a comprehensive taxonomy of counterfeit defects and a novel algorithm for calculation of the Counterfeit Defect Coverage (CDC) metric, defined as the confidence of detecting a counterfeit component given a set of tests have been performed. These risk management approaches provide a quantitative approach to addressing anti-counterfeit and supply chain risk management, but do not address the other nine outlined areas of concern. Comparable risk management approaches are required for the remaining areas of concern.

Although many CPS security vulnerabilities may propose a complicated or expensive solution, many easy or relatively cheap solutions can be implemented by individuals to further prevent system compromise as identified in **Table 5.2**. Protection of each cyber entry point of the vehicle is paramount when the system is found to have inadequate defenses. Practicing security habits such as using a steering wheel lock (e.g., The Club 1000), installing a OBD port lock, or regular checks of the computer infotainment software version, diagnostic logs, and engine bay scans (e.g., wire harness scan) are some easy but admittedly less convenient prevention techniques [37]. Another practice would include turning off Bluetooth or wireless capabilities when not in use or even using physical accountability and secure storage of key fobs while avoiding unnecessary remote transmission in public places (e.g., lock doors using central door lock button instead of key fob after parking). Understandably, it can be difficult in today's fast-paced culture, but one of the best security practices is simply the user's familiarity with the CPSS and vigilance of diversions or anomalies. However, this type of risk mitigation requires user diligence to be effective and is not a true solution. Instead, these are user preventative responses to the inadequacies found in existing CPS solutions. This type of approach is not suitable as a long-term solution since user intervention is required to be effective, which is likely not sustainable. Until designed-in, trusted technical solutions

TABLE 5.2 Examples of cyberattack vectors

Point of introduction	Expression	Impact	Examples
Boot Process Vulnerabilities	Chain-of-trust exploitation	CPS unprotected relying on boot security features	• Based on identified vulnerabilities, adversary is able to introduce new hardware or software that allows unauthorized access • Weak key management compromises the personal data of thousands of users • A hacker gains unauthorized access to stored personal information such as home address, credit card information, etc. • Hacking of telematics system for denial of service to engine start for ransomware
Hardware Implementation Exploitation	Test/debugging port exploitation, reflashing external memory, timing attacks, etc.	CPS exposed through downgraded security features and security software/firmware bypass	• Adversary is able to reverse engineer critical safety measures and identify vulnerabilities for an entire fleet of vehicles • A hacker inserts false data into vehicle system to cause ECUs to fail
Chip-Level Exploitation	Onboard keys and secure protocols exploitation through semi-invasive and invasive chip-level integrated circuit exploitation	CPS compromised through exposure of sensitive encryption/decryption keys, protocols, and trusted boot sequences; CPS only as secure as the weakest point in any one of its parts	• Hackers target certain tainted hardware obtained from the broker market • Counterfeit parts fail earlier than intended, causing quality and reliability issues • Counterfeit parts include Trojans, allowing hackers to gain access to the system • Material passes through untrusted part of the supply chain and is altered by an adversary
Encryption and Cryptographic Hash Function Implementation	Security key and associated password or sensitive information exploitation by side-channel attacks, information based cryptanalysis methods, and the exploitation of improper hash implementation	CPS compromised through reduced key robustness caused by cryptographically weak hash and resulting encryption algorithm	• Hackers are able to guess passwords that do not meet basic hygiene best practices • Adversary is able to reverse engineer critical safety measures and identify vulnerabilities for an entire fleet of vehicles
Remote Access	Remote status, sensitive information, or "command and control" exploitation through improperly secured communication channels or over-the-air (OTA) access	CPS information and control exposed through administrative access bypassing user controls through unsecured or under secured communication channels including OTA (e.g., RF, Wi-Fi, Bluetooth, cellular, etc.)	• Remotely disable braking system of fleet through zero-day vulnerability in telematics systems • Hacking of keyless entry, remote start of car for theft • A hacker remotely gains access to the electronic control module and displays false dashboard readings • Hackers activate dormant malicious code that software allows remote access, causing malfunctions and safety issues • A hacker gaining access to active monitoring could gain control of vehicle functions

TABLE 5.2 *(Continued)* Examples of cyberattack vectors

Point of introduction	Expression	Impact	Examples
Software Exploitation	Scalable exploitations of software vulnerabilities similar to embedded systems and general computing	CPS similarly vulnerable to software-level attacks as traditional embedded systems and general computing with typically reduced update capabilities for mitigating software patches	• A hacker gains unauthorized access to stored personal information such as home address, credit card information, etc. • Hackers exploit unpatched vulnerabilities from unsupported, obsolete software or firmware settings • Vehicle does not have capability to address a zero-day vulnerability that is discovered until it is brought in for service

© SAE International

start replacing the resulting security advantages of these habits, these techniques can be quick and cheap prevention behaviors against CPSS vulnerabilities.

A solution example in response to the cyberattack vectors identified in Table 5.2 may include elements represented in **Figure 5.6**, as identified in a recent report by the U.S. Government Accountability Office [2]. For example, the ECU message encryption may require the integration of FlexRay or Ethernet messages with the capabilities of 254 bytes

FIGURE 5.6 Example of layered defenses designed to address CPS areas of concern and requirements flowdown, adapted from [2].

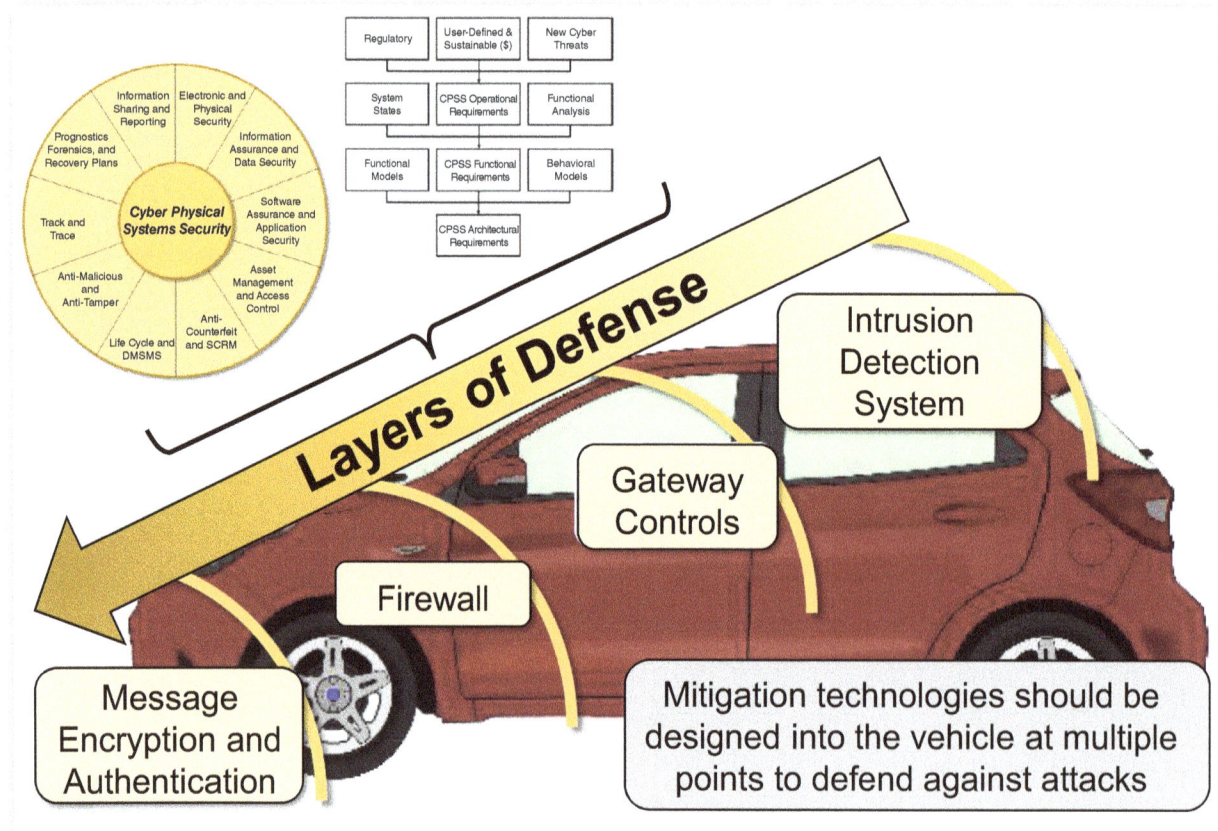

© SAE International

and 1500 bytes, respectively. This expanded level of communication could support the needed hash encryption required to mitigate cyberattacks of the safety-critical systems. Firewalls and gateway controls furthermore mitigate unauthorized access to the vehicle's CPS whether through software network firewalls or physical isolation between access points (e.g., telematics systems, radio systems, navigation systems, etc.) and critical communications. On-board design features for intrusion detection and prevention capability with performance monitoring in the form of system-on-a-chip (SoC) could identify and respond to a proficient attacker. Additionally, a supply chain risk screening test of the microelectronic components for embedded malware and hardware Trojans prior to system manufacturing can identify early malicious introduction in the systems life cycle. Software malware and virus intrusion detection can further supplement these other mitigations. These suggestions are merely an example of some features an implemented CPSS could include. These features should be informed by the areas of concern and the systems engineering requirements flowdown described above.

5.3.3 Verification and Validation

As demonstrated above, CPS are increasingly vulnerable to external attack, and it is critical that we are able to design systems that are resistant to such attack. However, current methodologies do not allow us to guarantee with any level of confidence that a particular systems design is secure. For example, traditional systems engineering processes have introduced verification as part of "Design V" or "V-model" [38] as shown in **Figure 5.7**. Along the left side of the V, designers progressively decompose and define a problem from requirements to architecture to design to implementation. Along the right side of the V, we do systematic testing to verify the design and architecture. Ideally, this is an iterative process done at multiple steps of the design process. We have modified this traditional V-model in Figure 5.7 to adapt it for secure systems engineering. At the high level, as we do system functional requirements, we must also determine security requirements including risk analysis. As we decompose the system, security requirements become more refined with threat modeling (understanding the possible threats on the system) and attack surface analysis (determining vulnerabilities at component interfaces). The verification

FIGURE 5.7   The design V-model for systems engineering adapted for security.

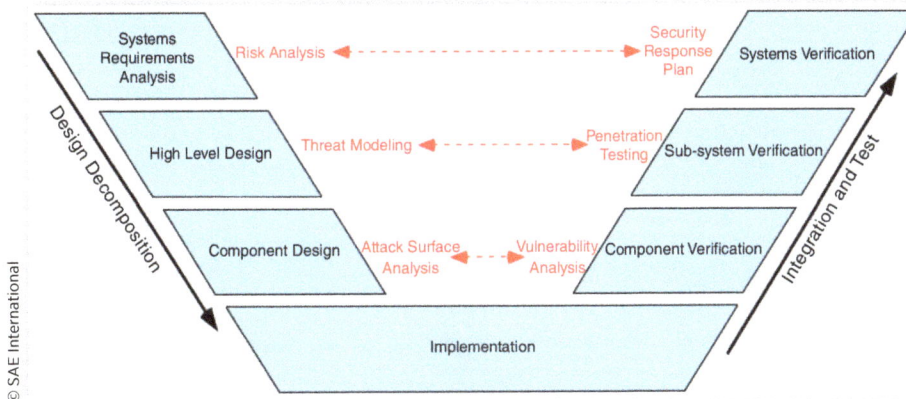

side requires active security probing such as penetration testing and a system deployment strategy to plan for security responses. NIST has developed a cybersecurity framework that discusses how organizations should address cyber risks through a process of identify, protect, detect, respond, and recover [39]. These processes only address the top level of the V-security requirements analysis and security response plan. There are no well-defined standard processes that address the lower levels of the V.

At these lower levels, the use of models could help and the automotive industry has begun using models as part of their functional systems engineering process [40]. However, in the context of security, devising comprehensive tests to assess security vulnerabilities is difficult. Models must be expanded in the security domain with analytical and formal models that can be verified with mathematical proofs that offer some confidence in a system's resilience against attacks. Formal techniques have been successfully applied in a number of functional domains and are beginning to be used to verify security properties as well [41, 42]. The difficult part is defining security properties and security requirements and determining the impacts of design and implementation decisions on these security requirements.

5.4 Conclusions and Recommended Next Steps

Research is needed to design and build real-world models and ranges supporting experimentation and validation for embedded malware, hardware Trojans, and CPS security. Mechanisms to run untested programs and tests with inputs from unverified or untrusted third parties or suppliers, without risk to the host CPS or user, are needed to ensure resilient systems. Security measures, if appropriately implemented, should not need to be clandestine. Ideally, iron-clad security measures could be published without risk to the system. Rigorous reviews conducted in cyber ranges with security enhancements to address identified vulnerabilities are an evolving process that is best managed using real-world models in a sandbox environment.

Operational CPSS modeling tools are needed to enable cost-effective, risk-based cyber resilience requirements. Standard work that identifies industry best practice for evaluating risk in each of the ten areas of concern previously discussed in this chapter and systems engineering approaches to resolve gaps are needed to support modeling tools. Standard work that includes metrics to evaluate the security coverage of the system will enable better understanding of the risk that can be used in the modeling tools and in support of solutions used to mitigate identified risk. The modeling tools can take credit for systems engineering solutions used to address identified risks, while balancing the consequences of occurrence for identified risks and costs needed to implement solutions. The definition of the operational requirements, which involves understanding the operating environment in which the system exists, can aid in the identification of relevant external factors and stressors that may be disruptive to the critical functionality of the system. It is important to consider the external environment from physical, information, cognitive, and social dimensions [43], which serves as the foundation for a risk assessment.

Research is needed for detection tools for embedded malware and hardware Trojans. The SAE G-19A has formed a Tampered subgroup. The purpose of the subgroup is to generate methods and metrics for the characterization, detection, and reporting of unauthorized features and design vulnerabilities that may lead to malicious tampering in microelectronic parts. The result of these solutions should enable a buyer of electronic parts to make informed risk-based decisions about the use of that part in their application.

SAE G-19A Tampered Subgroup Committee deliverables include:

1. Advancing the knowledge of how vulnerabilities and unauthorized features are introduced and exploited in electronic parts

2. Developing a detailed taxonomy of vulnerabilities and unauthorized features associated with electronic parts

3. Developing cost-effective test methods capable of detecting vulnerabilities and unauthorized features associated with electronic parts

4. Establishing and standardizing methods for detecting the presence of vulnerabilities and unauthorized features in electronic parts that could be introduced at any point in the component life cycle

5. Developing an AS6171 Test Method X, titled "Techniques for Suspect/ Counterfeit Tampered Device Detection by Various Test Methods"

Some of the information in this subject area will need to continue to be controlled by government classification programs and will need to be deferred to authorized U.S. agencies to ensure compliance to ITAR and EAR regulations and compliance with international laws. The committee will address only vulnerabilities that are currently known and published in international white papers or other venues available to the general public. The work of the G-19A Tampered subgroup is ongoing as of this publication. However, more research in this area is needed that includes laboratory analysis and peer review of new test methods.

Further research is needed to move from a risk-based view toward a resilience-based approach. Recognizing that a fully prevention-based perspective will eventually result in a successful attack, resilience focuses on the ability of the system (e.g., the vehicle, the supply chain) to minimize the effects of failures which will inevitably occur and quickly "bounce back" to a desirable, functional state [44]. Where risk assessment is difficult in low-likelihood/high-consequence events, resilience instead is focused on designing systems that recover quickly from disruptions, and learn and adapt to new operating environments [45, 46]. Resilience has been characterized by multiple capabilities, such as robustness (withstand disruptive forces), redundancy (satisfy functional requirements with substitutable system elements), resourcefulness (effectively leverage resources to diagnose and solve problems), and rapidity (recover quickly from a disruption) [47]. Modern CPS need a means to maintain awareness, detect evolving cyber threats, and to rapidly respond and recover. Fielded hardware will need periodic assessments that include updates to the system. Research for user assessment toolsets is needed to maintain sustainable trust and agility in a resilient trusted supply chain for CPS.

The "Functions" in the NIST cybersecurity framework (identify, protect, detect, respond, and recover [39]) are complementary with other quality decision-making tools such as Deming's Plan-Do-Study-Act (PDSA) cycle [48]. The first NIST function, "Identify," involves developing "the organizational understanding to manage cybersecurity risk," including understanding the business context and allocation of necessary resources. This is consistent with "Plan," which includes forming objectives, forming hypotheses, and planning the execution of the project. Next, the NIST functions of "Protect," "Detect," "Respond," and "Recover" broadly fit within the "Do" step in the PDSA cycle. The NIST functions represent the development and implementation of the protective system safeguards and procedures, detection of cybersecurity threats, executing proper responsive countermeasures to detected threats, and quickly recovering functionality in the event of a successful attack. This is all related to the activities in the "Do" step, which is focused on carrying out the plan and documenting the results. The final two steps of the PSDA cycle, "Study" (where data is analyzed and compared to predictions) and "Act" (where changes and adaptations are made for the next cycle iteration), collectively map onto NIST sub-functions (called "categories") of "Respond: Improvements" and "Recover: Improvements," which specifically identify the need to incorporate lessons learned and update response and recovery strategies. Viewed together, the NIST functions can be used within a PDSA cycle for continuous improvement and resilience building within the organization and CPS.

Support to emerging system-on-a-chip architectures is needed for designed-in cyber resiliency and security. System-on-a-chip architectures can be designed to validate software, hardware, and firmware settings. They can also be used to monitor security of the system through prognostic and diagnostic tools built into the chip. System-on-a-chip architectures can also be used to enhance encryption and validate updates and changes to the system.

Support and further development of emerging track and trace authentication taggants are needed. While botanical DNA solutions and RFID technologies assist in track and trace and authentication of devices, these technologies have their limitations. New optical photon-counting security tagging and verification of integrated circuits (IC) using optically encoded QR codes might present a low-cost mechanism for authentication of ICs and could be used to store data on the device(s) being verified [49].

The IT industry's use of penetration testing and code reviews should be adopted in the security evaluations of CPS. Black hat teams, peer review, and evaluation tools analyzing the system and software ensure resilient designs built to withstand attacks. Engineering teams should collaborate with researchers and independent security firms to test vehicles' digital security and identify CPS security vulnerabilities and solutions to resolve them.

Designs should incorporate domain separation for in-vehicle networks and safety-critical systems. Domain separation of safety-critical systems from the network improves the security and integrity of the vehicle's safety systems. If it is not possible to isolate the safety systems from the in-vehicle networks, firewalls and encryption techniques should be incorporated into the design of safety-critical systems to ensure systems are not compromised and function as intended.

Systems designs should implement a layered approach to security. This includes various authentication and security measures. For instance, passwords can be significantly enhanced by requiring alpha-numeric text and biometric inputs for authentication.

Encryption, domain separation, and firewalls between critical systems enhance security. A layered approach significantly enhances the security of a CPS.

There is a need to develop over-the-air update capabilities, particularly for systems that are already connected to the Internet. Systems designers can't anticipate all vulnerabilities of complex CPS. Updates to system software and firmware are needed as zero-day vulnerabilities are discovered and patches to these issues are developed. Over-the-air updates ensure resiliency and rapid recovery of vulnerable systems.

Finally, organizations should hire dedicated staff and high-level managerial positions focused on CPS security. The cost of an attack and the impact to the organization's brand and reputation far outweighs the cost of a dedicated staff addressing the issue. While no organization can be guaranteed to be hack-free with or without dedicated staff, the problem isn't going away. Consumers value their privacy and security and will not tolerate organizations who ignore the issue. A dedicated staff can focus on the problem and develop solutions that will bring value to systems that are designed for security.

In summary, modern cyber-enabled features such as adaptive cruise control, collision warning systems, built-in navigation, Bluetooth systems, and braking systems are all vulnerable points of attack that makes the case for a systems engineering approach to CPS security to bolster safety, enable resilience, and protect consumers. As technology evolves and the line continues to blur between physical products and electronic hardware, software, and firmware, strategies to ensure these systems are secure should be a significant priority. Much more effort is needed to address security of vehicles and CPS at large. The example vulnerabilities listed in this chapter are not unique to vehicles, and the same risk analysis and systems engineering approach could be applied to any CPS in any industry, including aircraft, buildings, and the critical infrastructure that supports our everyday way of life. This chapter should be a call for action which demonstrates the need for government, industry, and academic collaboration to promote research and advance the state of the art in CPS security. Efforts such as industry standard work to help codify efforts to address vulnerabilities and development of college curriculum to train the future workforce are needed. The time for action is long overdue. Industry, government, and academia cannot wait for a disaster to occur before taking action to address some of the world's most critical challenges around safety and security.

References

1. 42 U.S. Code § 7521(a)(6), "Emission Standards for New Motor Vehicles or New Motor Vehicle Engines."

2. U.S. Government Accountability Office, "Vehicle Cybersecurity. DOT and Industry Have Efforts Under Way, but DOT Needs to Define Its Role in Responding to a Real-World Attack," GAO-16-350, U.S. Government Accountability Office, Washington, DC, 2016.

3. Kelley Blue Book, "Kelley Blue Book Insights Data," *RSA Conference 2016*, San Francisco, CA, February 29-March 4, 2016.

4. Burakova, Y., Hass, B., Millar, L., and Weimerskirch, A., "Truck Hacking: An Experimental Analysis of the SAE J1939 Standard," *10th USENIX Workshop on Offensive Technologies (WOOT 16)*, Austin, TX, 2016.

5. Jafarnejad, S., Codeca, L., Bronzi, W., Frank, R. et al., "A Car Hacking Experiment: When Connectivity Meets Vulnerability," *2015 IEEE Globecom Workshops (GC Workshops)*, San Diego, CA, 2015.

6. Bennett, J., "Thieves Go High-Tech to Steal Cars," *The Wall Street Journal*, http://www.wsj.com/articles/thieves-go-high-tech-to-steal-cars-1467744606, accessed September 22, 2016.

7. Greenberg, A., "Hackers Remotely Kill a Jeep on the Highway—With Me in It," *Wired Magazine*, https://www.wired.com/2015/07/hackers-remotely-kill-jeep-highway, accessed September 22, 2016.

8. Breeden, J., "Why Car-Hacking Could Threaten the Federal Government," NextGov, http://www.nextgov.com/cybersecurity/2016/05/vehicle-vulnerabilities-could-force-unwanted-pit-stops/128674, accessed September 22, 2016.

9. Rubin, A.J., Blaise, L., Nossiter, A., and Breeden, A., "France Says Truck Attacker Was Tunisia Native with Record of Petty Crime," *New York Times*, http://www.nytimes.com/2016/07/16/world/europe/attack-nice-bastille-day.html, accessed September 22, 2016.

10. Suciu, P., "Ransomware: The Next Big Automotive Cybersecurity Threat?," Car and Driver, http://blog.caranddriver.com/ransomware-the-next-big-automotive-cybersecurity-threat/, accessed February 26, 2017.

11. Markoff, J., "Want to Buy a Self-Driving Car? Big-Rig Trucks May Come First," *New York Times*, http://www.nytimes.com/2016/05/17/technology/want-to-buy-a-self-driving-car-trucks-may-come-first.html, accessed September 22, 2016.

12. Klinedinst, D., and King, C., *On Board Diagnostics: Risks and Vulnerabilities of the Connected Vehicle*, (Pittsburgh: Software Engineering Institute, Carnegie Mellon University, 2016).

13. Eisenstein, P.A., "Automakers Turn to Hackers for Help in Improving Cybersecurity," *NBC News*, http://www.nbcnews.com/tech/tech-news/automakers-turn-hackers-help-improving-cybersecurity-n610271, accessed February 26, 2017.

14. Hoffenberg, S., and Rommel, C., *Automotive Cybersecurity: Meeting the High-Stakes Challenge*, (Natick: VDC Research, 2015).

15. Koscher, K., Czeskis, A., Roesner, F., Patel, S. et al., "Experimental Security Analysis of a Modern Automobile," *2010 IEEE Symposium on Security and Privacy*, Berkeley/Oakland, CA, 2010, 447-462.

16. Dawson, C., "Car's Data Transmitter Can Be Hacked to Take Control, Firm Says," *The Wall Street Journal*, https://www.wsj.com/articles/cars-data-transmitter-can-be-hacked-to-take-control-firm-says-1492088402, accessed April 19, 2017.

17. National Institute of Standards and Technology, "Framework for Cyber-Physical Systems. Release 1.0." Cyber Physical Systems Public Working Group, NIST, Gaithersburg, MD, 2016.

18. NITRD, "Cyber Physical Systems Vision Statement," Networking and Information Technology Research and Development Program, Arlington, VA, 2015.

19. Smirnov, A., Kashevnik, A., Shilov, N., Makklya, A. et al., "Context-Aware Service Composition in Cyber Physical Human System for Transportation Safety," *2013 13th International Conference on ITS Telecommunications (ITST)*, Tampere, Finland, 2013, 139-144.

20. Xiong, G., Zhu, F., Liu, X., Dong, X. et al., "Cyber-Physical-Social System in Intelligent Transportation," *IEEE/CAA Journal of Automatica Sinica* 2, no. 3 (2014):320-333, doi:10.1109/JAS.2015.7152667.12.

21. INCOSE, INCOSE Systems Engineering Handbook, Version 3.2.2, INCOSE-TP-2003-002-03.2.2, International Council on Systems Engineering (INCOSE), San Diego, CA, 2012.

22. DiMase, D., Collier, Z.A., Heffner, K., and Linkov, I., "Systems Engineering Framework for Cyber Physical Security and Resilience," *Environment Systems & Decisions* 35, no. 2 (2015):291-300, doi:10.1007/s10669-015-9540-y.

23. Rouf, I., Miller, R., Mustafa, H., Taylor, T. et al., "Security and Privacy Vulnerabilities of In-Car Wireless Networks: A Tire Pressure Monitoring System Case Study," *USENIX Security Conference*, Washington, DC, 2010, 323-338.

24. Villasenor, J., *Compromised by Design? Securing the Defense Electronics Supply Chain*, (Washington: Brookings Institution, 2013).

25. Gaffey, C., "German Missiles 'Hacked by Foreign Source," *Newsweek*, http://europe.newsweek.com/german-missiles-hacked-by-foreign-source-329980, accessed March 30, 2017.

26. Defense Microelectronics Activity (DMEA), "Trusted Accreditation," http://www.dmea.osd.mil/trustedic.html, accessed September 22, 2016.

27. Valasek, C., and Miller, C. "Remote Exploitation of an Unaltered Passenger Vehicle," *IOActive*, Seattle, WA, 2015.

28. Hotten, R., "Volkswagen: The Scandal Explained," *BBC News*, http://www.bbc.com/news/business-34324772, accessed March 30, 2017.

29. Kaplan, J., "Verizon's Hum Gives Your Old Car a New Brain," Digital Trends, http://www.digitaltrends.com/cars/verizons-hum-gives-your-old-car-a-new-brain/, accessed October 23, 2016.

30. Al-Kuwari, S. and Wolthusen, S.D., "On the Feasibility of Carrying Out Live Real-Time Forensics for Modern Intelligent Vehicles," in Lai, X. et al. (Eds.), *E-Forensics 2010 LNICST*, Vol. 56, Springer-Verlag, 2011, 207-223.

31. Nilsson, D.K. and Larson, U.E., "Combining Physical and Digital Evidence in Vehicle Environments," *IEEE Third International Workshop on Systematic Approaches to Digital Forensic Engineering*, Berkeley, CA, 2008, 10-14.

32. DiMase, D., Collier, Z.A., Carlson, J., Gray, R.B. et al., "Traceability and Risk Analysis Strategies for Addressing Counterfeit Electronics in Supply Chains for Complex Systems," *Risk Analysis* 36, no. 10 (2016):1834-1843, doi:10.1111/risa.12536.

33. Lintern, G. *Tutorial: Work Domain Analysis*, (Victoria: Cognitive Systems Design, 2016).

34. Kaplan, S. and Garrick, B.J., "On the Quantitative Definition of Risk," *Risk Analysis* 1, no. 1 (1981):11-27, doi:10.1111/j.1539-6924.1981.tb01350.x.

35. Collier, Z.A., Walters, S., DiMase, D., Keisler, J.M. et al., "A Semi-Quantitative Risk Assessment Standard for Counterfeit Electronics Detection," *SAE Int. J. Aerospace* 7, no. 1 (2014):171-181, doi:10.4271/2014-01-9002.

36. Guin, U., DiMase, D., and Tehranipoor, M., "A Comprehensive Framework for Counterfeit Defect Coverage Analysis and Detection Assessment," *Journal of Electronic Testing* 30, no. 1 (2014):25-40, doi:10.1007/s10836-013-5428-2.

37. Threat Brief, "Cybersecurity and Hacking Concerns for Today's Cars and Tomorrow's Driverless Vehicles," http://threatbrief.com/cybersecurity-hacking-concerns-todays-cars-tomorrows-driverless-vehicles, accessed March 22, 2017.

38. MITRE Corporation, "Verification and Validation," MITRE Systems Engineering Guide, https://www.mitre.org/publications/systems-engineering-guide/se-lifecycle-building-blocks/test-and-evaluation/verification-and-validation, accessed October 23, 2016.

39. National Institute of Standards and Technology, "Framework for Improving Critical Infrastructure Cybersecurity, Version 1.0," NIST, Gaithersburg, MD, 2014.

40. Davey, C., "Automotive Software Systems Complexity: Challenges and Opportunities," *INCOSE International MBSE Workshop*, Jacksonville, FL, 2013.

41. Sanwal, M.U., and Hasman, O., "Formal Verification of Cyber-Physical Systems: Coping with Continuous Elements," *Computational Science and Its Applications—ICCSA 2013*, Ho Chi Minh City, Vietnam, 2013, 358-371.

42. Klein, G., Andronick, J., Elphinstone, K., Heiser, G. et al., "seL4: Formal Verification of an Operating-System Kernel," *Communications of the ACM* 53, no. 6 (2010):107-115, doi:10.1145/1743546.1743574.

43. Linkov, I., Eisenberg, D.A., Plourde, K., Seager, T.P. et al., "Resilience Metrics for Cyber Systems," *Environment Systems & Decisions* 33, no. 4 (2013):471-476, doi:10.1007/s10669-013-9485-y.

44. Collier, Z.A., DiMase, D., Heffner, K., and Linkov, I., "Building a Trusted and Agile Supply Chain Network for Electronic Hardware," *20th International Command and Control Research and Technology Symposium (ICCRTS)*, Annapolis, MD, June 16–19, 2015.

45. Park, J., Seager, T.P., Rao, P.S.C., Convertino, M. et al., "Integrating Risk and Resilience Approaches to Catastrophe Management in Engineering Systems," *Risk Analysis* 33, no. 3 (2013):356-367, doi:10.1111/j.1539-6924.2012.01885.x.

46. Fiksel, J., Polyviou, M., Croxton, K.L., and Pettit, T.J., "From Risk to Resilience: Learning to Deal with Disruption," *MIT Sloan Management Review* 56, no. 2 (2015):1-8.

47. Tierney, K. and Bruneau, M., "Conceptualizing and Measuring Resilience: A Key to Disaster Loss Reduction," *TR News* 250 (2017):14-17.

48. Moen, R.D. and Norman, C.L., "Circling Back: Clearing Up Myths about the Deming Cycle and Seeing How It Keeps Evolving," *Quality Progress* 2010 (November 2010): 22-28.

49. Markman, A., Javidi, B., and Tehranipoor, M., "Photon-Counting Security Tagging and Verification Using Optically Encoded QR Codes," *IEEE Photonics Journal* 6, no. 1 (2014):1-9, doi:10.1109/JPHOT.2013.2294625.

About the Authors

Daniel DiMase is the Director of Compliance and Quality at Honeywell International Inc., working on the Quality Systems & Regulatory Compliance team for the Aerospace Strategic Business Group. Daniel has primary responsibility for mitigating the severe and growing threat that counterfeit parts pose to Honeywell Aerospace's highly diversified business operations. He guides the implementation of internal policies, procedures, and tools to mitigate counterfeit and cyber physical systems security risks related to product assurance, safety, and security, and manages customer and regulatory concerns in those areas. Daniel co-authored numerous peer-reviewed technical publications on supply chain risk management, counterfeit avoidance and detection, and cyber physical systems security. He is a long-term collaborator with industry, academia, and government teams commissioned with advancing the state of the art and scientific knowledge in related topics.

Daniel DiMase is an active member of SAE International's G-19 Counterfeit Electronic Parts Document Development group. He is the SAE Chairman Emeritus of the Test Laboratory Standards Development committee, co-chairman of the Distributor Process Rating committee, and participates in other committees and working groups addressing counterfeit avoidance and detection and cyber physical systems security. He is a proven contributor to the executive committee of the Aerospace Industry Association's Counterfeit Parts Integrated Projects Team. He provides ideation in his role as a member of the Department of Homeland Security's Customs and Border Protection Advisory Committee on Commercial Operations in the Intellectual Property Rights subcommittee and on the Government-Industry Data Exchange Program (GIDEP) Industry Advisory Group. Daniel received a special recognition award at the DMSMS and Standardization 2011 Conference for his industry leadership role in mitigating counterfeit parts. He also received the 2017 Newman Award for Supply Chain Excellence by the National Defense Industrial Association (NDIA) Manufacturing Division. In addition, he received the SAE 2017 Arch T. Colwell Cooperative Engineering Medal for his instrumental role in formulating solutions to the evolving counterfeit parts problem.

Daniel DiMase has over 28 years of industry experience, previously serving in leadership positions as president of SemiXchange, Inc. and ERAI. He is a results-oriented leader with expertise in business growth, supply-chain, operations and finance, international logistics, global sourcing, risk management, and strategic planning. He has a Six-Sigma Green Belt Certificate from Bryant University. He received his Bachelor of Science degree in Electrical Engineering from The University of Rhode Island and an Executive MBA from Northeastern University.

Zachary A. Collier earned his PhD in 2018 from the Department of Systems and Information Engineering at the University of Virginia. He earned a master of engineering management from Duke University and a bachelor of science in mechanical engineering from Florida State University. Collier's work experience includes the U.S. Army Engineer Research and Development Center, where as a member of the Risk and Decision Science Team he developed and implemented strategic decision-making frameworks for government and industry clients on topics such as supply chain risk management, energy security, technology acquisition, and critical infrastructure. He has also held engineering consulting positions in the area of traffic accident reconstruction. His research is focused on assessing and managing enterprise risks in support of long-term strategic planning, decision modeling, and project management. He is a member of SAE's G-19A Standards Development Committee, responsible for the publication of the AS6171 Test Methods Standard for Suspect/Counterfeit, Electrical, Electronic, and Electromechanical Parts.

Prof. John A. Chandy is a Professor and the Associate Head of the Electrical and Computer Engineering Department at the University of Connecticut. Prof. Chandy is also Co-Director of the Connecticut Cybersecurity Center; Interim Director of the UConn Center for Hardware Assurance, Security, and Engineering; and Co-Director of the Comcast Center for Cybersecurity Innovation. Prior to joining UConn, he had executive and engineering positions in software companies working particularly in the areas of clustered storage architectures, tools for the online delivery of psychotherapy and soft-skills training, distributed architectures, and unstructured data representation. His current research areas are in high-performance storage systems, reconfigurable computing, embedded systems security, distributed systems software and architectures, and multiple-valued logic. Dr. Chandy earned Ph.D. and M.S. degrees in electrical engineering from the University of Illinois in 1996 and 1993, respectively, and a S.B. in electrical engineering from the Massachusetts Institute of Technology in 1989.

Bronn Pav is an aerospace electronic systems engineer with nearly a decade of experience including Cyber-Physical Systems Security (CPSS) technology development and integration. His background originated in systems engineering for DoD programs with experience in cryptographic technology, U.S. weapon, satellite, and aircraft systems. His experience includes employment with GE Aviation, Honeywell Defense and Space, and collaboration with the National Labs including the National Nuclear Security Administration (NNSA). His roles include subject matter expertise in malicious detection and avoidance of microelectronics vulnerabilities and supply chain trust as part of systems security engineering. He has also acted as program technical lead, test lead, and counterfeit mitigation lead during systems security prototype development, manufacturing, and validation. Pav has evaluated multiple platforms and advised U.S. government prime contractors in recent supply chain trust regulations, processes, requirements, and associated electronic designs for avoidance and detection of malicious vulnerabilities. Bronn is an active participant in the SAE International's G-19 Counterfeit Electronic Parts Document Development group and leads the G-19A Tampered Detection Test Methodology subcommittee for methods in detecting tampered microelectronic parts for Trojans and malware. He holds a BSEE in electrical engineering from the University of South Florida with graduate work from Liberty University. Bronn is also certified with a green belt in Design for Six Sigma (DFSS). His industry recognition includes Honeywell Outstanding Engineer for 2016, NASA ISC group achievement award for his work on the Exploration Flight Test 1 of the Orion capsule, and a contributor to the X-47b, which is a recipient of the NAA Robert J. Collier Trophy.

Kenneth Heffner received his Ph.D. degree in chemistry from the University of South Florida, Tampa, FL. He is currently an Engineering Senior Fellow for Honeywell Aerospace in Clearwater, FL, supporting Honeywell's Aerospace business units. He is the technology leader for Honeywell's new Systems Security Engineering business unit. His research includes sensors for inertial navigation systems, autonomous thin film instrumental analysis, high-density vertically integrated microsystems, high-performance computing, and embedded secure microelectronics systems. Dr. Heffner holds 16 U.S. patents. He is also a certified Design for Six Sigma black belt for hardware design.

Steve Walters leads the Honeywell Aerospace Antennas and Space Electronics Reliability Engineering group and the Honeywell Clearwater site Failure Analysis Lab. Steve has been actively involved with the identification and mitigation of counterfeit components since 2009. His primary area of focus has been the development of counterfeit part detection methodologies, which included the development of the Honeywell counterfeit avoidance testing process and the development and implementation of a third-party test labs certification and proficiency testing methodology. Steve has also been active in the G-19A Test Laboratory Standards Development Committee. In this role, Steve led the risk assessment and mitigation subgroup responsible for risk model design and development. Steve holds a BSEE in electrical engineering from the University of South Florida. He is also Design for Six Sigma (DFSS) green belt certified.

CHAPTER 6

"When Trucks Stop, America Stops"

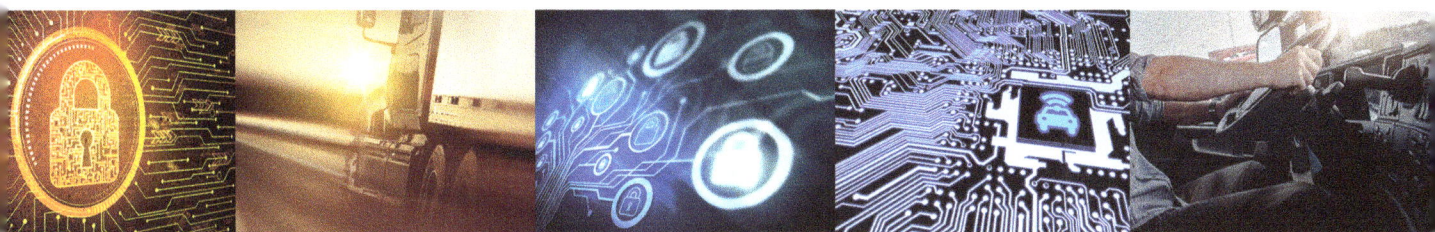

Commercial truck traffic is vital to our nation's economic prosperity and plays a significant role in mitigating adverse economic effects during a national or regional emergency. Our economy depends on trucks to deliver ten billion tons of virtually every commodity consumed—or nearly 70 percent of all freight transported annually in the U.S. In the U.S. alone, this accounts for $671 billion worth of goods transported by truck. Add $295 billion in truck trade with Canada and $195.6 billion in truck trade with Mexico and it becomes apparent that any disruption in truck traffic will lead to rapid economic instability.

The unimpeded flow of trucks is critical to the safety and well-being of all Americans. However, it is entirely possible that well-intended public officials may instinctively halt or severely restrict truck traffic in response to an incident of national or regional significance.

Recent history has shown us the consequences that result from a major disruption in truck travel. Immediately following the 9/11 terrorist attacks, significant truck delays at the Canadian border crossings shut down several auto manufacturing plants in Michigan because just-in-time parts were not delivered. The economic cost to these companies was enormous. Following Hurricane Katrina, trucks loaded with emergency goods were rerouted, creating lengthy delays in delivering urgently needed supplies to the stricken areas.

Although in the face of an elevated threat level, a terrorist attack, or a pandemic, halting truck traffic may appear to be the best defense, it actually puts citizens at risk. Officials at every level of government must recognize that a decision to halt or severely curb truck traffic following a national or regional emergency will produce unintended health and economic consequences not only for the community they seek to protect, but for the entire nation.

The American Trucking Associations researched seven key consumer industries to quantify the potential consequences of restricting or halting truck traffic in response to a national or regional emergency. This report details the findings.

The Food Industry

Every day, Americans purchase billions of dollars of groceries. Most of these goods are brought to market via daily truck deliveries.

- Significant shortages will occur in as little as three days, especially for perishable items following a national emergency and a ban on truck traffic. Minor shortages will occur within one to two days. At convenience stores and other small retailers with less inventory, shortages will occur much sooner.

- Consumer fear and panic will exacerbate shortages. The forecast of a winter storm quickly exhausts basic commodities at grocery stores and supermarkets. It takes retailers up to three days to recover from these runs on supplies. News of a truck stoppage—whether on the local level, state or regional level, or nationwide—will spur hoarding and drastic increases in consumer purchases of essential goods. Shortages will materialize quickly and could lead to civil unrest.

- Supplies of clean drinking water will run dry in two to four weeks. According to the American Water Works Association, Americans drink more than one billion glasses of tap water per day. For safety and security reasons, most water supply plants maintain a larger inventory of supplies than the typical business. However, the amount of chemical storage varies significantly and is site specific. According to the Chlorine Institute, most water treatment facilities receive chlorine in cylinders (150 pounds and one ton cylinders) that are delivered by motor carriers. On average, trucks deliver purification chemicals to water supply plants every seven to 14 days. Without these chemicals, water cannot be purified and made safe for drinking. Without truck deliveries of purification chemicals, water supply plants will run out of drinkable water in 14 to 28 days. Once the water supply is drained, water will be deemed safe for drinking only when boiled. Lack of clean drinking water will lead to increased gastrointestinal and other illnesses, further taxing an already weakened healthcare system.

Healthcare

Both healthcare providers and consumers rely on regular delivery of medications and healthcare supplies to hospitals, pharmacies, nursing homes and other healthcare facilities. Trucks deliver nearly all of these supplies. Al Cook, former president of the Materials Management Association and current member of the Medical Materials Coordinating Group, which is advising the U.S. Department of Health and Human Resources on emergency preparedness, describes over-the-road commercial transportation as "life and death to being able to care for sick people."

- Without truck transportation, patient care within the truck stoppage zone will be immediately jeopardized. According to Cook, many hospitals have moved to a just-in-time inventory system. In fact, some work from a low-unit-of-measure system. This means that essential basic supplies, such as syringes and catheters, are not ordered until the supplies are depleted. These systems depend on trucks to deliver needed supplies within hours of order placement. Internal redistribution of supplies in hospitals could forestall a crisis for a short time; however, in a matter of hours, hospitals would be unable to supply critical patient care.

- If an incident of national significance produces mass injuries, truck transportation is the key to delivering urgently needed medical supplies necessary to save lives. According to Cook, there are not enough medical supplies in any local area to support a large scale medical emergency. The Medical Materials Coordinating Group has worked with U.S. Health and Human Services to develop contingency plans that will coordinate national redistribution of essential medical supplies during a national emergency.

These contingency plans ensure that affected areas receive adequate medical supplies to support the crisis while also maintaining adequate supplies for the basic medical needs of the larger community. Cook states that the medical redistribution program relies on trucks as the primary mode of transport to carry out the expedient redistribution of supplies, and ties the success of the program to the ability of trucks to access medical facilities and warehouses during an emergency situation.

- Hospitals and nursing homes will exhaust food supplies in as little as 24 hours. Hospitals and nursing homes receive daily truck deliveries of food for patients. The International Food Distributors Association notes that because these facilities lack significant warehousing capabilities, a truck stoppage will result in food shortages within 24 to 48 hours, particularly among perishable items.

- Pharmacy stocks of prescription drugs will be depleted quickly. Although pharmacies typically carry several weeks inventory of many drugs, seasonal pharmaceuticals, such as antibiotics and flu-shots during winter months, have faster turnover rates. According to the National Association of Chain Drug Stores, most of the nation's 55,000 drug stores receive daily merchandise deliveries by truck.

- Hospitals and other diagnostic and treatment facilities will exhaust supplies of radiopharmaceuticals and oxygen. Radiopharmaceutical supplies for cancer treatment and diagnostic services, which have an effective life of only a few hours, will become unusable. Hospital size and storage capacity determine the amount of oxygen a facility can maintain; however, in general, hospitals will exhaust oxygen supplies within seven to ten days.

Transportation

The impact of a truck stoppage would not be limited to highway transportation but would affect all modes of transportation. Trucks are the fundamental unit within the transportation sector. Trucks transport just about all cargo to and from air and rail terminals and maritime ports. Trucks also deliver fuel to the majority of rail yards. The Air Transport Association estimates that trucks account for approximately 80 percent of the fuel deliveries to the nation's airports. Truck transport is the mechanism for fuel delivery to service stations and truck stops.

- Service station fuel supplies will start to run out in just one to two days. According to the Service Station Dealers of America, the nation's busiest fuel stations sell between 200,000 and 300,000 gallons per month. These stations require multiple deliveries every day to meet this demand. An average service station requires a delivery every 2.4 days. Based on these statistics, the busiest service stations could run out of fuel within hours of a truck stoppage, with the remaining stations following within one to two days. Researchers predict that automobile travel will cease within one week if fuel deliveries are halted.

- Air, rail and maritime transportation will be disrupted. Airlines and air cargo operations will be grounded due to the lack of supplies for operations. Railroads will cease all truck trailer-on-flat-car or piggy-back and container operations. Rail freight will pile up at rail terminals since intermodal trucks provide the first and last mile in intermodal moves. Smaller railroads (Classes 2 and 3) will stop operation due to lack of truck-supplied fuel and will create significant congestion on feeder lines to the large (Class 1) railroads. If the truck stoppage occurred in a coastal region, inbound and outbound container ships, which rely on intermodal truck transportation, will sit idle at the maritime ports.

- A fuel shortage will create secondary effects. Without access to automobile travel, people will be unable to get to work causing labor shortages and increased economic damage. Without cars, many people cannot access grocery stores, banks, doctors, and other daily needs. Public bus systems will cease to operate as well, preventing many disabled and elderly people from accessing these necessities. Without fuel, police, fire, rescue and other public service vehicles will be paralyzed, further jeopardizing public safety. U.S. Mail and other package delivery operations will cease. Within two days, garbage will start to pile up in urban and suburban areas due to a lack of regular pickups, creating a health hazard.

Waste Removal

The Environmental Protection Agency estimates that Americans generate more than 236 million tons of municipal or household waste annually. This does not take into account manufacturing, medical, or other types of commercial waste.

- Within days of a truck stoppage, Americans will be literally buried in garbage with serious health and environmental consequences. Further, without fuel deliveries, many waste processing facilities will be unable to operate equipment such as backhoes and incinerators.

- Uncollected and deteriorating waste products create rich breeding grounds for microorganisms, insects, and other vermin. Hazardous materials and medical waste will introduce toxins as well as infectious diseases into living environments. Urban areas will, of course, be significantly impacted within just a couple of days. But rural and agricultural areas as well as food processing plants will be impacted as well. Without waste removal and treatment, food wastes, such as slaughtered animal byproducts and overripe fruits and vegetables will create extremely toxic conditions.

- Beyond the health risks, a lack of waste removal creates pollution hazards. Biological pathogens, hazardous chemicals and even radioactive materials will be released into the soil, water, and atmosphere. And, the sheer volume of uncollected waste could block water run-off and drainage leading to pooling and flooding.

The Retail Sector

A disruption of truck deliveries to retail outlets will have serious effects on both consumers and retailers.

- Replenishment of goods will be disrupted. Many of the nation's leading retailers rely on just-in-time delivery to keep inventory levels as low as possible. Similar to the low-unit-of-measure hospital inventory system, these stores rely on frequent deliveries to replenish basic goods. Often, delivery of a shipment is not triggered until the current inventory is nearly depleted. Without truck deliveries, retailers will be unable to restock goods, including consumer basics such as bottled water, canned goods, and paper products.

- Consumer behavior during emergencies triples the rate of inventory turn-over. Since many large retail outlets typically keep inventories as lean as possible, problems often arise quickly during truck transportation slowdowns that occur from crises such as hurricanes.

In a hurricane situation, supplies that would normally last a few days, such as water, powdered milk, and canned meat, typically disappear within one day. Given these inventory rates, this means that perishable goods could be depleted in a matter of days and non-perishables in just a few days. Runs on food and non-food staples during hurricanes, and even before big winter storms, provide a good example of how fast some retail inventories can be depleted during panic buying. The same quick depletion of inventories could occur if trucks stopped making deliveries for any reason.

Manufacturing

In recent years, manufacturers in the United States have shifted increasingly to just-in-time manufacturing. Aimed at improving efficiency, just-in-time manufacturing reduces the need for extensive warehousing of manufacturing components because parts and components are delivered to the production line just hours before assembly. As a result, manufacturers are extremely sensitive to disruptions in the supply chain.

- Just-in-time manufacturers will shut down assembly lines within hours. Major American manufacturers, ranging from computer manufacturers such as Dell and Compaq to major automakers such as GM and Ford, rely on just-in-time manufacturing. Without truck deliveries, component shortages and manufacturing delays will develop within hours. If assembly lines are forced to shut down, manufacturers will incur significant disruption costs and thousands of employees will be put out of work. Just-in-time manufacturers also rely on trucks to transport goods to the market within hours of assembly. A truck stoppage will cripple deliveries to retailers.

Banking & Finance

Even with today's high-tech electronic exchange of currency and information, trucks play a critical role in transporting hard copies of financial documents and currency. Disruption of truck deliveries to banks and ATMs will paralyze the banking industry, affecting both consumers and businesses. The bottom-line: cash is still heavily used as legal tender.

- ATM and branch bank cash resources will be exhausted quickly. In today's fast paced, high-technology economy, consumers access cash 24/7 from 370,000 ATMs nationwide. JP Morgan Chase, the nation's second largest consumer bank, replenishes its 6,600 ATMs via armored truck delivery every two to three days. Given the increase in ATM activity that occurs before and after any type of crisis, ATMs would run out of cash much sooner.

- Small and medium-size businesses will lose access to cash. Banks provide daily cash and coin deliveries to thousands of small and medium-size businesses via armored trucks. According to JP Morgan Chase, without these daily deliveries and collections, the ability of businesses to carry out normal commercial transactions will be disrupted and eventually cease.

- Regular bank functions will cease. Bank branches transfer paper documents for every transaction, via daily truck service, to a central location for processing. Unable to conduct transactions at a central location, banks will be unable to process deposits, checks, and other standard bank transactions, bringing the American banking system to a halt.

Other Effects

In addition to the effects on the key industries outlined in this paper, a ripple effect could significantly disrupt a variety of services and activities beyond the affected area, extending into communities nationwide.

For example, the Department of Defense (DoD) supply chain includes 1,100 shipping points inside the United States, connects to airports and seaports, and is the supply lifeline to warfighters deployed globally. Nearly all DoD freight involves truck movement and all trucks with DoD freight are subject to orders by local law enforcement. Stopping trucks with DoD freight would ultimately cripple the Department of Defense in ways no adversary has been able to achieve.

A truck stoppage in the Great Lakes region will close auto manufacturing, steel mills, and other major industries, not only putting thousands of workers out of work, but also disrupting the flow of automobiles and steel products to the rest of the nation. A stoppage at harvest time in or around any of the nation's agricultural regions will cut off the transport of fresh foods, such as fruits, vegetables, and grains. Not only would Americans be deprived of fresh foods, the economic impact to the farming industry would be devastating.

Consider a truck stoppage in the states around Washington, D.C. The federal government will be slowed down within a week and could grind to a halt within two or three weeks. Moreover, a truck stoppage in a major metropolitan area will result in rapid depletion of food, bottled water, and critical medical supplies, potentially leading to civil unrest as citizens compete for basic necessities.

Conclusion

As demonstrated by the analysis of just a few key industries, restricting or shutting down all truck operations in response to a natural disaster, elevated threat level, terrorist attack, or pandemic will have a swift and devastating impact on the food, healthcare, transportation, waste removal, retail, manufacturing, and financial sectors.

Members of the trucking industry must educate government officials at the local, state, and federal levels about the dire consequences of a truck stoppage. At first glance, halting all truck travel appears to be a powerful tool to neutralize a terrorist threat or protect citizens from a pandemic. However, this is a solution that could be worse than the problem.

Instead, we must urge governments at all levels to develop contingency and action plans that use sophisticated techniques to isolate and respond to a threat. Working together, ATA believes that a solution can be found that avoids the ruinous effects that will be brought about by a freeze on truck travel.

Case Study: The Effect of Border Delays on Auto Manufacturers Following September 11th

The auto industry is one of the manufacturing industries that has extensively integrated just-in-time inventory techniques. According to *Supply Chain Management Review*, the auto industry has saved $1 billion over the past decade through just-in-time techniques. Every day, auto manufacturing plants located along the Canadian border in Michigan receive numerous truck deliveries from auto parts plants in Canada. These parts will be assembled into autos and shipped within hours of delivery.

While the just-in-time manufacturing process is efficient, the experience of auto manufacturers during the days following September 11th showed the sensitivity of assembly lines to disruptions in the components and parts supply chain. Immediately following September 11th, stringent security measures (and some closings) at border crossings created delays ranging from 12 to 36 hours at certain checkpoints, including the crossing between Detroit, Michigan, and Windsor, Ontario. Approximately 7,400 trucks carrying commerce worth half-a-billion dollars flows across this checkpoint every day.

The border crossing delays caused shutdowns in operations at assembly plants operated by Ford, General Motors, DaimlerChrysler AG, Toyota Motor Sales, and American Honda Motor Company. Mark Nantais, head of the Canadian Vehicle Manufacturer's Association, estimates that these shutdowns cost auto manufacturers $1.5 million per hour.

In the first week following September 11th, Ford's production fell by 16,000 vehicles due to component shortages and GM's production fell by 10,000 vehicles. In the following months, according to a study by the Center for Automotive Research in Ann Arbor, Michigan, continued delays slowed manufacturing speed, causing losses of approximately $60,000 per hour for assembly plants.

Despite the disruptions caused by the border crossing delays, the just-in-time method is so important to corporate competitiveness that the model remains in place today. As a result, manufacturers continue to be sensitive to supply disruptions. A disruption as large and significant as a truck stoppage—whether local, regional or nationwide—will have a similar impact on auto and numerous manufacturing industries, with enormous costs stemming from reduced sales, lost production, and lost employment.

A Timeline Showing the Deterioration of Major Industries Following a Truck Stoppage

The first 24 hours	• Delivery of medical supplies to the affected area will cease.
	• Hospitals will run out of basic supplies such as syringes and catheters within hours. Radiopharmaceuticals will deteriorate and become unusable.
	• Service stations will begin to run out of fuel.
	• Manufacturers using just-in-time manufacturing will develop component shortages.
	• U.S. mail and other package delivery will cease.
Within one day	• Food shortages will begin to develop.
	• Automobile fuel availability and delivery will dwindle, leading to skyrocketing prices and long lines at the gas pumps.
	• Without manufacturing components and trucks for product delivery, assembly lines will shut down, putting thousands out of work.
Within two to three days	• Food shortages will escalate, especially in the face of hoarding and consumer panic.
	• Supplies of essentials—such as bottled water, powdered milk, and canned meat—at major retailers will disappear.
	• ATMs will run out of cash and banks will be unable to process transactions.
	• Service stations will completely run out of fuel for autos and trucks.
	• Garbage will start piling up in urban and suburban areas.
	• Container ships will sit idle in ports and rail transport will be disrupted, eventually coming to a standstill.

Within a week	• Automobile travel will cease due to the lack of fuel. Without autos and busses, many people will not be able to get to work, shop for groceries, or access medical care.
	• Hospitals will begin to exhaust oxygen supplies.
Within two weeks	• The nation's clean water supply will begin to run dry.
Within four weeks	• The nation will exhaust its clean water supply and water will be safe for drinking only after boiling. As a result, gastrointestinal illnesses will increase, further taxing an already weakened health care system.

This timeline presents only the primary effects of a freeze on truck travel. Secondary effects must be considered as well, such as inability to maintain telecommunications service, reduced law enforcement, increased crime, increased illness and injury, higher death rates, and likely, civil unrest.

Prepared by the American Trucking Associations (July 14, 2006).

On the Digital Forensics of Heavy Truck Electronic Control Modules

James Johnson, Jeremy Daily, and Andrew Kongs
The University of Tulsa

Concepts of forensic soundness as they are currently understood in the field of digital forensics are related to the digital data on heavy vehicle electronic control modules (ECMs). An assessment for forensic soundness addresses (1) the integrity of the data, (2) the meaning of the data, (3) the processes for detecting or predicting errors, (4) transparency of the operation, and (5) the expertise of the practitioners. The integrity of the data can be verified using cryptographic hash functions. Interpreting and understanding the meaning of the data is based on standards or manufacturer software. Comparison of interpreted ECM data to external reference measurements is reviewed from the current literature. Meaning is also extracted from interpreting hexadecimal data based on the J1939 and J1587 standards. Error detection and mitigation strategies are discussed in the form of sensor simulators to eliminate artificial fault codes. A transparent process for data gathering and handling is discussed.

The needs for improved techniques are motivated through examples of manipulated data and an analysis of potential opportunities that exist to alter the data. As an example, a step-by-step process of changing the records of a DDEC Reports .XTR file is provided. A detailed examination of resetting the ECM clock is also presented, which motivates a design of a hardware write-blocking device.

Some recommendation to provide more forensically sound records of ECM data is outlined. The strategy records and hashes network traffic to create a verification technique for later use. The data is then used in a replay algorithm so the diagnostic software can recreate the information from the forensic copy of the network traffic. Finally, application of the digital forensics beyond accident reconstruction is noted.

7.1 Introduction

Each year, many thousands of heavy vehicles are involved in traffic accidents; in 2015, 328,000 trucks were involved in accidents [1]. Litigation connected with these crashes can ensue, and, in recent years, this litigation has come to depend more and more heavily on electronic event data recorded on the vehicle's electronic control modules (ECMs). Much like passenger vehicle event data recorders (EDRs), this evidence includes speed records and other event data that can help accident reconstructionists determine what transpired during the event. Since the digital data from these devices may end up as evidence in court it, should be secure.

Modern heavy vehicle ECMs store incident data that may be valuable to accident reconstructionists in corroborating or clarifying physical evidence. Practices for extracting this information, however, are immature relative to other areas of digital forensics and require great care on the part of investigators in order to properly preserve and present evidence.

Due to the large financial stakes associated with litigating truck crash cases, there may be sufficient financial motive for bad actors to exploit unsecure digital forensic data from heavy vehicles. There are a few circumstances that make this prospect feasible, namely, (1) the digital forensic data on a truck is short lived and extended preservation of original data (i.e., securely storing the vehicle until litigation is over) is not likely. While some data is stored in nonvolatile storage mediums, like flash or EEPROM, the operation of the vehicle will likely create new events that are not attributable to the crash in question. Without extended storage of the original sources, the investigations rely on the copies of the data that get generated during the initial investigation. This means there is little chance of a verification of the original data after the years it takes for a case to adjudicate. As such, the application of cybersecurity principles to harvest and maintain digital forensic data is a worthy undertaking.

The art of digital forensic recovery has been focused mainly on event data regarding operation of the vehicle. The forensic context related to cyber exploits running on the embedded truck systems is beyond the scope of this chapter.

7.1.1 Motivation

Like other digital evidence, heavy vehicle digital event data should be held to a standard of forensic soundness, which is a series of principles that ensure that forensic evidence is handled in a secure manner that guards against unintentional, and intentional, alteration or misinterpretation. Currently, evidence is extracted from heavy truck systems using software that was not originally designed to be forensic software, and is not particularly suited to the task. While evidence can be collected using this data in a forensically sound manner, doing so requires diligence on the part of the investigator, and mistakes can result in evidence being dismissed. Additionally, the current methods by which data are stored are not particularly resistant to tampering; encryption, when used, is weak and no mechanism for integrity checking is implemented. To date, there has not been any literature that links the concepts developed in investigation practice of personal computers to the digital evidence obtained from heavy-duty ECMs. Some of the solutions presented in this paper could help first responders gather more crash data without the fear of spoiling or losing the data.

7.1.2 Paper Organization

First, the paper will review digital forensics concepts that apply to any kind of digital evidence, including heavy vehicle event data recorder (HVEDR) data. The first concept reviewed is forensic soundness, a term for the legal fitness of evidence. We then give a brief introduction to cryptographic methods for ensuring the integrity of digital evidence. We then discuss the format in which HVEDR data are transmitted and stored. The final item in the background section highlights some potential sources of error in existing evidence extraction processes and current methods by which those risks are mitigated.

Next is a discussion of trustworthiness of data extracted by current methods, including trust levels of vehicle network data and data generated by diagnostic software. Some previously unpublished methods by which data created by current forensic processes may be altered to modify or destroy crucial HVEDR evidence data is included. This highlights areas in which current data may not be trustworthy and provides a proof of concept of threats to evidence integrity.

Finally, we outline recommended practices to bring the trustworthiness of HVEDR data up to currently accepted standards for digital forensic soundness. We also introduce preliminary proof-of-concept implementations of these recommendations, including a cryptographic process for detecting data alteration and a novel method of storing and replaying HVEDR data.

7.2 Digital Forensic Concepts

A heavy vehicle ECM is a specialized process control computer that may have data of interest to an investigator. While ECMs may not have keyboards and monitors, they still are computers and have central processing units, memory, storage, and a means of networking or communicating with external devices. As such, many of the principles from computer forensics can be applied to heavy vehicle ECMs.

A Special Report issued by the National Institute for Justice [2] on Electronic Crime Scene Investigation highlights the following basic forensic principles applied to dealing with digital evidence:

- Any process or procedure of collecting, transporting, or storing of digital evidence should not incur any changes to the evidence.
- Only specifically trained experts should examine digital evidence.
- Transparency during the operations of acquisition, transportation, and storage of the evidence should be maintained.

These basic tenets lay the foundation for the idea of forensic soundness. It is important to understand the term "forensically sound" as it relates to digital evidence. Many authors and professional organizations have attempted to rigorously define this concept, including the National Institute for Standards and Technology (NIST) [3], law enforcement entities such as the National Institute for Justice (NIJ) [2], and academic bodies like the International Organization on Computer Evidence [4]. The methods of extracting, analyzing, and presenting digital evidence are forensically sound if they perform the task in a manner such that the results can be used in legal proceedings with a high degree of confidence in their admissibility. The process of extracting, interpreting, and presenting evidence will be referred to in this paper as the "forensic process."

In addition to the NIJ report, McKemmish [5] enumerates the following components of forensic soundness:

- **Meaning** is a term that denotes confidence in the interpretation of extracted evidence data.
- **Error Detection** denotes processes for detecting or predicting errors in the forensic process.
- **Transparency** means the forensic process is documented, known, and verifiable.
- **Expertise** is required for those investigators examining digital data.

Anyone who designs a forensic process that deals with heavy truck ECMs should consider these requirements and design them into the solution. Another tenet of forensic soundness is the principle of tamper resistance as pointed out by Casey [6]. If it is infeasible to prevent tampering, then the ability to detect tampering should ensure soundness of data. This section of the paper continues with discussions of tamper resistance (encryption and hashing), data meaning (decoding hex data), error detection (in terms of fault elimination), and transparency. It is beyond the scope of this technical paper to discuss the expertise required for handling digital forensic data in general.

7.2.1 Data Integrity

The primary tool for ensuring data integrity and detecting tampering in digital forensics is a cryptographic hash algorithm. A cryptographic hash algorithm is an algorithm which generates a fixed-length fingerprint of the input data, in such a way that small changes in the data result in very large changes in the generated hash. This functionality is similar to cyclic redundancy checks (CRCs) used in networks to ensure data integrity.

Because the hash is generally much smaller than the data being hashed, the pigeon-hole principle dictates that multiple input data strings must result in the same hash when run through a cryptographic hashing algorithm. A situation in which two inputs have the same hash is known as a hash collision. One important property of cryptographic hash algorithms is that generating collisions is computationally infeasible, meaning that for practical purposes, it is impossible (i.e., the computation time is measured in hundreds of years). This is important to integrity checking because if the algorithm did not have this property, someone could modify the source data in such a way that the change was undetectable by the hashing process. CRC collisions, by contrast, can be generated fairly easily. It is worth noting that the SAE standard for HVEDR data, J2728, recommends a 32-bit CRC for integrity checking [7].

Though an in-depth discussion of cryptographic hashing algorithms is beyond the scope of this paper, the interested reader can find a reasonably complete discussion in [8]. As of this writing, the SHA-256 [9] hashing algorithm is an industry standard choice for cryptographic hashing as a collision generation is still infeasible. Hashing does not encrypt data, so the original data remains unchanged.

A cryptographic hash function is an algorithm that generates a "fingerprint" of a given input. This fingerprint is generated in such a way that if a single bit of the input changes, at least half the bits in the output file are altered [8]. As such, they are commonly used to confirm that data have not been altered. A forensically sound tool should use a hash of the extracted truck evidence data to ensure that it has not been altered. This hash can be calculated and distributed to or stored by a third party immediately after the download takes place to ensure the data is authentic. This practice would eliminate any gaps in time in which someone could manipulate a digital record. Furthermore, hashing immediately at the time of extraction eliminates the need for the download to be witnessed in person or videotaped.

7.2.2 Meaning of the Digital Data from ECMs

If digital data are only represented as binary (or hexadecimal) listings, its meaning is lost. Therefore, it is paramount to an investigation to interpret the digital data in a meaningful way. Within the context of heavy vehicle ECM data, the tools that interpret the data are typically written by the original equipment manufacturers (OEMs) in the form of diagnostic software. Some OEMs provide other data reporting tools, like DDEC Reports and Cummins PowerSpec, to interpret the data, which can be useful for reconstructing crashes. However, these programs do not implement some of the safeguards used to ensure forensic soundness that are found in other areas of digital forensics.

7.2.2.1 **Standards-Based Meaning:** Most of the research presented in the literature so far has focused on comparing interpreted values to external instrumentation to verify the correctness of those values. Arguably, this is all that is necessary for an expert to establish the meaning of the data. However, the interpretation of the HVEDR record is still left up to the OEM software. There are, however, standards that help interpret digital data into engineering units; the units specified in communications standards such as

J1587 [10] provide clues as to the semantics of digital data. Furthermore, standards-based tools can be used to gather standardized data like odometer reading, engine hours, VIN, and component information to supplement other information in the HVDER record.

The SAE J1587 standard [10] defines the data interchange between microcomputer systems in heavy-duty vehicle applications. The physical network for J1587 is specified by the J1708 standard [11]. This protocol specification includes standard formats and interpretations for messages regarding the vehicle's operation.

The J1939-71 [12] standard serves a similar purpose to the J1587 standard, defining the J1939 application layer. The standard defines message structure, data format, and engineering units for various parameters. For example, the wheel-based vehicle speed message is defined as a 2-byte value with a resolution of 1/256 km/h per bit. Therefore, it is reasonable to expect that speed data will be stored internally in km/h, the field will be 2 bytes wide, and the 2-byte integer will be the wheel-based vehicle speed multiplied by 256.

The American Trucking Association's Technology & Maintenance Council (TMC) Recommended Practice number 1210 (RP1210) [13] provides a methodology for a personal computer to communicate with J1587 and J1939 compliant networks, as illustrated in **Figure 7.1**.

Currently, nearly all OEM-developed applications are capable of using an RP1210 compliant device to communicate with vehicles. As such, any new forensic software or device designed to communicate with these types of networks should support RP1210. Some compatible applications, such as DG Diagnostics, allow for recording of all messages passed between the Vehicle Data-link Adapter and the Vendor Application through the RP1210 API. This RP1210 recording or log file can serve as a rudimentary forensic record of the communications between an ECU and Vendor Application. As discussed earlier, performing a cryptographic hash of the log file after recording can be used to show that the log file has not been altered.

The J2728 standard for HVEDR data [7] specifies data sources that a HVEDR should record. Not only does it specify data points; it also specifies the preferred J1587 or J1939 messages that the EDR should sample to obtain those records. Referring to the J2728 standard for this information and cross-referencing, the messages to the relevant vehicle networking standards will allow the investigator to determine which electronic device is responsible for computing the data and sending the message. The standards can also be used to understand the sensor that the device used to obtain that data. This can enhance the investigator's understanding of the physical phenomena that the EDR data represent.

FIGURE 7.1   RP1210 information flow chart.

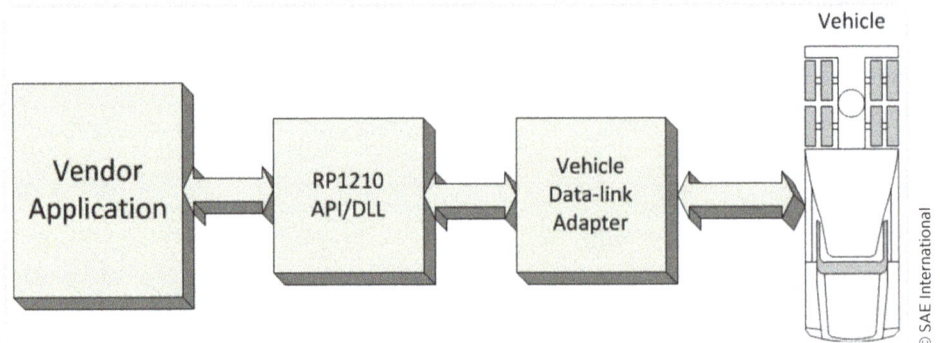

Not all ECMs conform to the J2728 standard, so using standards-based meaning should not be done without verification.

The J1939-73 standard for diagnostic data communications in heavy vehicles [14] contains specifications for diagnostic trouble codes used to indicate system or component faults, as well as direct memory access procedures that occur over J1939. The diagnostic specifications are useful for understanding the freeze frame data that may be attributed to a crash. Knowledge of the memory access procedure is also useful for evaluating how commercial tools access evidence data.

The networking protocols (J1708/J1587 and J1939) enumerate and define the various data that are transmitted over the in-vehicle network, which provides clues as to the nature of the data present in the ECM. For example, road speed data on a J1708 network only has resolution to the nearest 0.5 mph. The value reported by the ECM may have the same resolution.

7.2.2.2 **Proprietary Meaning:** Most new heavy vehicles contain some event data recording capability. These data are accessed from software supplied by the ECM manufacturer, as shown in **Table 7.1.**

While it may be tempting to trust the interpretation of the OEM software, research has shown that the interpretation of the data may be at issue. For example, in some Cummins Sudden Deceleration events, speed data is reported in the report at one sample per second; however, after comparing to external reference measurements, it is discovered that the data actually is reported at approximately 0.2 s intervals (5 Hz) [15]. Similarly, Austin and Farrell [16] showed that many snapshot records from Caterpillar are reported on intervals that are approximately 0.5 s apart instead of every second as represented in the OEM software.

Establishing and verifying meaning for all data elements in the digital records is a formidable task. The current body of literature has made strides to evaluate many aspects of these EDR records, but more work remains to be done in this area.

When distilled into some basic elements, the meanings from heavy vehicle digital data representing time history traces (like speed and RPM) can be understood in two dimensions: time and value. Since the ECMs on heavy vehicles are designed to control some process that takes time, the timing of events is often recorded. The other aspect of data is its value. In other words, the question investigators want answered by digital ECM data is, "when did this value occur and what is the accuracy of this value?" As such, there are two dimensions (time and value) to evaluate when looking at establishing the meaning of the data in an ECM record.

Most of the literature regarding HVEDR data is dedicated to these timing and value accuracy studies. Recently, Plant et al. [17] studied the timing and synchronization of data from DDEC V control modules. Other notable research on the meaning of EDR data includes Steiner et al. on the Mack V-MAC III [18], Reust on the DDEC IV [19], van Nooten et al. on the Caterpillar and Detroit Diesel Modules [20], Drew et al. on Caterpillar ECM data [21],

TABLE 7.1 A list of manufacturers and the software used to access the data on the ECM

ECM Family	Software
Caterpillar	Caterpillar Electronic Technician (CatET)
Cummins	Cummins PowerSpec
	Cummins Insite
Detroit Diesel	DDEC Reports
	Detroit Diesel Diagnostic Link (DDDL)
Mack	Mack and Volvo Proprietary
Mercedes	DDEC Reports
	Detroit Diesel Diagnostic Link (DDDL)
Navistar	ServiceMaxx
Paccar	Cummins PowerSpec
	Cummins Insite
	or
	Paccar Davie
Volvo	Mack and Volvo Proprietary

© SAE International

and Reust et al. on DDEC series modules [22]. Messerschmidt et al. [23] explored the effect of power loss for ECM data, which is useful in explaining so-called anomalies in the EDR data.

7.2.2.3 Daily Engine Usage from DDEC Reports:

Most of the aforementioned papers explored speed and brake timing and history. As such, researchers are beginning to focus on some of the other historical data in HVEDRs (e.g., [24]). The premise of the research was to evaluate the timing of the transitions from different states of the vehicle according to the graphs shown in **Figure 7.17** in the appendix. The table entry for this graph from DDEC Reports is shown in line with the text in **Figure 7.2** and repeated in the appendix as **Figure 7.18** with a large format.

Examining the digital record in the XTR file may provide some insights into what numbers are recorded and whether or not the transitions from off to idle to drive correspond to stored data.

After opening the .XTR file produced by DDEC Reports in a hex editor, as shown in **Figure 7.3**, one can find that the daily engine usage logs start on byte 1956 in the XTR file and each day takes up 36 bytes. Figure 7.3 shows the 36 bytes corresponding to the data in Figure 7.2 highlighted in blue. The interpretation of these hex values is shown in **Table 7.2** in the appendix. The daily engine usage log has a pattern in the .XTR file that is repeated 30 times to get the last 30 days.

In a DDEC IV and V (and maybe some others), timestamps are stored as a 32-bit number (0-44294967295) representing the number of seconds from an epoch corresponding to January 1, 1985, at midnight (UTC). Since each machine interpreting these values may be different, the epoch must be calculated before interpreting the number. For example, the time passed from the Windows epoch may be different than the time since the J1587 epoch. This difference is accounted for by subtracting off the PC's epoch from the J1587 epoch. Implementing this with the Python time library [25] is as follows:

```
import time
tZero=time.mktime((1985,1,1,0,0,0,0,1,-1)) - time.timezone
```

FIGURE 7.2   Example data from DDEC Reports regarding daily engine usage.

DDEC® Reports – Daily Engine Usage

Print Date: 8/21/2013 11:08 AM   Date Range: 01/18/07 To 01/07/00 (EST)
University of Tulsa
800 S. Tucker Dr   Vehicle ID: TIB DDEC4
Tulsa, OK 74104   Driver ID:
(918)631-3056   Engine S/N: 06R0499534

Date:	1/18/2007	
Start Time:	00:00:00	EST
Odometer:	1006109.00	mi
Distance:	548.80	mi
Fuel:	95.25	gal
Fuel Economy:	5.76	mpg
Average Speed:	59.54	mph

Total(hh:mm)	09:13	06:00	08:47
Hour(EST)	Drive(min)	Idle(min)	Off(min)
00:00-02:00	0	120	0
02:00-04:00	0	120	0
04:00-06:00	96	24	0
06:00-08:00	104	16	0
08:00-10:00	110	10	0
10:00-12:00	54	66	0
12:00-14:00	120	0	0
14:00-16:00	69	4	47
16:00-18:00	0	0	120
18:00-20:00	0	0	120
20:00-22:00	0	0	120
22:00-24:00	0	0	120

© SAE International

FIGURE 7.3   The first record of the daily engine usage log shown in a hex editor.

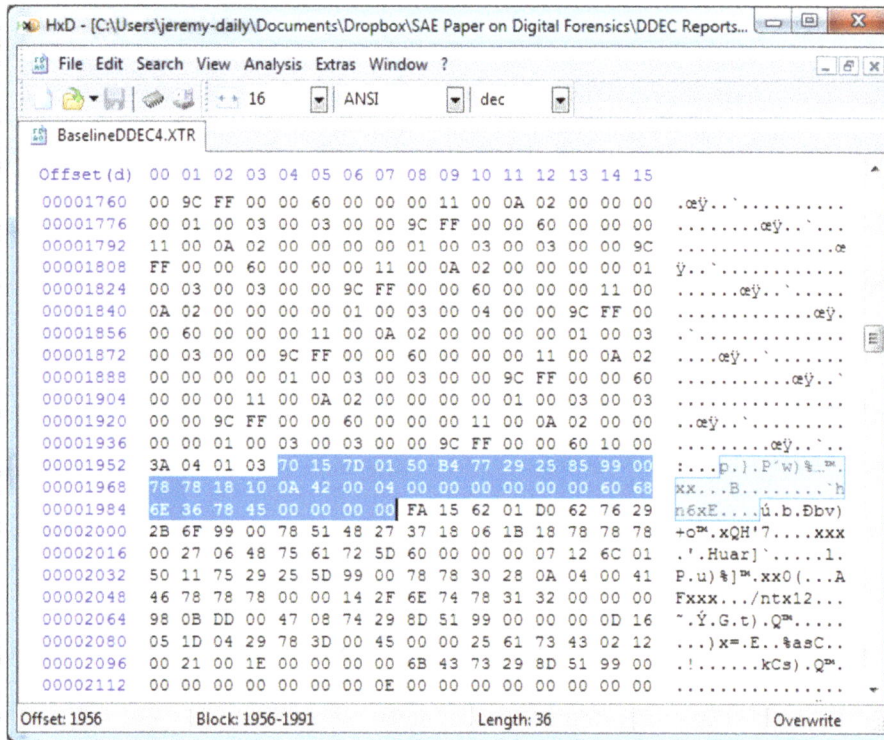

TABLE 7.2 Interpretation of the data shown in Figures 7.17 and 7.18

Bytes sequence	Hex value (s)	Decimal	LSB value	Meaning	Value
0-1	70 15	5488	0.1 mile	Distance	548.8 miles
2-3	7D 01	381	0.25 gal	Fuel	95.25 gal
4-7	50 B4 77 29	695710800	1 s from epoch	Start Time (see discussion in text)	Wednesday, 17 January 2007 at 23:00:00 Central Standard Time
8-11	25 85 99 00	10061093	0.1 mile	Odometer	1006109.3 miles
12-23	78 78 18 10 0A 42 00 04 00 00 00 00	120 120 24 16 10 66 0 4 0 0 0 0	1 min	Idle time	Same as decimal
24-35	00 00 60 68 6E 36 78 45 00 00 00 00	0 0 96 104 54 120 69 0 0 0 0	1 min	Drive time	Same as decimal

which yields a value of tZero = 473385600 s. The `time.timezone` value includes the PC's time zone into the `tZero`. For Central Standard Time, `time.timezone` = 21600 s or 6 h. Interpreting the 32-bit number (Intel format) can be done with Python's struct library [26] as follows:

```
import struct
TSbytes = b'\x50\xB4\x77\x29'
TS = struct.unpack('<L',TSbytes)[0]
```

where b'\x50\xB4\x77\x29' is the byte string corresponding to the time stamp in the .XTR file. Adding TS to tZero (the epoch from J1587) gives a value of 1,169,096,400 s from the PC's epoch of midnight of January 1, 1970 UTC. Converting these seconds into a usable timestamp uses the following Python command:

```
timeString = time.strftime("%A, %d %b %Y at %H:%M:%S %Z", time.
localtime(TS+tZero))
```

which produces a result of "Wednesday, 17 Jan 2007 at 23:00:00 Central Standard Time." This matches the instance from DDEC Reports since the ECM was programmed to work in the Eastern Time Zone.

The odometer reading is stored as a 32-bit unsigned integer representing the number of tenths of a mile passed. This is consistent with J1587. The next 12 entries are for the idle time, and the following 12 bytes are for the drive times as broken down in the number of minutes per 2 hour block. This completes the pattern and these are the only recorded data. The key observation is that the XTR file does not contain time stamps regarding the transition from one state to another. This means the graphic shown in Figure 7.17 is generated by DDEC Reports on the PC with an algorithm only based on the drive and idle times. The off time, fuel economy, and average speed are all calculated parameters in DDEC Reports. Therefore, determining the exact timing of when a driver transitioned from off to idle to drive requires more evidence than can be provided by DDEC Reports. The data also does not account for the number of transitions between drive, idle, and off.

Examining the digital record at the byte level and matching the hex data to the DDEC Reports enables investigators to know what data were recorded and what data are calculated. A deeper examination of the RP1210 log files show that parts of the J1708 network traffic is contained within the XTR file. Therefore, the XTR file is a representation of the communications from the ECM.

The example shown herein is for a Detroit Diesel DDEC IV interpreted by DDEC Reports version 8.02. Other manufacturers interpret and save their data in plain text formats, like HTML or XML, which can be edited.

7.2.3 Error Detection and Mitigation

Boggess et al. [27] demonstrated that another issue with evidence preservation is the creation of spurious fault codes during bench downloads. Following this work, Plant, Austin, and Smith [28] showed how many data elements remain unchanged when an ECM is removed from a vehicle. Ideally, incident data are extracted through the diagnostic port in the cab of the vehicle. The ECM resides in the vehicle in question, and so the diagnostic trouble codes that it records reflect the actual condition of the vehicle. However, the in-vehicle network may be damaged during the crash, making the diagnostic port unusable for this purpose.

This necessitates removal of the ECM from the vehicle and reading it directly, a process commonly known as a bench download. The simplest method of performing a bench download, connecting to the ECM using an engine programming harness, and performing the download as usual with a diagnostic tool, can often yield spurious fault codes because the ECM fails to detect various temperature sensors, pressure sensors, and so forth.

Investigators have developed several methods for mitigating this problem. If available, a solution is to use a donor vehicle that is nearly identical to the vehicle involved in the incident. The potential difficulties of obtaining a surrogate vehicle, or difficulty in the case of non-fleet vehicles, are self-evident. Another method involves purchasing all of the components that the ECM expects and connecting them to the ECM. Doing this for each ECM could be prohibitively expensive.

7.2.4 Establishing Transparency and Trust

Regardless of the meaning of the digital data, it is necessary to present data in its final form in such a way that is transparent of its handling to establish trustworthiness. According to the transparency principle of forensic soundness, actions taken by an investigator should be available for later examination. Additionally, any error conditions encountered by the software should be recorded so that the legal weight of the evidence may be accurately considered. Journals are log files generated by forensic software to meet these requirements and should at least be an option in any forensic solution.

One example of a possible attack on the integrity of current data formats is deleting audit trails in Cummins Insite's EIF files. EIF files are simply XML-formatted plain text files compressed into encrypted ZIP files. While the cryptographic security of ZIP file formats has historically been suspect [29], in this case, the encryption strength was irrelevant, as the encryption passphrase was pulled from memory when Insite opened a file. With the passphrase, one can simply decompress the text file, alter the data using a program like notepad, and re-encrypt it.

7.2.4.1 Baseline of Trust: In order to make a full accounting of potential sources of error in any forensic process, it is important to examine the assumptions that the process requires. Standard methods of truck evidence extraction assume that the ECM is reporting data faithfully, that the vehicle network is accurately transmitting the communication to and from the ECM. It also assumes the vehicle diagnostic adapter (VDA) is faithfully sending data from the vehicle network to the RP1210 library that is called by the OEM software.

While verifying the correct internal functioning of heavy truck ECMs is beyond the scope of this paper, the fidelity of data transmitted across the various networks (CAN bus, serial bus, and USB) can be easily verified. Both protocols make use of error-correcting codes that detect transmission errors with very high reliability; for example, the CAN CRC that is defined in the CAN standard process detects all errors from 1 to 5 bits in the data frame, and only fails to detect 0.00018% of 6-bit errors in a CAN frame [30]. This represents a level of certainty that would take many years to experimentally verify because the likelihood of having an undetected error is so low.

Assuming the ECM is performing correctly, the programmed parameters (like tire size) are known, and data flowing through the VDA maintains its integrity, then the next level of trust would be the OEM software and the native file formats they produce.

These files, however, cannot be trusted because they can be altered without detection. Therefore, the baseline of trust for digital forensic purposes is the vehicle network traffic, which can be accessed with a network data logger or with a verbose log file generated by the RP1210 library for the VDA. Details of why native files cannot be trusted are shown with an example from DDEC Reports in the following sections.

7.2.4.1.1 **Data Manipulation.** Currently, truck evidence data are stored in a variety of file formats. These include standard data formats, such as HTML or PDF, and proprietary formats used by the various engine maintenance software products. However, these formats are vulnerable to alteration. Through examination of the files and the programs that generate them, it was determined that:

- Cummins PowerSpec Reports are stored natively as HTML files that can be manipulated with a text editor.

- Cummins EIF files are password-protected ZIP archives that contain data formatted using XML, a text-based format for organizing data similar to HTML. The same password is shared between all EIF files so knowing the password allows someone to modify any EIF file by decompressing, modifying, and recompressing the data. While the cryptosystem used by ZIP files is known to have some flaws [26], one of the authors simply recovered the password by reading it from memory with a debugging program while Insite opened an EIF file.

- DDEC Reports XTR files are a raw record of the data extracted from the ECM via J1708 or J1939. Editing the binary data can be done with a hex editor. Location of the bytes containing the speed data in the hard brake records and daily engine usage log is presented below.

- DDDL Drumroll files are obfuscated XML documents. The obfuscation scheme is a fairly simple XOR scheme that can be deciphered using methods discussed in [28]. Knowing this scheme allows someone to decrypt, modify, and re-encrypt Drumroll files at will. The DDDL software is implemented using Microsoft's .NET framework, which means obtaining the source code is not difficult.

- CAT warranty reports are stored as plaintext XML files that can be altered with any standard text editor.

7.2.4.1.2 **DDEC Reports XTR File Manipulation Example.** A DDEC IV module was downloaded using DDEC Reports 8.02, and the .XTR file was automatically saved in C:\Detroit Diesel\DDEC Reports\Diagnostic\DATA PAGES. This file was renamed to BaselineDDEC4.XTR and the DDEC reports software produced a report that was 39 pages long. For the sake of demonstration, hard brake #1 is considered. The report contains a graphical representation of the data along with a data table. The graph represents typical data from a DDEC IV ECM and is shown in **Figure 7.4**.

The report in Figure 7.4 shows the incident odometer is 1311407.0 mi. The authors determined empirically that this data element is in the XTR file and is encoded as a 4-byte (32-bit) unsigned number in Intel format (little endian, i.e., the least significant byte is at the lowest address) where the least significant bit (LSB) represents 0.1 miles. This comports with SAE J1587 standard for distance quantities. After searching through the .XTR file,

FIGURE 7.4 Hard brake #1 graph from DDEC Reports representing typical DDEC ECM data.

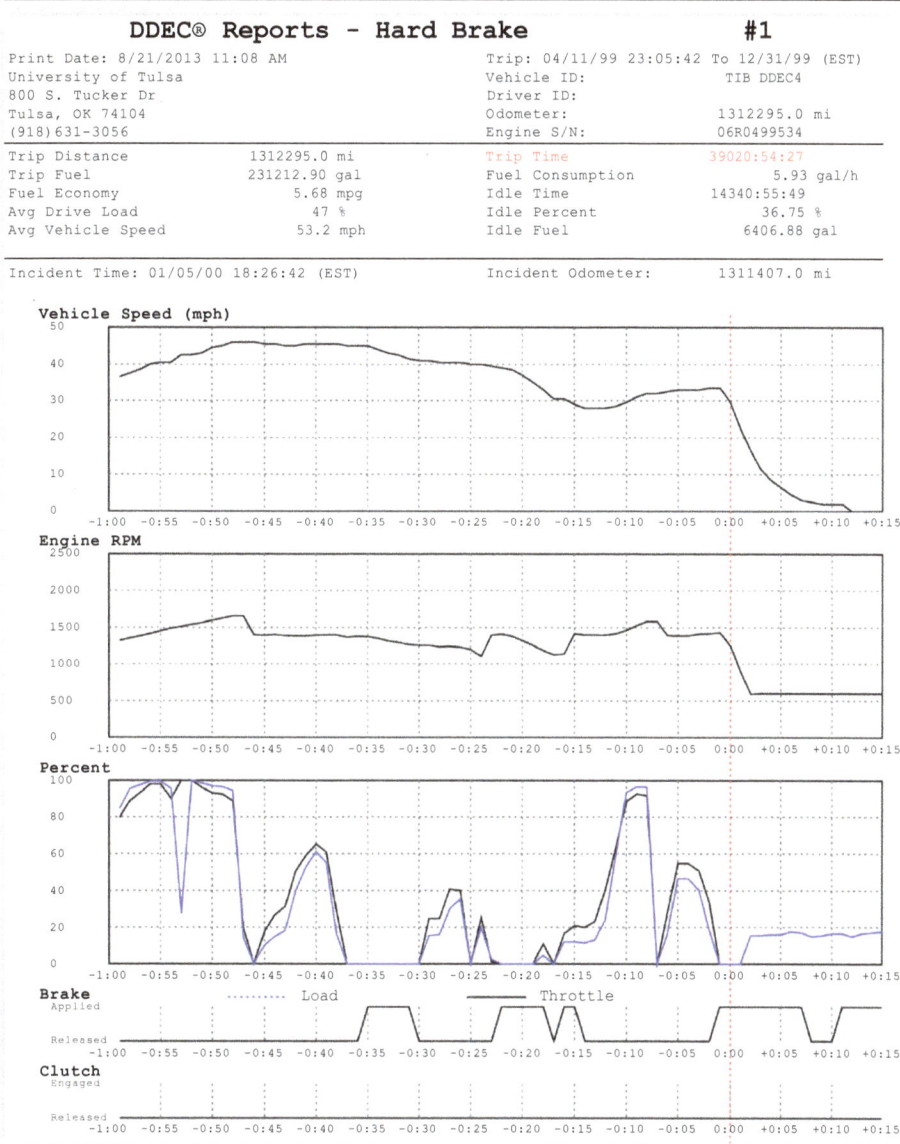

```
         DDEC® Reports - Hard Brake                    #1
Print Date: 8/21/2013 11:08 AM          Trip: 04/11/99 23:05:42 To 12/31/99 (EST)
University of Tulsa                      Vehicle ID:           TIB DDEC4
800 S. Tucker Dr                         Driver ID:
Tulsa, OK 74104                          Odometer:             1312295.0 mi
(918)631-3056                            Engine S/N:           06R0499534

Trip Distance        1312295.0 mi        Trip Time             39020:54:27
Trip Fuel           231212.90 gal        Fuel Consumption          5.93 gal/h
Fuel Economy             5.68 mpg        Idle Time             14340:55:49
Avg Drive Load             47 %          Idle Percent              36.75 %
Avg Vehicle Speed       53.2 mph         Idle Fuel              6406.88 gal

Incident Time: 01/05/00 18:26:42 (EST)         Incident Odometer:    1311407.0 mi
```

it was found that the hard brake #1 incident odometer is stored in block 4060-4063 and is encoded as 0xD5 0x1A 0xC8 0x00, as shown in **Figure 7.5**. Decoding this hex value involves byte swapping, converting to decimal and multiplying by 0.1 miles, or 0x00C81AD5 = 13114069 * 0.1 = 1,311,406.9 miles. DDEC Reports apparently rounds the number, so precise interpretation of the XTR data can resolve distances to the nearest 10th of a mile.

Immediately after the incident odometer, there are 5 bytes that are not represented in DDEC Reports output. Following those 5 bytes, the hard brake tabulated data begins in 6-byte chunks starting with byte 4069. The time is not included in the XTR file. The 6 bytes representing each line in order are as follows: road speed (1 byte),

FIGURE 7.5 Data representing the incident odometer reading for hard brake #1 in a .XTR file.

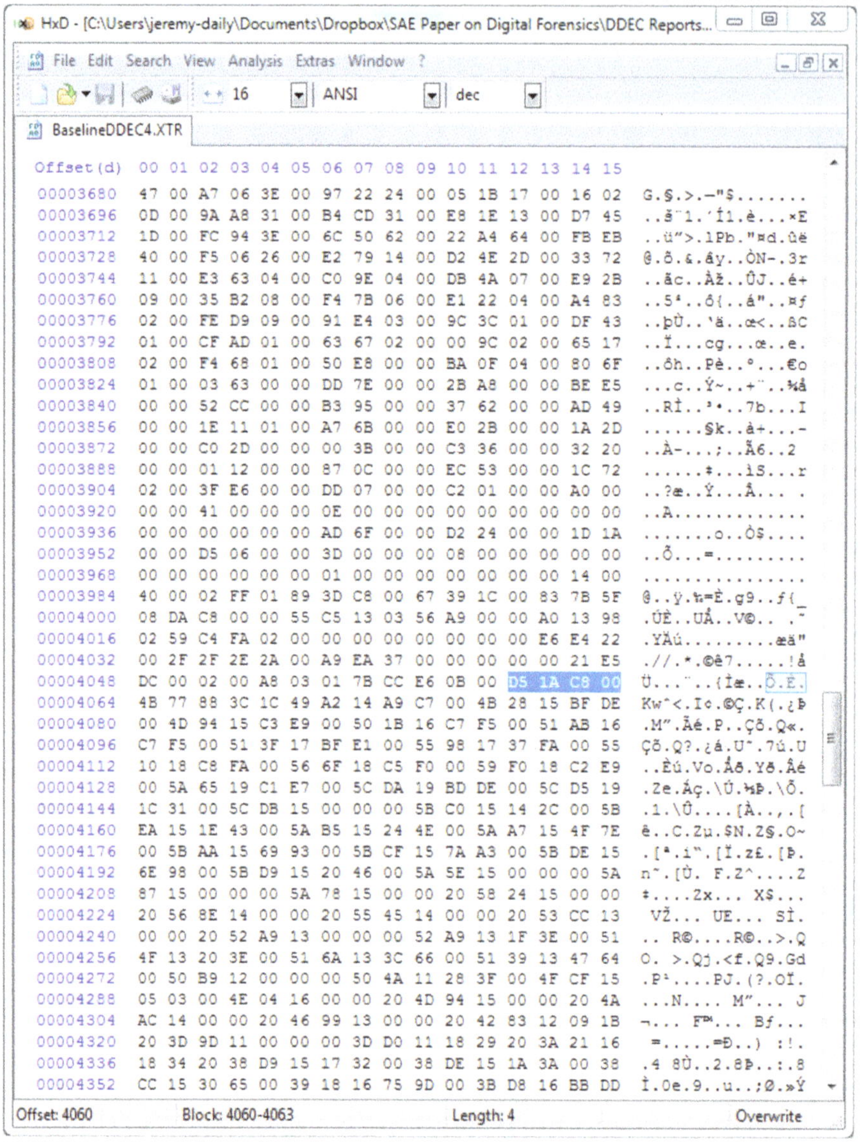

© SAE International

FIGURE 7.6 Binary bit-mapped meaning in DDEC Reports switch data.

Position	7	6	5	4	3	2	1	0
Value	128	64	32	16	8	4	2	1
Bit	X	X	X	0	0	0	0	X

© SAE International

engine speed (2 bytes), engine load (1 byte), throttle (1 byte), and switch status (1 byte). The first line of the data table is shown decoded in **Error! Reference source not found**. The binary bit-mapped code for the switch status is shown in **Figure 7.6**.

To demonstrate how to manipulate the hard brake record, the speed data was changed to 0xFF in a hex editor and saved as a new file. The new file has a speed record that now starts with 127.5 mph. The graphics corresponding to these changes are shown in **Figures 7.7 and 7.8**.

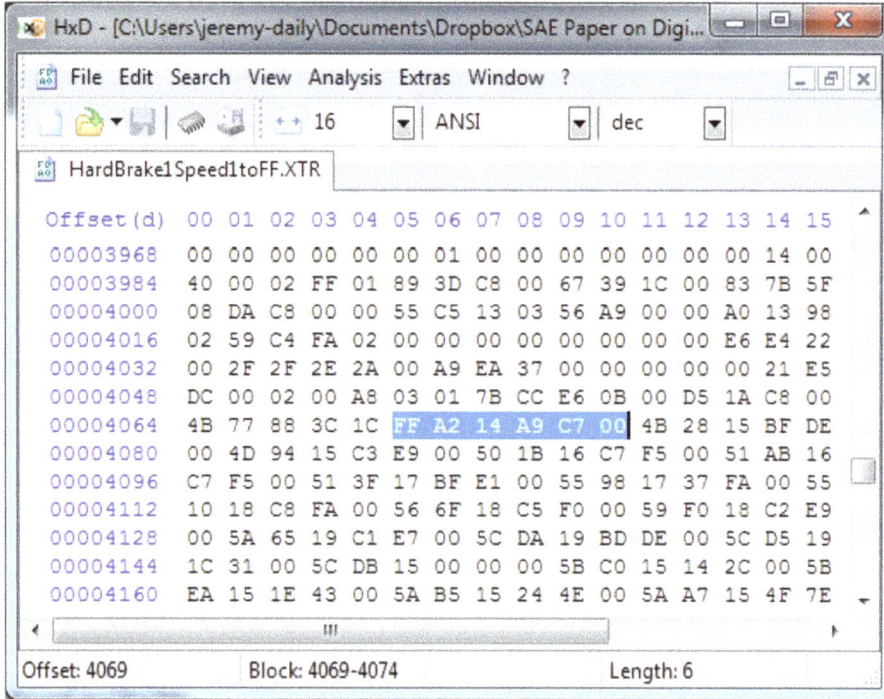

FIGURE 7.7 Hex editor showing the change of speed for the first entry in hard brake #1 to 0xFF.

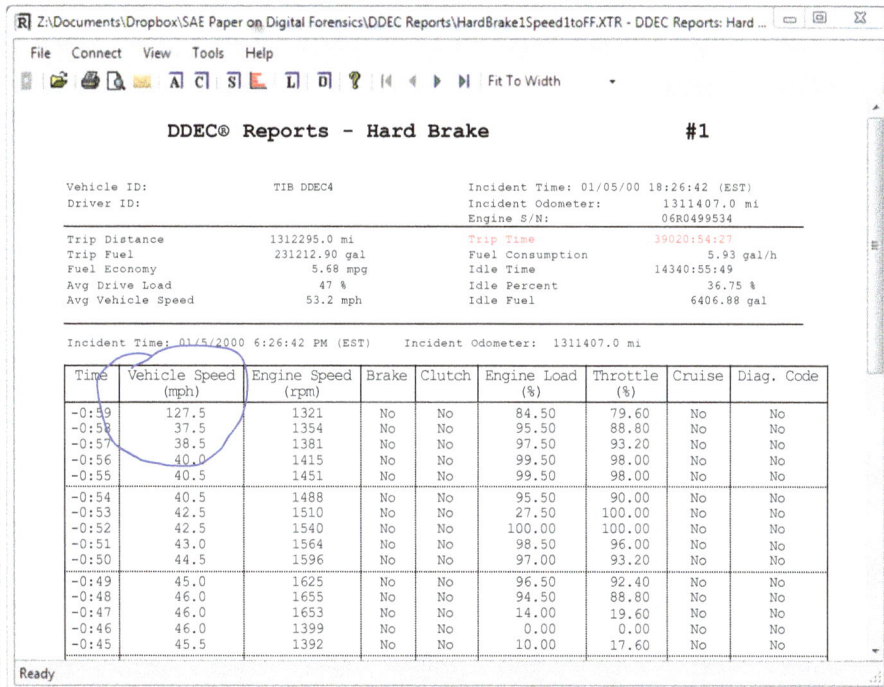

FIGURE 7.8 The hard brake record reflecting the change implemented from Figure 7.7.

FIGURE 7.9

Manipulating the data in a hex editor for the first two entries of the Last Stop Record.

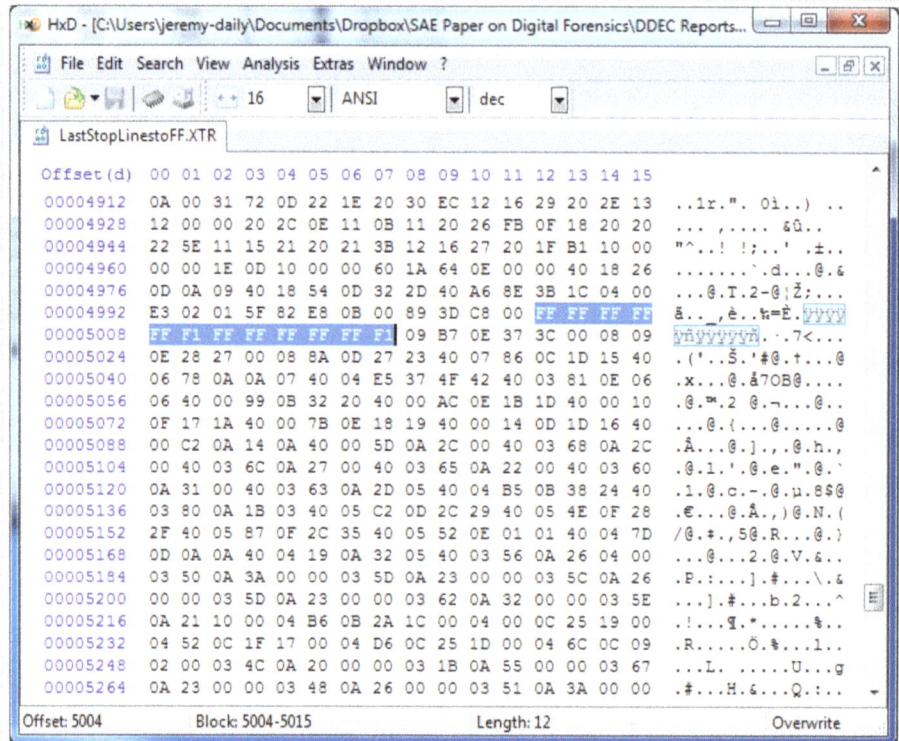

In a similar fashion, hard brake #2 and the Last stop record can be changed by replacing the contents in appropriate bytes. The tabulated data for hard brake #2 starts at byte 4536 and the last stop record starts at byte 5010. **Figures 7.9 and 7.10** show the changed hex data and corresponding table entries in DDEC Reports, respectively.

With an understanding of the data locations and meaning, any record can be generated by manipulating an XTR file. Since all printed reports come from the XTR, those can all be replaced by subsequent report production. There is no algorithmic mechanism in place to reveal manipulation (or corruption) of the data after it has been downloaded. Being able to open a modified XTR file in DDEC reports means there is no checksum, hash, or other test for modified data within the file. As such, a potentially contested file will have to have been downloaded and distributed within a short enough time period to eliminate the possibility of someone having the time to tamper with the data. This precludes investigators from performing a verifiable download at a crash scene without the presence the other parties. This could be particularly challenging for law enforcement when dealing with pressures to clear the scene and open the roads again.

7.2.4.2 **ECM Time Stamps:** Some of the data on heavy truck ECMs is volatile, meaning that it requires a continuous supply of electrical power to avoid being deleted. One important example is the on-board system clock, which is important to investigators because it correlates with the timestamps associated with incident records. An exemplar circuit, shown in **Figure 7.11**, is based around the EM Micro V3020, as used in several

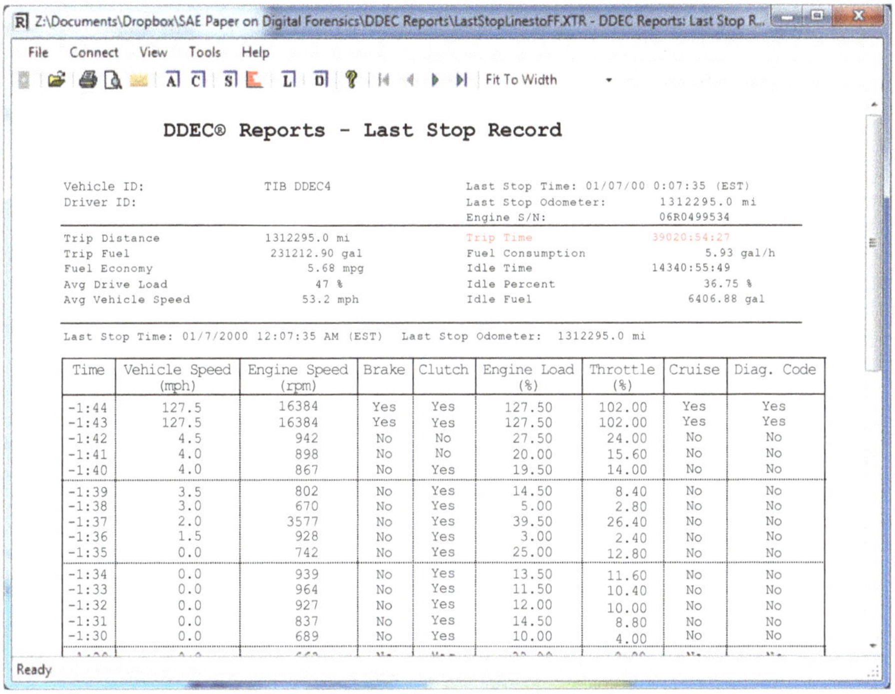

Z:\Documents\Dropbox\SAE Paper on Digital Forensics\DDEC Reports\LastStopLinestoFF.XTR - DDEC Reports: Last Stop R...

File Connect View Tools Help

Fit To Width

DDEC® Reports - Last Stop Record

Vehicle ID:	TIB DDEC4	Last Stop Time:	01/07/00 0:07:35 (EST)
Driver ID:		Last Stop Odometer:	1312295.0 mi
		Engine S/N:	06R0499534
Trip Distance	1312295.0 mi	Trip Time	39020:54:27
Trip Fuel	231212.90 gal	Fuel Consumption	5.93 gal/h
Fuel Economy	5.68 mpg	Idle Time	14340:55:49
Avg Drive Load	47 %	Idle Percent	36.75 %
Avg Vehicle Speed	53.2 mph	Idle Fuel	6406.88 gal

Last Stop Time: 01/7/2000 12:07:35 AM (EST) Last Stop Odometer: 1312295.0 mi

Time	Vehicle Speed (mph)	Engine Speed (rpm)	Brake	Clutch	Engine Load (%)	Throttle (%)	Cruise	Diag. Code
-1:44	127.5	16384	Yes	Yes	127.50	102.00	Yes	Yes
-1:43	127.5	16384	Yes	Yes	127.50	102.00	Yes	Yes
-1:42	4.5	942	No	No	27.50	24.00	No	No
-1:41	4.0	898	No	No	20.00	15.60	No	No
-1:40	4.0	867	No	Yes	19.50	14.00	No	No
-1:39	3.5	802	No	Yes	14.50	8.40	No	No
-1:38	3.0	670	No	Yes	5.00	2.80	No	No
-1:37	2.0	3577	No	Yes	39.50	26.40	No	No
-1:36	1.5	928	No	Yes	3.00	2.40	No	No
-1:35	0.0	742	No	Yes	25.00	12.80	No	No
-1:34	0.0	939	No	Yes	13.50	11.60	No	No
-1:33	0.0	964	No	Yes	11.50	10.40	No	No
-1:32	0.0	927	No	Yes	12.00	10.00	No	No
-1:31	0.0	837	No	Yes	14.50	8.80	No	No
-1:30	0.0	689	No	Yes	10.00	4.00	No	No

Ready

FIGURE 7.10 Reflection of the first two lines in the Last Stop Record showing their maximum values corresponding to 0xFF from Figure 7.9.

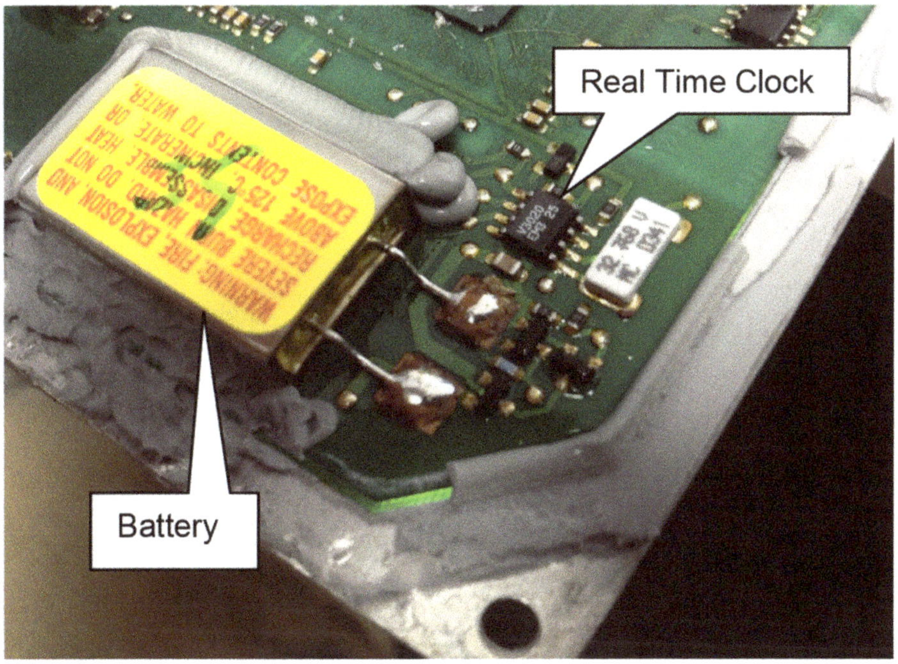

FIGURE 7.11 Real-time clock circuit from a DDEC V ECM. The battery is under the yellow sticker.

Detroit Diesel ECMs. It is important that the state of the ECM's on-board system clock be preserved because the system clock is not necessarily synchronized with standard times and the offset from the standard time must be recorded at the time of investigation. Often this is done by taking a screenshot of the diagnostic software displaying the ECM time and the computer clock showing the PC time on the same screen.

While the batteries which power the system clocks on heavy truck ECMs are quite large and would take at least several years without external power being applied to fully discharge, some older ECMs have time clocks that have run out of battery storage. When this happens, the clock usually is set to its epoch. Some systems use the Unix epoch of January 1, 1970 at midnight. Other systems, like DDEC IV and V use the epoch defined in the J1587 PID for the date and time of January 1, 1985 at midnight [14]. These times are set for Universal Coordinated Time (UTC). Therefore, if a DDEC Reports timestamp shows the ECM time as December 31, 1984, then this likely indicates that the PC program is interpreting numbers that are zeros or close to zero. If the clock was reset when power was removed due to the backup battery being drained, then the time counter may also restart at zero. The clock actually works by counting the number of seconds from the epoch (t = 0).

Since time is important in an investigation to tie digital records to events, resetting the clock eliminates the ability for perform a verifiable analysis regarding the time. Therefore, changing the time should be avoided during a download. The clock on an ECM gets set by sending it a command through the vehicle network. The command for a DDEC IV and DDEC V is the J1587 Data Link Escape command (PID 254). For the DDEC IV and DDEC V, the clock is read by using the J1587 PIDs 251 (0xFB) for the clock and PID 252 (0xFC) for the date. For example, to request the date using J1587 over the J1708 network, the following command is sent from the host computer through a VDA:

0xB6 0x00 0xFC

where 0xB6 corresponds to the MID of the VDA, 0x00 is the request PID, and 0xFC is the date PID. The ECM would then respond with a message that says, for example,

0x80 0xFC 0x03 0x65 0x08 0x1C

where 0x80 is the MID of the ECM (128), 0xFC is the PID for the date (252), 0x03 says there are three more bytes in the message, 0x65 represents the number of quarter days (i.e., 101*0.25 = 25.25), 0x08 represents the month (August), and 0x1C is the number of years since 1985 (i.e., 1985 + 28 = 2013). The day can be calculated as CEILING(a/4) according to J1578 [10]. Similarly, if the time is requested, the following message would be crafted and sent:

0xB6 0x00 0xFC

which differs from the date request by only the PID being requested. An example response over the network could be

0x80 0xFB 0x03 0x70 0x13 0x02

where 0x80 is the MID of the ECM (128), 0xFB is the PID for the time (251), 0x03 says there are three more bytes in the message, 0x70 represents the number of quarter seconds (i.e. 112*0.25 = 28 s), 0x13 represents the minutes (19), and 0x02 represents the hours. So interpreting the time and date according to the J1587 standard [10] gives

August 25, 2013 at 2:19:28 AM. Understanding how dates and times are read over the network is important to set a date and time over the network. Using the same date and time encoding scheme, the 6 bytes to represent time and date could be a b c d e f. The message to send to the ECM to set the clock uses the Data Link Escape PID (254) and is crafted as follows:

0xB6 0xFE 0x80 0xC7 0x06 a b c d e f

where 0xB6 corresponds to the MID of the VDA, 0xFE is the Data Link Escape PID, 0x80 is the MID for the engine, 0xC7 is the code to reset the clock, 0x06 is the number of bytes to follow, a b c are the time characters, and d e f are the date characters.

Interestingly, the data stored in the XTR file corresponding to dates and times does not follow these standards. However, DDEC Reports does use the J1587 [10] and J1708 [11] standards to read and reset the clock for DDEC IV and DDEC V modules. Understanding how the clock gets read and reset is important in the design of forensic tools such as a write-blocking device.

Some Caterpillar ECMs report records that are off by a day. Based on the way the date PID is interpreted, it could be easy for a program to interpret the day code in the J1708 message using a truncation algorithm, which would produce data that is off by a day. This observation is untested and remains for future study.

7.2.4.3 Current Strategies to Establish Transparency and Trust: The RP1210 recommended practice from the American Trucking Association's Technology & Maintenance Council [14] governs the Windows API that diagnostic programs use to access diagnostic link connector devices. RP1210 Log files can usually be set to contain a record of all the network traffic. Some device manufacturers include debug logging capability in their driver software, presumably to facilitate development of programs that utilize their RP1210 interface. This capability is also useful for maintaining logs of the evidence extraction process, as they can log all function calls to the RP1210 interface. This provides another level of transparency, since the ECM extraction process can be recreated from these logs.

Based on current technologies, a method of ensuring manipulation free reports requires exposure of the process to the other party during the actual extraction. Alternatively, a neutral third party can be used to extract the data. The use of video captures, immediate file distribution, and experts witnessing the data extractions are ways to maintain transparency during the forensic process. Sometimes these options are not available for at-scene investigators who need to clear a roadway.

7.3 Recommendations for Digital Evidence Extraction from Heavy Vehicles

In this section, procedures and devices are described that can enable more forensically sound digital data extraction from heavy vehicle ECMs.

7.3.1 Sensor Simulators

Heavy vehicle ECMs connect to many sensors and other systems when installed in a fully functioning vehicle. When an ECM or is removed from the system, many of the components the device would normally expect are missing. For instance, during a bench download of an ECM that has been removed from a vehicle, the absence of these sensors can create new fault codes, cause spurious sensor input and overwrite potentially valuable information, especially in some current Caterpillar ECMs [28]. Previous work has shown how a passive system could simulate many of these sensors to achieve a simulated vehicle in the key-on engine-off (KOEO) state [27].

While the passive sensor systems explained in the literature will work on some ECMs, many ECMs require active signals to be fault-free. These signals may include the accelerator pedal position, throttle position, and J1939 messages. The J1939 messages emulate the ABS controller and body controller and prevent associated faults.

For demonstration purposes, a bench-top system was built based on a Navistar MaxxForce 13 engine, as shown in **Figure 7.12**. The system was equipped with an instrument cluster to visualize live J1939 network data. For the active signal generation, a National Instruments CompactRIO was used.

A fault-free state can be created for forensic imaging as verified by an exemplar ECM. Once verified, the subject ECM can be installed in the simulator for a fault-free download. It should be emphasized that the software determines what fault conditions exist and there can be different versions of the software loaded on any given ECM. Therefore, it is important to test an exemplar ECM that contains the same software as the suspect

FIGURE 7.12 A sensor simulator system with active and passive devices.

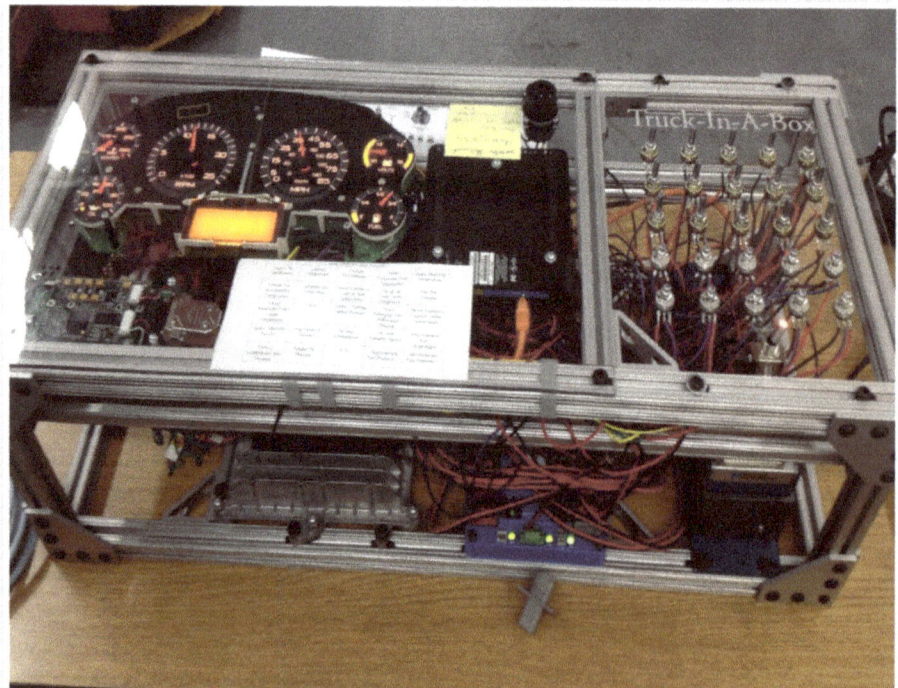

© SAE International

ECM to ensure a fault-free condition exists. The correct software can often be loaded onto an ECM by a dealer or maintenance shop if the VIN is known. This procedure was proposed in Reference [27].

7.3.2 Write Blockers

In extracting digital evidence from hard drives, it is standard practice to use a device that blocks all write signals to the evidence drive under examination in order to preserve the evidence. Such devices are, appropriately, known as "write blockers" [15].

A similar solution for a heavy truck evidence extraction write blocker would mediate the network communication between the diagnostic link adapter and the vehicle ECM. Any communication recognized as a command to modify data on the engine would be ignored, and not forwarded on to the ECM.

For example, the ECM clock reset command for a DDEC IV or DDEC V is accessed through the J1587 PID for Data Link Escape. A write blocker, consisting of two J1708 transceivers and control logic capable of parsing J1587 messages, would sit between the diagnostic link connector and the ECM, mediating traffic between the two. The control logic would examine the clock reset message and fail to forward the message, thus preventing the message from altering the data. For example, if the write blocker received a message from the VDA on the J1708 network that contained 0xFE 0x80 0xC7 0x06 in the 2nd-5th positions, then the message would be discarded, thus preventing the ECM from seeing a request to set the clock. A flow chart of the functionality of a write-blocking device and how it interacts with the software and VDA are shown in **Figure 7.13**.

7.3.3 Authentication Algorithms

Hard disk images (bit-for-bit copies of the data on hard drives) are typically verified using cryptographic hashes. When the image is created, the image is hashed and the hash is stored as part of the documentation of chain of custody. The difficulty with ECMs and embedded devices in general, is there is no hard drive and getting a bit-for-bit copy of stored data will require opening the case and reading the memory devices directly, which can be destructive to the physical ECM. Furthermore, Reference [31] indicates that hashes can be used to verify that the extraction process has not altered the source device. As some data alteration in standard ECM extractions is unavoidable (notably the ECM running time parameter, which is updated while the ECM is powered on), cryptographic hashing does not lend itself to source evidence verification in this application.

Therefore, the network traffic, representing the next level of trusted data, can be preserved and hashed. The presence of vehicle network traffic is common to all manufacturers and represents the most authentic data source. The network traffic represents a recording of the data that was exchanged between the devices and a replay of that data can be used to generate reports with the meaning of the data. The network traffic log is static and lends itself to authentication through cryptographic hashing. Additional documentation and external reference information can also be included along with the network log. It is recommended that a common algorithm such as SHA-256 [8] is used to hash the data. All parties in the case should record a copy of the hash in a secure location. If all parties are unknown at the time of an extraction, then the hash value should be stored by a neutral

FIGURE 7.13 Flowchart of a write blocker functionality.

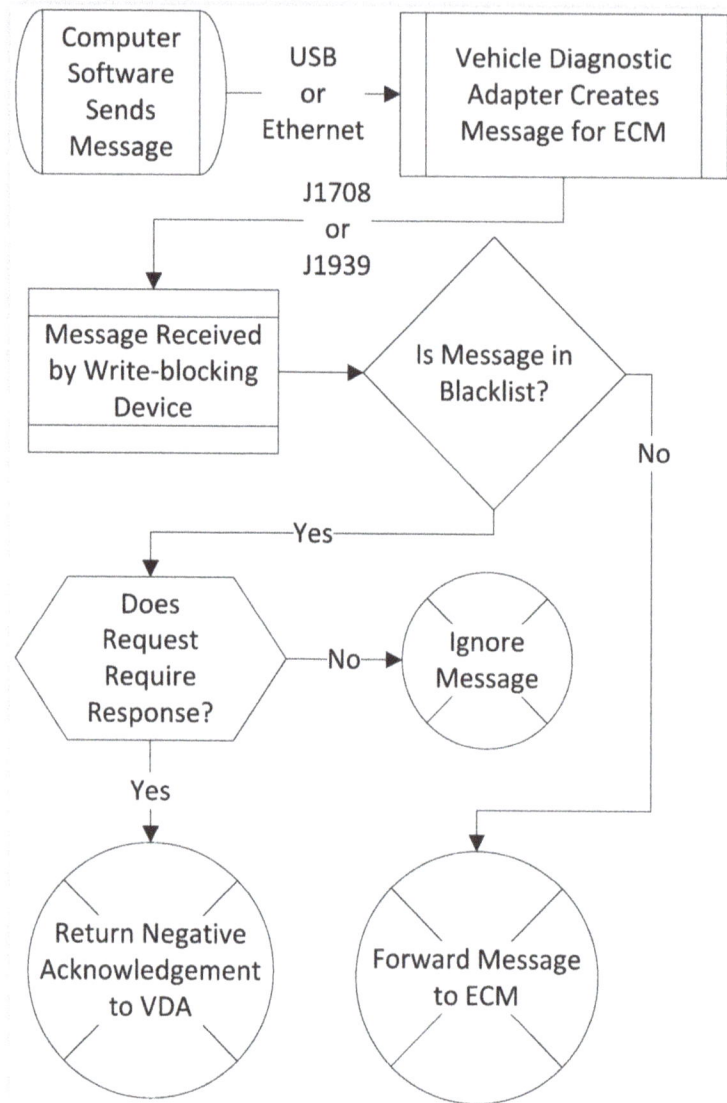

© SAE International

third party. If there is ever any doubt about the authenticity of evidence data after it has been collected, the hash of the file in question may be calculated and compared to the recorded hash. This concept is suggested through the use of a verification file in the J2728 standard.

A diagram of a suggested storage and authentication mechanism is given in **Figure 7.14**. The forensic VDA contains a set of public keys, with the associated private keys residing with a trusted third party. All relevant data are compiled into a report, which is then hashed. The report data is encrypted with a strong symmetric key algorithm, using a randomly-generated 128-bit key. This key is then encrypted with one of the public keys; the encrypted key is then stored alongside the public key and the encrypted report. The symmetric encryption algorithm is used due to the increased security of these algorithms [8].

FIGURE 7.14 Process diagram for forensic preservation that prevents tampering and provides verification through a hash.

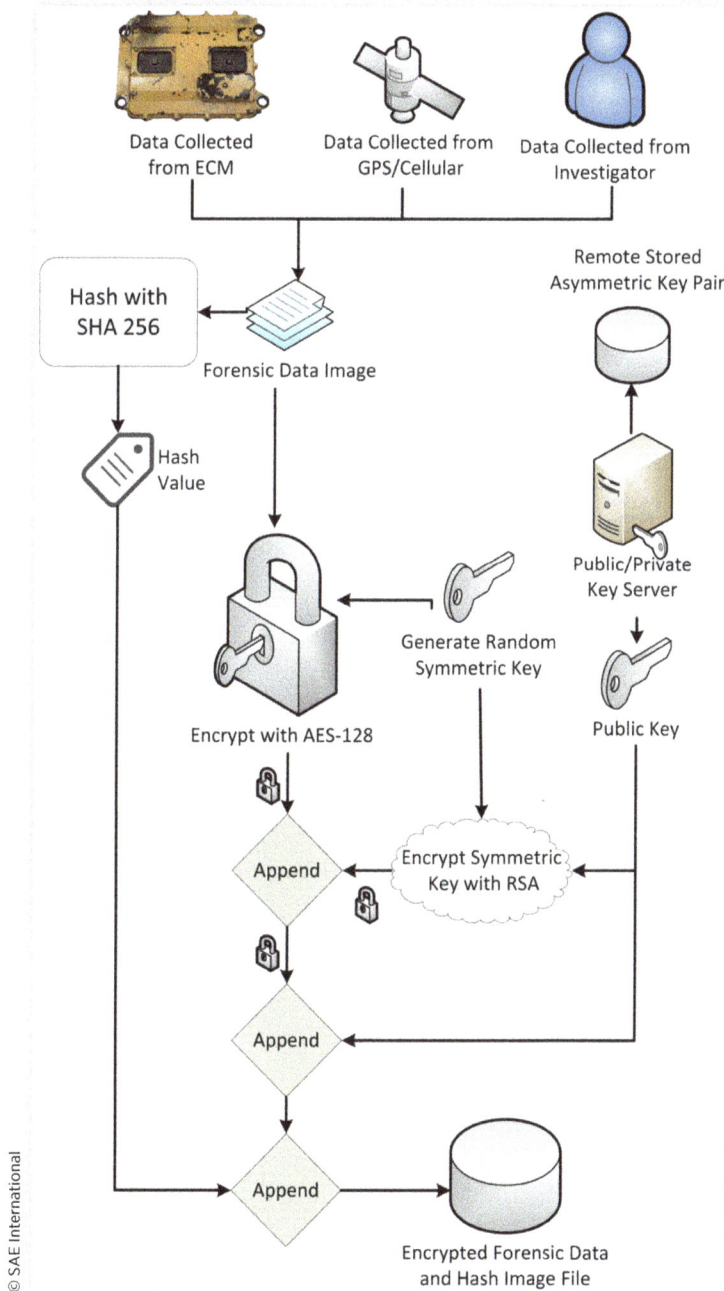

To decrypt, the appropriate private key is obtained from the trusted third party. The symmetric key is decrypted and used to decrypt the report. The report may then be hashed to verify its integrity.

Because the data are strongly encrypted, they cannot be altered or tampered with in a meaningful way without detection. This means that if even 1 bit is modified in the encrypted file, then the entire contents of the decrypted file will be meaningless and the hash values will not match. Because of this ability to detect tampering, if the files are decrypted and the hash values match, then there was no manipulation of the file and the data is forensically sound.

7.3.4 Forensic Replay Mechanism

The network traffic from the ECM data extraction can be treated as an evidence image. A device or program that is aware of the protocol used by the diagnostic program to extract data can impersonate an engine control module and consistently replay the same data every time for better repeatability in analysis. A flow chart of the design of a replay system is shown in **Figure 7.15**. The implementation challenge in the replay mechanism is parsing the different periodic and aperiodic messages. Also, the timing of the requests and responses was recorded, but may not be able to be reproduced if human interaction with the diagnostics is needed.

Having a verifiable replay mechanism with authenticated data eliminates the need to have the original ECM as the source of the data. This has a practical impact, since the ECM could be put back into service, the vehicle could be driven, and the burden for storage space could be reduced.

FIGURE 7.15   Flowchart describing the data flow for a network traffic replay system.

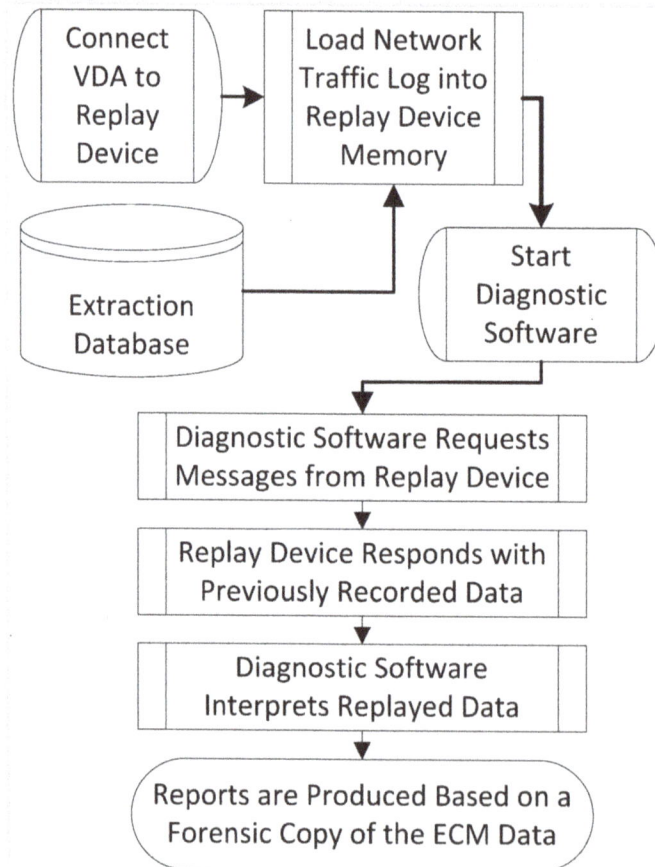

© SAE International

7.3.5 Journal Preservation

In order to satisfy the forensic tool requirements outlined in the digital forensics literature, it is recommended that any vehicle data extraction tool maintain log files documenting the actions of the investigator performing the extraction and records any errors encountered during the evidence extraction process. These logs should be stored alongside the crash data and network traffic itself, and handled with the same care, including hashing for integrity. Knowledge of the process by which the evidence was extracted is vital to authenticating its trustworthiness.

7.3.6 Chip Level Forensics

Forensics tools are used to access lower levels of abstraction in data and present them in a human-readable format [6]. In typical ECM data extractions, the data is presented by the ECM over a networking interface. Thus, the network interface acts as a layer of abstraction over the raw data recorded on the flash memory on board the ECM.

This layer of abstraction can be removed by accessing the raw data on the chip directly, as has been practiced in forensic analysis of cellular phones for some time [32]. Methods for accessing this data include using a debug interface such as the Joint Test Action Group (JTAG) port to read data from storage while it is on the circuit board, and removing data-bearing chips from the board so that they may be read with a purpose-built device. In some cases, removing flash or other storage devices from the ECM may render them unusable or significantly more difficult to decipher, due to algorithms used to arrange data on flash memory devices for reliability (wear-leveling) [33].

Frequently, a combination of techniques may be required for extraction to mitigate any risk of altering the data. For example, a DDEC V ECM contains at least four identifiable data-bearing components: two mass-storage flash devices, a microprocessor with integrated storage memory, and a real-time clock. The two mass-storage flash devices may be removed and imaged using a commonly available chip-reader. The main processor and real-time clock data would require extraction through JTAG [30] or some other form of in-circuit interrogation.

Chip removal or in-circuit data extraction from ECMs for forensic purposes can be challenging in itself. Most heavy truck ECMs are designed with robust environmental protection. Protections can include conformal coatings for anti-corrosion and water damage protection, flexible printed circuits for vibration tolerance, and metal enclosures sealed in a manner that was never intended to be opened. **Figure 7.16** is a Navistar ECM having its metal case milled away to expose data-bearing components.

Currently, interpreting raw information from these data-bearing components requires either proprietary knowledge or a significant amount of reverse engineering. The easiest strategy is to use a surrogate ECM and transplant the data-bearing chips and perform a download accordingly.

FIGURE 7.16 Milling around the edges of an ECM to expose the data-bearing components.

© SAE International

7.3.7 Beyond Crash Reconstruction

Recent awareness of information security regarding over the road vehicles has prompted researchers to explore potential vulnerabilities. Researchers from University of California San Diego and the University of Washington published two papers outlining several network-based vulnerabilities on a modern sedan [31, 32]. Following that research, notable demonstrations of vulnerabilities on a Toyota Prius and Ford Escape were published by Miller and Valasek [36]. While these efforts were sanctioned and performed under controlled environments, they do expose the possibility of nefarious activity involving vehicle ECMs. As such, detecting these activities may require the use of more advanced digital forensic techniques. Larson and Nilsson [37] presented ideas related to systematic approaches to understanding the interrelationships between digital and physical crimes in vehicles.

As an example, Bully Dog [38] makes ECM programmers that change truck engine tuning parameters, usually to achieve greater power output or remove speed governors. Investigators should be aware of these devices, as engine parameters that are not with the OEM specification can affect crash reconstruction calculations based on supposed engine performance parameters. Additionally, user changeable parameters, such as tire size and gearing can alter the speed calculated by the ECM. These figures are part of the digital record, but verifying them requires physical inspection.

7.4 **Summary/Conclusions**

This paper has discussed several principles of digital forensics, and applied them to the handling of incident data residing on heavy vehicle ECMs. The concepts of tamper resistance, meaning, error detection, transparency and trust were discussed in detail. A demonstration of manipulating a DDEC Reports .XTR file shows that native file formats should not be used as a baseline of trust.

While it is possible to extract, store, and present evidence data in a forensically sound manner using current methods, doing so requires great care and cooperation on the part of the investigators. This is often not practical when vehicles are driven from the scene, thus altering the data. Furthermore, all interested parties may not have been identified at the time of the data extraction. This represents a problem for any investigator using current practices without witnesses and may be problematic for first responders.

Details regarding the interpretation of the daily engine usage, ECM Time, hard brake, and Last Stop records were discussed in such a way as to bring meaning to raw hex data that is stored in the .XTR file produced by DDEC Reports. This level of investigation explains that the timing of transitions shown in some of the graphical representations of the daily engine usage is arbitrary as no time measure exists in the .XTR file. Therefore, attributing a specific time for the transition from off to idle or idle to drive based on the graph is inappropriate. Also, understanding the way time is represented in binary form can also help interpret data that represent the epoch and can provide sound explanations for dates that show up in the past. Common years that indicate a clock is set near its epoch are 1984 for the J1587 epoch or 1969 for the Unix epoch.

In addition to discussing the principles of forensic soundness, methods to make the forensic processes more reliable and sound were proposed. These include using journal logs to record investigators' actions, hashing and encrypting evidence files to ensure integrity, blocking message traffic that can alter a digital record, storing, hashing and replaying diagnostic network traffic to minimize alteration of the source device, and performing forensic extractions at lower levels of abstraction such as direct chip reading.

The fact that the data files can be easily altered should call their trustworthiness into question, but the data on the network may be trusted implicitly. Data integrity in adverse environments is a primary design consideration in the design of vehicle networks, so system designers have implemented extensive protections against accidental alteration of network data. Archiving the network data, hashing the data, and encrypting the data to prevent undetected tampering is the forensically sound solution proposed that should work with all manufacturers that use the vehicle network to transmit the data. The hash value should be distributed immediately after the data extraction is complete to the other parties or a neutral third party.

As vehicle and systems designers strive to harden the cybersecurity posture of heavy vehicles, the forensics associated with driving events will need to expand to cyber events. Building mechanisms for easy and meaningful data extraction that is forensically sound will be necessary to uncover cyber-attacks in the future.

Definitions/Abbreviations

AES - NIST Advanced Encryption Standard—algorithm based on Rijndael cipher

API - Application Programming Interface

CRC - Cyclic Redundancy Check

ECM - Electronic (or Engine) Control Module

ECU - Electronic Control Unit

EDR - Event Data Recorder

HVEDR - Heavy Vehicle Event Data Recorder

JTAG - Joint Test Action Group—Refers to IEEE 1149.1 Test Access Port/Boundary Scan

MID - Message Identifier (see J1587)

NIST - National Institute for Standards and Technology

PC - Personal Computer

PID - Parameter Identifier (see J1587)

RSA - Public-Key Cryptography system authored by Rivest, Shamir, and Adleman

SHA - Secure Hash Algorithm—family of cryptographic functions published by NIST

TMC - Technology & Maintenance Council of the American Trucking Association

UTC - Coordinated Universal Time, which is the successor of Greenwich Mean Time

VDA - Vehicle Diagnostic Adapter

References

1. FMCSA, https://www.fmcsa.dot.gov/safety/data-and-statistics/large-truck-and-bus-crash-facts-2015.

2. NIJ Special Report, "Electronic Crime Scene Investigation: A Guide for First Responders," Published April 2008, https://www.ncjrs.gov/pdffiles1/nij/219941.pdf, accessed 26 December 2013.

3. National Institute of Standards and Technology, "Hardware Write Blocker Device (HWB) Specification," Published May 19, 2004, http://www.cftt.nist.gov/HWB-v2-post-19-may-04.pdf, accessed December 26, 2013.

4. International Organization on Computer Evidence, *Guidelines for Best Practice in the Forensic Examination of Digital Technology*, (Digital Evidence Standards Working Group, 2002).

5. McKemmish, R., "When Is Digital Evidence Forensically Sound?" in Ray, I. and Shenoi, S., eds., *Advances in Digital Forensics IV*, (Boston: Springer, 2008), 3-15, doi:10.1007/978-0-387-84927-0_1.

6. Casey, E., "Error, Uncertainty, and Loss in Digital Evidence," *International Journal of Digital Evidence* 1, no. 2 (2002): 1-45.

7. SAE International Surface Vehicle Recommended Practice, "Heavy Vehicle Event Data Recorder (HVEDR) Standard – Tier 1," SAE Standard J2728, Rev. June 2010.

8. Schneier, B., *Applied Cryptography*, (New York: John Wiley & Sons, 1996).

9. Anonymous, "Secure Hash Standard (SHS), Federal Information Processing Standards Publication FIPS Pub 180-4," National Institute for Standards and Technology, Gaithersburg, MD, March 2012, v + 30pp, http://csrc.nist.gov/publications/fips/fips180-4/fips-180-4.pdf; http://csrc.nist.gov/publications/PubsFIPS.html#fips180-4.

10. SAE International Surface Vehicle Recommended Practice, "Electronic Data Interchange Between Microcomputer Systems in Heavy-Duty Vehicle Applications," SAE Standard J1587, Rev. July 2007.

11. SAE International Surface Vehicle Recommended Practice, "Serial Data Communications Between Microcomputer Systems in Heavy Duty Vehicle Applications," SAE Standard J1708, Rev. October 2008.

12. SAE International Surface Vehicle Recommended Practice, "Vehicle Application Layer," SAE Standard J1939-71, Rev. February 2010.

13. TMC Recommended Practice, "Windows Communication API," TMC RP 1210, Rev. June 2007.

14. SAE International Surface Vehicle Recommended Practice, "Application Layer – Diagnostics," SAE Standard J1939-73, Rev. February 2010.

15. Bortolin, R., van Nooten, S., Scodeller, M., Alvar, D. et al., "Validating Speed Data from Cummins Engine Sudden Deceleration Data Reports," *SAE Int. J. Passeng. Cars - Mech. Syst.* 2(1): 970-982, 2009, doi:10.4271/2009-01-0876.

16. Austin, T. and Farrell, M., "An Examination of Snapshot Data in Caterpillar Electronic Control Modules," *SAE Int. J. Passeng. Cars - Mech. Syst.* 4(1): 611-635, 2011, doi:10.4271/2011-01-0807.

17. Plant, D., Cheek, T., Austin, T., Steiner, J. et al., "Timing and Synchronization of the Event Data Recorded by the Electronic Control Modules of Commercial Motor Vehicles - DDEC V," *SAE Int. J. Commer. Veh.* 6(1): 209-228, 2013, doi:10.4271/2013-01-1267.

18. Steiner, J., Cheek, T., and Hinkson, S., "Data Sources and Analysis of a Heavy Vehicle Event Data Recorder - V-MAC III," *SAE Int. J. Commer. Veh.* 2(1): 49-57, 2009, doi:10.4271/2009-01-0881.

19. Reust, T., "The Accuracy of Speed Captured by Commercial Vehicle Event Data Recorders," SAE Technical Paper 2004-01-1199, 2004, doi:10.4271/2004-01-1199.

20. van Nooten, S. and Hrycay, J., "The Application and Reliability of Commercial Vehicle Event Data Recorders for Accident Investigation and Analysis," SAE Technical Paper 2005-01-1177, 2005, doi:10.4271/2005-01-1177.

21. Drew, K., van Nooten, S., Bortolin, R., Gervais, J. et al., "The Reliability of Snapshot Data from Caterpillar Engines for Accident Investigation and Analysis," SAE Technical Paper 2008-01-2708, 2008, doi:10.4271/2008-01-2708.

22. Reust, T., Morgan, J., and Smith, P., "Method to Determine Vehicle Speed during ABS Brake Events Using Heavy Vehicle Event Data Recorder Speed," *SAE Int. J. Passeng. Cars - Mech. Syst.* 3(1): 644-652, 2010, doi:10.4271/2010-01-0999.

23. Messerschmidt, W., Austin, T., Smith, B., Cheek, T. et al., "Simulating the Effect of Collision-Related Power Loss on the Event Data Recorders of Heavy Trucks," SAE Technical Paper 2010-01-1004, 2010, doi:10.4271/2010-01-1004.

24. Messerschmidt, W., "DDEC Reports Version 8.02: Analysis of Daily Engine Usage Data," Presented at *the Annual Illinois Association of Technical Accident Investigators Conference*, East Peoria, IL, September 18–20, 2013.

25. Python Software Foundation, "time - Time Access and Conversions," Python 3.3.3 Documentation, http://docs.python.org/3.3/library/time.html, accessed January 13, 2014.

26. Python Software Foundation, "struct - Interpret Bytes as Packed Binary Data," Python 3.3.3 Documentation, http://docs.python.org/3.3/library/struct.html, accessed January 13, 2014.

27. Boggess, B., Dunn, A., Morr, D., Martin, T. et al., "A New Passive Interface to Simulate On-Vehicle Systems for Direct-to-Module (DTM) Engine Control Module (ECM) Data Recovery," SAE Technical Paper 2010-01-1994, 2010, doi:10.4271/2010-01-1994.

28. Plant, D., Austin, T., and Smith, B., "Data Extraction Methods and Their Effects on the Retention of Event Data Contained in the Electronic Control Modules of Detroit Diesel and Mercedes-Benz Engines," *SAE Int. J. Passeng. Cars - Mech. Syst.* 4(1): 636-647, 2011, doi:10.4271/2011-01-0808.

29. Kohno, T., "Attacking and Repairing the WinZip Encryption Scheme," *Proceedings of the 11th ACM Conference on Computer and Communications Security*, Washington, DC, 2004. doi:10.1145/1030083.1030095.

30. Koopman, P. and Chakravarty, T., "Cyclic Redundancy Code (CRC) Polynomial Selection for Embedded Networks," *The International Conference on Dependable Systems and Networks*, Florence, Italy, 2004

31. Carrier, B., *File System Forensic Analysis*, (Upper Saddle River, Addison Wesley, 2005), ISBN:0-321-26817-2.

32. Carrier, B., "Defining Forensic Examination and Analysis Tools Using Abstraction Layers," *International Journal of Digital Evidence* 1(4): 1-12, 2003.

33. Breeuwsma, M., de Jongh M., Klaver C., van der Knijff R. et al., "Forensic Data Recovery from Flash Memory," *Small Scale Digital Device Forensics Journal* 1(1): 1-17, 2007.

34. Checkoway, S., McCoy, D., Kantor, B., Anderson, D. et al, "Comprehensive Experimental Analyses of Automotive Attack Surfaces," *Usenix Security*, San Diego, CA, 2011.

35. Koscher, K., Czeskis, A., Roesner, F., Patel, S. et al., "Experimental Security Analysis of a Modern Automobile," *IEEE Symposium on Security and Privacy*, Oakland, TN, 2010.

36. Miller, C., Valasek, C., "Adventures in Automotive Networks and Control Units," Published August 2013, http://illmatics.com/car_hacking.pdf, accessed October 13, 2013.

37. Larson, U. and Nilsson, D., "Securing Vehicles Against Cyber Attacks," *Proceedings of the 4th Annual Workshop on Cyber Security and Information Intelligence Challenges Ahead*, Oak Ridge, TN, 2008.

38. Bully Dog, "Bully Dog Semi-Truck," http://bigrig.bullydog.com, accessed October 16, 2013.

Contact Information

The corresponding author is Dr. Jeremy Daily, e-mail: jeremy-daily@utulsa.edu, PH: (918) 631-3056.

Acknowledgments

Portions of this project were supported by Award No. 2010-DN-BX-K215 awarded by the National Institute of Justice, Office of Justice Programs, U.S. Department of Justice. The opinions, findings, and conclusions or recommendations expressed in this document are those of the author(s) and do not necessarily reflect those of the Department of Justice.

A. Appendix

FIGURE 7.17 Excerpt from a DDEC Reports generated daily engine usage log.

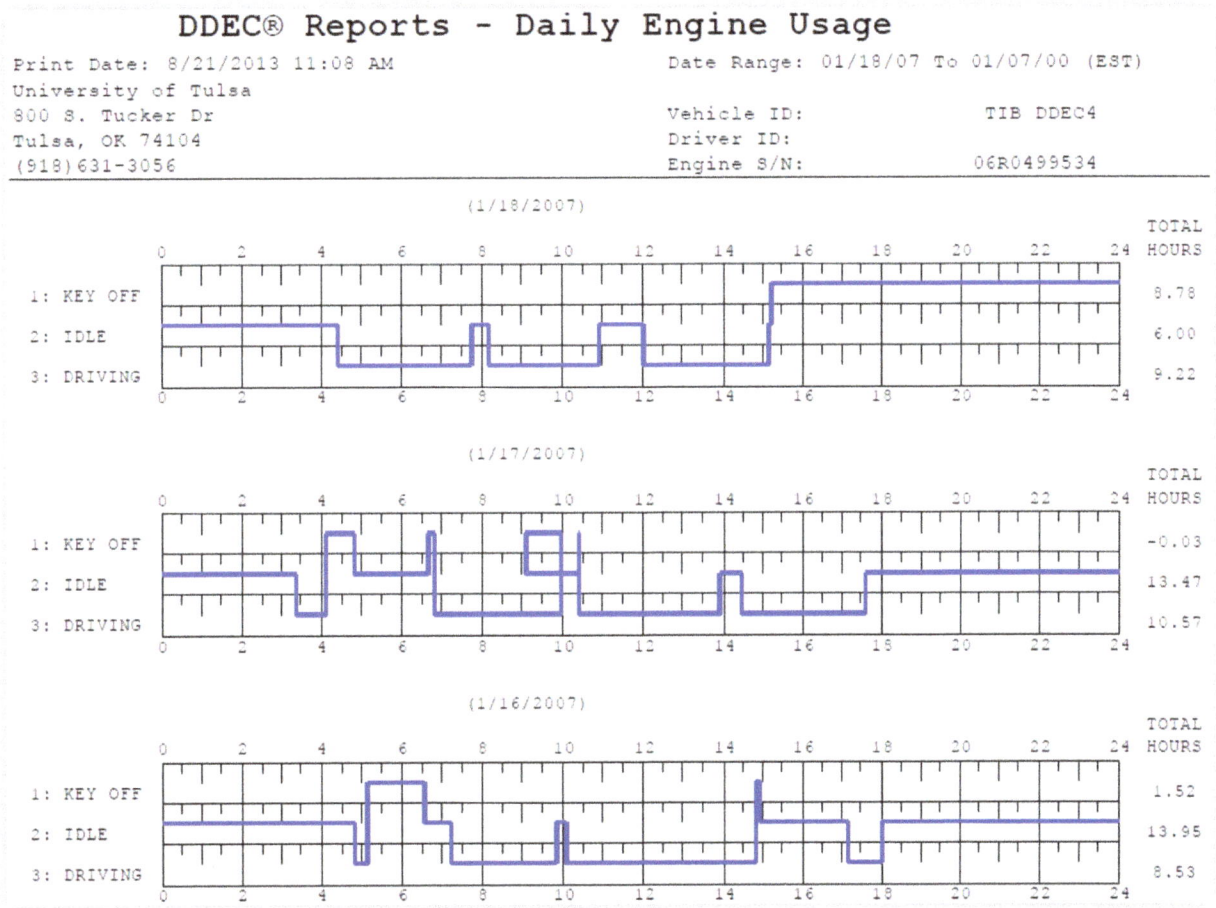

DDEC® Reports - Daily Engine Usage

Print Date: 8/21/2013 11:08 AM
University of Tulsa
800 S. Tucker Dr
Tulsa, OK 74104
(918)631-3056

Date Range: 01/18/07 To 01/07/00 (EST)

Vehicle ID: TIB DDEC4
Driver ID:
Engine S/N: 06R0499534

FIGURE 7.18 Table entry from DDEC Reports corresponding to the first line of the graph in Figure 7.17.

DDEC® Reports - Daily Engine Usage

Print Date: 8/21/2013 11:08 AM
University of Tulsa
800 S. Tucker Dr
Tulsa, OK 74104
(918)631-3056

Vehicle ID: TIB DDEC4
Driver ID:
Engine S/N: 06R0499534

Date:	1/18/2007
Start Time:	00:00:00 EST
Odometer:	1006109.00 mi
Distance:	548.80 mi
Fuel:	95.25 gal
Fuel Economy:	5.76 mpg
Average Speed:	59.54 mph

Total(hh:mm)	09:13	06:00	08:47
Hour(EST)	Drive(min)	Idle(min)	Off(min)
00:00-02:00	0	120	0
02:00-04:00	0	120	0
04:00-06:00	96	24	0
06:00-08:00	104	16	0
08:00-10:00	110	10	0
10:00-12:00	54	66	0
12:00-14:00	120	0	0
14:00-16:00	69	4	47
16:00-18:00	0	0	120
18:00-20:00	0	0	120
20:00-22:00	0	0	120
22:00-24:00	0	0	120

© SAE International

About the Author

Jeremy Daily is researching traffic crash reconstruction, vehicle digital forensics, and commercial vehicle cybersecurity. He teaches automotive design, machine dynamics, and finite element analysis. A couple years ago, Jeremy was able to transition some of the University research on heavy vehicle digital forensics to practice by starting a technology company, Synercon Technologies.

CHAPTER 8

Telematics Cybersecurity and Governance

Glenn Atkinson

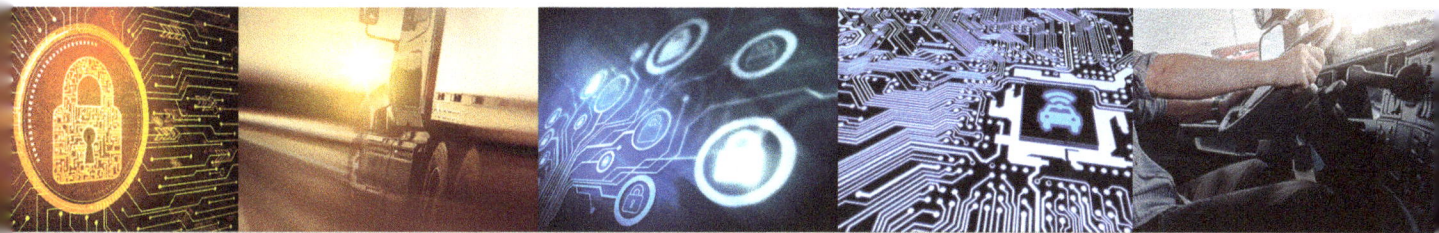

8.1 Background: Author

During my formative years during and after high school, I worked at a garage servicing cars and I also owned, serviced, and drove motorcycles. I attended civil engineering, mechanical engineering, and eventually completed a project based master of engineering. A variety of work terms in a wide range of industries from oil and gas to municipal infrastructure, mining, forestry, and manufacturing. A consistent theme for me, almost from the beginning, was to learn about tools and processes to better identify how things might go wrong early in the design phase and use these to create more resilient designs that work in the wide variety of upset conditions that could occur.

It was during some graduate work in the early 1990s while attending the University of Southern California that I studied "system safety" under Professor Herc Roland. I learned design analysis tools and methods developed by Bell labs, among others, to proactively identify hazards in the nuclear and aerospace industries. It's very harmful to have a nuclear meltdown or, likewise, to have a nuclear-armed ICBM missile detonate over friendly (or even unfriendly if unintended) territory. As a result, great effort was taken in developing analytical and behavioral methods (system safety) to proactively identify hazards in these designs and make them as fault tolerant as possible. The system safety programs for these high-risk industries are massive, bureaucratic behemoths. It makes sense when the stakes are so high to have commensurate effort expended to control risk to an acceptable level.

My task, at ONRSA (Ontario Natural Resource Safety Association), was to reconfigure these methods in a form that would be applicable to the mining and other natural resource industries. I facilitated many proactive hazard analysis projects involving a wide cross section of workers, managers, and consultants over the next 5 years. This was one of my first tastes of the power of collaboration.

In the early 1990s, mining automation and the proposed use of remotely operated underground equipment was envisioned by some forward-looking large mining companies. Automation promised big advantages in safety and some reduced ventilation requirements and improved overall efficiency. The MOL (Ministry of Labor) in Ontario had safety concerns about remote-controlled large underground vehicles moving within mines where other miners worked who were on foot or in light vehicles. Technology companies with the expertise and drive to facilitate the remote technology were aggressively marketing the benefits. I was approached and accepted a project to facilitate a collaborative four-party functional safety analysis among the MOL, INCO, Mintronics, and McGill University.

The study was not the win in itself. The win came from the willingness and indeed enthusiastic support for a detailed proactive analysis of what could go wrong and what controls would be recommended to address identified risks. Through the process, all parties understood each other's concerns, worked towards constructive solutions, and a couple of years later, then PM (Prime Minister—for those of us who aren't Canadian) Chretien operated an underground mobile equipment in Sudbury from the surface. I briefly share this story to provide insight into development of my early modus operandi to value open, transparent collaboration to solve tough problems among a variety of stakeholders.

Later, as director of product safety for a large global hydraulic press manufacturer and chair of the related ANSI safety standards committee, I again learned first-hand the real life benefit of emphasizing a proactive collaborative approach to product safety. In this case, one key objective was to harmonize the requirements between Europe and North America. We made a lot of progress. However, there was one nut that we could not crack. North America included a requirement for a mechanical device as the third independent operator gate interlock. Europe used advanced fail-safe circuitry and two operator gate interlocks. It was the product liability issues from legacy machines in North America that delayed harmonization and points to the huge benefit to design in global, harmonized security requirements for the connected vehicle from the start and to avoid the swamp of legacy issues rooted in different jurisdictions that are difficult to change and harmonize later.

8.2 **Collaboration**

My observations from within a variety of industries over the past 30 years impressed to me that meaningful accomplishments in functional safety occurred as a result of transparency and collaboration, not individual genius in isolation. No efficient business has all of its departments and subsidiaries working in independent silos that pass on WIP (work in progress) till it becomes the final product. A problem in one area has its roots in others upstream and its effects in others downstream. We view this analogous to vehicle cybersecurity. A meaningful solution can only come from collaboration among interested stakeholders.

Benefit of collaboration within a company is sometimes easier understood as all of the participants (employees) have their paycheck drawn from the same account. A contribution from one employee in their area benefits the company and helps everyone. Companies normally provide positive reinforcement for internal teamwork as it leads to improved business processes and profit. Internally collaborative companies are not uncommon and generally invigorating productive companies.

In some cases, leaders and other individuals within companies believe that working collaboratively with other companies (their direct competitors in some cases) will disadvantage their interests by sharing insight with competitors. True in some cases, false in others. Understanding the difference, and collaborating within industry where appropriate, is critical in this case of advancing cybersecurity of the connected vehicle. As well, cybersecurity collaboration with all stakeholders across industry is required to optimize BOTH the size and growth of the overall industry and each specific company's share of this industry.

This idea applies in spades with the connected vehicle. However, there are a few significant issues that could inhibit broad adoption of the connected vehicle, beyond its entrenched place in the commercial fleet: they are functional safety, cybersecurity, and privacy governance. It seems quite intuitive that if industry-wide solutions are developed that adequately address these concerns, the entire connected vehicle industry grows, providing fertile grounds for all the businesses operating within it to the direct benefit of vehicle end users, the consumer. Likewise, if these concerns fester and erupt in nonconstructive activities, anywhere, with anyone or business, the entire industry suffers in the setback, and everyone's share of the industry will commensurately shrink.

Over the last few years, this seems particularly relevant in vehicle security. Fortunately, as readers of this SAE text can observe, the "connected vehicle" industry participants, from chip manufacturers to OEM, Tier 1, TSP (telematics service provider), fleet owners, consultants, related industry associations, and regulatory authorities, all understand this and there are scores of related simultaneous initiatives ongoing to advance connected vehicle security.

We caution that should industry participants build walls to limit or close access to the connected vehicle, this will effectively thwart future innovation and advancement of new value add propositions, and the industry will shrink from what it could have been. This is not a trivial obstacle and should be a concern to all, especially end users of vehicles as it's their future value via restricted innovation that would be sacrificed in a "closed" connected vehicle. Readers who wish to explore this argument are urged to read the blog "Securing the Future of Connected Mobility - Six Essential Recommendations for Fleet Management, Leasing and Rental Operations."

Those of us who participate as members of SAE standards development committees hear the standard antitrust boiler plate read to begin each meeting as the rules for engagement in SAE committee collaborative standards and guidelines development. No mention of sales volumes, pricing, and no comments that can be construed by others as self-serving, etc. This provides a very powerful construct for working collaboratively, and I believe it serves our companies and industry well to observe these same rules when working on the connected vehicle security issues generally.

There are technical challenges with connected vehicles for sure. However, we have many extremely bright and motivated minds that can develop technical solutions to defined problems. We engineers excel at solving this kind of problem. The real issue is to frame the challenge in a form that includes all stake holders' interests and facilitate the process to develop a solution collaboratively among all stakeholders, one that is end user beneficial and friendly. I discuss collaboration as an essential requirement to solve serious industry issues as my first point because I believe it is the overarching required construct to address vehicle security issues.

The recent "standing-up" of the Auto ISAC is such an example where OEMs and other stakeholders up and down the connected vehicle value chain share cybersecurity information and develop best practices. This collaborative approach recognizes this is the most effective approach to address the common security threat to the industry overall.

8.2.1 And So My Journey Begins

And, so my journey began with Geotab.

While relaxing during a well-earned sabbatical in 2014, I got a call from a prior colleague, now a board member involved with a vehicle telematics company, to inquire if I was interested in providing some guidance to them regarding product safety. It didn't take too long to realize this was an emerging industry, with immense opportunity to unlock value. It was also clear, proactive product stewardship was essential for the industry overall to reach its potential. I was interested, so the journey began for me with Geotab.

Geotab is a global leader in telematics, providing open platform fleet management solutions to businesses of all sizes. Geotab's intuitive, full-featured solutions help businesses better manage their drivers and vehicles by extracting accurate and actionable intelligence from real-time and historical trips data.

This chapter is not a detailed "how to" guide for telematics cybersecurity. Although basic related requirements are covered conceptually at least, I reference other guidelines, blogs, and white papers that cover the technical requirements more fully.

This chapter is written for connected vehicle end users, commercial vehicle fleet owners, and managers. We provide an overview of the issues and questions that compare the mostly existing frame of reference, laws, and governance mostly developed pre-cyberspace that relate well to the "classic" physical-only vehicle and compare this against the new reality of the commercial "connected" cyber-physical vehicle fleet. Engineers looking to dive straight into the technical solutions can focus on the references provided at the end of this chapter, and the extensive technical research provided within other chapters in this text.

I would like to thank Neil Cawse, Geotab's CEO for granting me a broad mandate and the autonomy and resources to oversee product safety, and in particular do what was required to begin the ongoing journey to raise the security level of the Geotab platform and develop and share best practices within the telematics and, more broadly, connected vehicle industry overall. When there are exploited telematics cybersecurity issues within the industry, it's problematic for everyone, not just those directly involved. It results in safety and economic loss issues, restricts growth of the industry, and constrains the ultimate value for end users. We've begun our journey; the ending is yet to be written.

8.2.2 Classic Electro-Hydraulic-Mechanical Vehicle

My first car was a beauty, a 1968 Mercury Montego MX, 2-door coupe, white, black vinyl roof 302 V8. She did require continuous tender loving care, as did every other vehicle of her era. The engine had to drive a lot more than the wheels! The engine was connected with pulleys and belts to turn the power steering pump, to assist in steering the wheels. The engine required the ignition points to be filed or replaced every 6 months or so, the distributor rotor and cover to be replaced/maintained regularly, and the carburetors (like all carbs) were finicky. Drum brakes eventually brought the cruiser to a stop. Although, it was not nearly as uneventful as rapid stops with today's vehicles. However, a person with a few years of study and application of same, could normally diagnose a vehicle problem and implement a solution to at least return the vehicle to running state. Emissions were high and fuel economy was low. But the love affair endured. I loved my first car.

As more vehicles entered our roadways, we began noticing something in the air becoming more prominent. It was something between smoke and fog, especially pronounced with temperature inversions and in valleys. It was quite undesirable. Vehicle emissions (and fuel consumption) became the focus.

Computers were in use in industrial SCADA (supervisory control and data acquisition) systems, at that time. They had been, for many years. A SCADA refers to an industrial computer system that monitors and controls a process. Initially, these systems were not connected to the Internet in an effort to prevent a "bad actor" from remotely upsetting critical industrial processes that could cause great harm. It seemed reasonable to employ a facsimile of the SCADA computer technology in vehicles to control the fuel burn in the cylinder and thus reduce emissions. As well, the power steering pump was exchanged for an electric servo drive motor. If you think about it, the pulley driving the old hydraulic pump was turning as long as the engine was running, even if the steering wheel was not required to turn. That was wasting energy, gasoline in this case. The electric servo motor to assist steering only required electricity (from the alternator also connected by pulley to engine) when the vehicle wheels were actually required to steer the vehicle. These improvements delivered a vehicle that not only ran better, smoother, and was more reliable, but the vehicle consumed less fuel and had lower emissions. It was the SCADA era of automobiles. The intercommunication of the vehicle computer systems was along a broadcast messaging network developed by Bosch called the CAN Bus. All sensor messages are broadcast on the CAN Bus, and every ECU (electronic control unit) can listen to each message. Sort of like the old country "party" telephone lines.

8.3 Connected Vehicles

Then in the 2000s, cell phones started to become more common. Cell tower coverage improved rapidly. The utility of the cell phone was so great its adoption has become almost universal today. Now cellular coverage is a virtual blanket over urban (and most of rural) North America. With good cellular coverage, OEMs (original equipment manufacturers) could offer cellular telematics and great features like remotely unlocking your car doors should you inadvertently lock your keys inside and offer roadside assistance in the event of a vehicle breakdown, without the driver requiring a cell phone.

Beginning in 1996, to comply with emission measuring requirements, an OBD (on-board diagnostic) port was mandated to be added under the dash, accessible close to the driver seat, and this OBD port was connected directly to the CAN Bus. Emissions diagnostic tools could be connected and indirectly measure the vehicle emissions. The next step was a small one. Commercial vehicle owners now had the potential to remotely observe the vehicle diagnostics data. All they needed was an OEM or a third-party cellular telematics device connected to the OBD port. A frenzy of discussions among commercial fleet owners and third-party telematics companies soon transformed 50 years of commercial fleet management best practices into the operating code embedded within the telematics device. Now commercial vehicles from Fortune 500 fleets to single small business fleets could benefit from using the best global fleet management practices. And, they have.

It is not an understatement that today, the commercial vehicle fleets are leading the way in demonstrating the value of remote access to their vehicles' data and routing location. There are literally hundreds of additional use cases that integrate third-party apps with vehicle data developed and many more on the doorstep of innovation.

8.4 Everything Was Coming and Going Along So Well...

I had been involved with Geotab for several months. I was beginning to understand the fleet management and telematics business and forming an overview of the technology, when we learned of a report in the press of a "hack" of a third-party telematics device in November 2014. The report circulated our offices and deep consternation followed. We began to prepare a communication document we could provide to our customers should they inquire with us after reading the same press article. Although the specifics of this hack would not have applied to our device, it didn't take too long to recognize we had only industry standard security of the time (some by obfuscation, rather than one-way algorithms). We were not happy with the explanation we could provide. Cybersecurity would envelope most of my and many others' efforts going forward.

Geotab's CEO, Neil Cawse, was 100% supportive that telematics security, more broadly connected car security, was a clear and present threat to commercial vehicle fleets, and all the advancements a connected fleet had brought to them. We had focused much attention on server security and Internet-facing applications. However, we had relied too much on cell service provider encryption that has been shown to be vulnerable. Our plan was to develop best practices for telematics platform cybersecurity, share with industry, and urge and motivate our industry to join the conversation and implement best practices.

Before we go further, it's worth taking a moment to think about some of the differences between the new environments of the cyber-physical "connected" vehicle and the "classical" vehicle that operated strictly in the physical space.

Cyberspace is everywhere, all the time. Because distance and physical matter is not relevant in the cyber domain, borders and rules that depend on physical matter do not apply. When we drive across national borders and see the controls in place for us and our vehicle, they all rely on a physical presence. A cyber adversary can conceivably reach across any border or ocean, from anywhere, and touch any IoT (Internet of Things), including a connected vehicle. One exception might be China. The main fiber-optic cables carrying Internet traffic under the ocean are all routed through a central Chinese clearing house. On one hand, this provides control of the Internet, on the other, is the lack of freedom of expression and innovation.

Although the Internet destroys physical national borders, some nations still implement different cyber rules in different nations. One example is the difference in privacy rules in Germany versus the United States. So telematics platform developers need to review and incorporate sufficient design configuration options that end users can configure systems that meet their nation's cyber requirements on privacy for example, securely. While on the topic of privacy, logically it follows there can only be consideration of privacy governance after security is established.

8.4.1 Anonymity on the Internet

The Internet is simply a giant switching system with no inherent authentication or authorization. It is relatively easy for a cyber adversary to "hide" in the darknet. The IP address is the address given every entity that has access to the Internet, like the mailing address used on envelopes with the postal service to get the letter to the right recipient. The darknet provides IP addresses that are routed and rerouted in a way that their true identity/origin can no longer be traced. This provides a level of anonymity for the cyber adversary quite different from the classic bank robber who had to be physically present in the bank to rob it.

Moore's law, that the computing power of the microprocessor will double every 18 months effectively means every cybersecurity control measure in place today, will soon/eventually be broken. Time is ticking on every implemented control measure today. As well, this means the microprocessors used in vehicles today will likely become obsolete long before the vehicle does. Some form of quick and easy upgrade for the connected commercial vehicle seems a reasonable requirement. Not just a secure OTA (over-the-air) update, but a wholesale upgrade of the hardware too would be nice.

Zero Day Exploits are exploits an adversary has found but has not yet deployed. They are a vulnerability that is guaranteed to be a successful exploit the first time it is used. There were multiple Zero Day Exploits used, for example, in the Stuxnet virus/worm, hinting at the sophistication required to implement such an attack. The vulnerability can lay dormant, then without warning, at a most inopportune time be exploited. Zero Day Exploits are a concern.

Cyberspace can cause real effects in the physical world. In the computer's early days, a computer bug or malware could corrupt files, but damage was limited to the cyber world. Stuxnet—the complex virus/worm that infected the Iranian SCADA system controlling nuclear centrifuges and caused them to rotate uncontrolled (although the malware was smart enough so computers indicated normal centrifuge rotation to the operator) and caused massive permanent loss of the physical system. Once this was demonstrated in real life, all cyber-physical systems (SCADA or not) were put on notice. Utilities, energy companies, chemical companies, and of course, the vehicle transportation system among others are on notice; cybersecurity can cause real effects in the physical world.

Cyber systems are sufficiently asymmetric so adversaries have a vast advantage over system designers. This means we need to anticipate the system will be exploited.

We need to plan for it. This means system designers have to go on the offensive. Systems need some forms of anomaly/intrusion detection to detect adversaries that are banging at their system or have breached parts of the system. A plan for recovery must be in place and tested to ensure it is there when required. This also means computer systems should be segmented so when a breach is detected, it can be contained, and the remainder of the system can continue operation.

There is also asymmetry in the costs and benefits of cybersecurity. This is a real issue, and if industry doesn't get it, regulators will move in. When Microsoft or Google develop code/applications that inevitably are found to have "bugs" in them, who is left to deal with the consequence? The end user of course, and there is currently no means to push that cost and inconvenience back to its originator. What incentive is there to spend the time and resources to develop "bug-free" code? If there are two software applications that perform a similar function, why would a consumer pay twice as much for one? The code writing enterprise who profits from the application sale is isolated from the customer who has to deal with suboptimal code development. This feedback loop does not reinforce good code development.

For our purposes, good telematics product stewardship requires a more holistic view of security, its costs, and benefits. We must as an industry, recognize that to optimize the value proposition of the connected vehicle to end users, security, and its related costs, must be developed and applied throughout the ecosystem. Because of the asymmetry of the costs of security versus the cost avoidance (end user), development and sharing of best practices is essential. This will not only ensure the current best ideas of industry are used but that they are also implemented throughout industry at as low a cost as possible. It is highly inefficient use of resources to develop these practices individually.

Security, as we will see, is an end-to-end process. Not only must the telematics provider provide a resilient, secured, platform; they must communicate the importance of good computer hygiene practices to other stakeholders up and down the connected vehicle value chain. Poor computer hygiene practices, wherever they occur, will frustrate all computer systems they interact with.

Hopefully by this point, you have not thrown your hands in the air and cried out "what's the point of trying?" Rather, we have to approach cybersecurity with a strong note of humility and recognize this is a journey, with milestones but not an end point. We need to deeply consider what the system is possible of doing, as well as what harm it could cause. We need to reflect on this reality as we develop the consensus within our company, our customers, and our industry to allocate resources, effort, and priority to developing our overall cybersecurity program.

About this time, we should also be recognizing that you cannot go it alone in building a secure and resilient telematics platform, or any other complex computer system. This is going to require a collaborative effort from all parties with a stake in the connected vehicle. From commercial fleet vehicle owners to telematics providers, OEMs, cellular carriers, computer chip manufacturers, industry associations, authorities and regulators, to universities. Around the world, there are scores of activities bringing these parties together with a common goal, to design a more secure and resilient connected vehicle. It seems broadly acknowledged this is a common goal, one that can be worked at collaboratively through industry associations while fully respecting antitrust rules.

For those new to the connected vehicle cybersecurity conversation, it is worth calling out the SAE (Society of Automotive Engineers), in particular, the <u>Vehicle Electrical System Security Committee</u>, that does an outstanding job of collecting the vehicle-related cyber events and discussing them. I strongly recommend experts join the SAE meetings in general and this committee in particular to learn the current cybersecurity issues enveloping the connected car.

8.5 The Geotab Story: Building a Telematics Platform Resilient to Cyber Threats

As noted above, an effective and quick means to becoming a part of the current vehicle security conversation is by enrolling as a regular guest of applicable SAE committee meetings. This was a good place to meet experts within the field and learn the current issues and range of projects underway to address them. Several Geotab engineers and specialists have become members of applicable SAE committees both to learn and contribute to advancing the current state of security and the connected car.

We learned of the need for a "Threat Analysis," and although the work of SAE J3061 was in progress, and too far along for us to become contributing members, we listened as guests and used other sources as a guide in developing our initial threat model. See Chapter 3 on SAE J3061. Later we "embraced" ISO 27001:2013 - Information security management systems requirements as a method to manage end user information security.

Next, we looked for a world-class expert in cybersecurity that included cellular communication. I had a family friend who was director of security for what had been the leading cellular handset company in the world. He brought a wealth and depth of knowledge and introduced me and others at Geotab to best practices of device security. We discussed "hackers'" possible motivation, how it could be rewarding for them, and to what extent we should anticipate their ability, resources, and determination to exploit. I refer to Chapter 2 for a deeper discussion on the motivation of hackers. In general, consider the value of what is being protected when considering an adversary motivation. The higher the value, the more resources are likely to be used to exploit them and the more corresponding emphasis and resources on security are required.

So when we are considering the "worst case" scenario resulting from a cyber attack, we have to consider any and all assets that could be compromised via or related to the system we are creating. Readers today have the benefit of SAE J3061 - Cyber security Guidebook for Cyber-Physical Vehicle Systems that was a work in progress only a couple of years ago. SAE J3061 includes multiple methods used to conduct cyber threat analysis to identify potential threats, and their related risks.

We, among others, always had a focus on Internet security, server security, redundancy, and security of Internet-facing applications. An initial threat analysis of our telematics platform indicated the OBD device and device cellular communication did not include equivalent security to that of our Internet and servers. This was a common issue in 2014 among OBD telematics devices, and it needed to change.

At Geotab, we are in the fortunate situation that we control the telematics platform from end-to-end. We design and manufacture the OBD device, design the device firmware, have exclusive control of any servers used, and design the software for outward (Internet)-facing applications. This allows us to use a "system" approach to design a telematics platform resilient to cyber threats.

8.6 Telematics Security: Vehicle to Server via Cellular Communication

8.6.1 Cybersecurity Best Practices

We've written this section so commercial vehicle fleet owners and managers understand the required security concepts and can have meaningful conversations with their TSP provider. We also refer readers to a Geotab blog by Alex Sukhov (ref) "15 Security Recommendations for Building a Telematics Platform Resilient to Cyber Threats."

Engineers and designers of telematics systems should be (or become) familiar with the technical material referenced in this chapter, and described in the accompanying chapters of this text. I'll leave it to the reader or system designer who wishes to dig a little deeper to refer to the references for more technical details (ref Alex S blog, IEEE paper, DOT-Volpe recommendations for gov vehicles, SAE J3169, SAE J3005-2, SAE J3138, NIST recommendations, and ISO 27001 among others).

8.6.2 Secrets

Before passing on a secret to a confidant, you look them in the eye to ensure they are who you think they are, then whisper the secret in their ear in a manner others can't eavesdrop.

8.6.3 Authentication

Are the sender and receiver of data communication really who they say they are? The same concept required to share a secret in someone's ear are the required steps for the connected vehicle (OBD device) to communicate data securely to the server. Both the vehicle and the server need to identify each other and ensure they are who they say they are. This process is called "Authentication."

There are different methods to complete authentication that can be used for different applications; however, they all share one element in common, the use of a mathematical function that is easy one way and difficult or impossible the other. Most math functions are reversible, addition, multiplication, for example. 2 + 3=? Or 5 = 3 +?; in either case (forward or backward), obtaining the answer is trivial. A one-way function is easy in one direction, very difficult or impossible in the other. For example, it is easy to combine two colors of paint to obtain a third color, but very difficult to then separate that back to the two specific colors. Easy to scramble eggs, impossible to then recreate the original eggs. In the 1970s, Diffie-Hellman developed the math for such a process. The math proof is beyond the scope of this chapter. For example, the telematics device has a "key" that could be a public key. The server has a "key" that is private and secure. Through application of a one-way function, the server can assure the telematics device is who they say they are in theory and if implemented well in practice too, at least until Moore's law provides quantum computers for the purpose of breaking these one-way math functions.

8.7 Cloning of Devices

We learned earlier that all security algorithms will eventually be defeated, so what if the "key" on the device is duplicated and reissued by another device? The best practice is to use a nonstatic key in the telematics device to reduce this risk and to individualize every telematics device to prevent an OBD device clone being able to communicate with the server. So if an adversary were to break into a vehicle, steal the device, and reverse engineer it exactly including the firmware, it would not be able to authenticate with the server, as each device is individualized and two of the same device would be detected.

8.8 Eavesdropping

Once the device and server know who each other are, they can begin data communication. However, an adversary could be eavesdropping on this communication. Therefore the data communication session is encrypted. One must be careful to use encryption code from libraries, and not try to develop one on their own. Public, well-used encryption libraries are used by many and exposed to adversaries who find vulnerabilities. These are discovered and patches sent out to fix the vulnerability. Over time, existing encryption libraries become more resilient, unless at some time a fatal vulnerability is discovered for which there is no patch, and the encryption protocol must be replaced. This occurs from time to time and requires constant monitoring of acceptable and unacceptable encryption protocols (ref: NIST encryption recommendations FIPS 1402). This is one example, of a theme we will develop, that collaboration to advance cybersecurity is essential on a number of fronts. Rust never rests, and neither do cyber adversaries.

8.9 Keep Embedded Code Secure

Every IoT device, including telematics devices, will need update patches to ensure recent security vulnerabilities are fixed and also used for other required functionality updates. If adequate precautions are not taken, it is often the process of updating firmware that could allow an adversary to inject malicious application/firmware that can then activate system elements which were not intended by the design.

The updated firmware requires digitally signing a "certificate" to indicate it comes from a reliable source before the telematics device will accept it.

8.10 Enable Hardware Code Protection

When the microcontroller supports the ability to read firmware code from the device, it should be disabled. This limits the adversary's ability to reverse engineer a device.

Notwithstanding the above, assume your code is public and accommodate for this in design of the telematics platform. Use cryptographically strong random numbers in security algorithms. There have been documented cases where random number generation was not "random" enough; encryption could be decrypted this way, for example.

8.11 Segregation

Use different security "keys" for different roles, for example, the same key should not be used for authentication as for the application signature used during firmware updates.

8.12 Disable Debug Features

Debug logic which contains security information should not be accessible in production software builds. Limit server access using practice of "least privilege." Only those required to have access should have access. Log-in and keystroke records for server access are an essential part of forensic analysis of suspicious account activity.

This highlights some important OBD device security considerations; however, it is not an exhaustive list.

8.12.1 Security Validation

When we completed the design and implementation of the OBD device security measures, they now needed to be validated. Earlier we mentioned how important collaboration is to the security effort. Always best to have a "fresh pair of eyes" that are independent from the developer to perform a "penetration" test of the device and device communication.

The penetration test is where the "researcher" acts as if they are an adversary looking for ways to compromise the intent of the application. Experience, tools, knowledge, and utilizing the most current versions of each are required for optimal penetration testing results.

There are at least a couple of approaches to penetration testing. Some believe it is best to have a "bug bounty" program. In this case, the connected vehicle/computer system provider develops a program that rewards "researchers" for finding and reporting "bugs" in the software application. This approach has the advantage that potentially thousands of "researchers" from around the world can be working away, providing blanket-like coverage. As well, depending on the particular program, most bug bounty programs only pay when "vulnerabilities/bugs" are found, so can be quite cost effective relative to the hours spent by many looking for them.

Another approach is to find an individual, or company, that has unique expertise in your specific type of application, execute an NDA (nondisclosure agreement) with same, and provide them the device, and/or code depending on the type of penetration testing being performed.

Black Box—no information given to the tester about the internal workings of the particular application/device. Good to detect vulnerabilities to brute force attacks. Trial and error approach to penetration testing.

White Box—full information and access given to both the source code and software architecture of the application/device. Faster than black box and can be more thorough. Can require development of logic to determine where to try to attack and uses more sophisticated/expensive tools.

Gray Box—a combination of the above, partial information and access is given to the "researcher."

For us, we wanted to find individual(s) that were recognized within the vehicle security industry as current experts in the field, had demonstrated work to verify this, and had taken a personal interest as a cyber activist for public good. The last point was important to us as collaboration within cybersecurity is a game of trust. We wanted to make sure we were working with ethical individuals and companies before handing over our device (and code in some cases) to a third party.

Coincidentally, I was attending the 2015 Escar security conference in Detroit. Although it was my first vehicle security conference, many others there seemed to know each other. We met an industry expert in cybersecurity at the 2015 Escar conference who was presenting a cybersecurity workshop. What impressed me was not just his CV of competence but also his interest in public service as a "researcher." It's one thing to discover vulnerabilities to cause damage or ransom of some kind; it's quite another to use this insight to warn others so they can take required actions and fix the problem before a related loss occurs. Seems to me that "Researchers" provide a significant public service. Later, I met another excellent candidate with a similar demonstration of taking personal time to discover vulnerabilities in a telematics device. We learned that not every telematics provider wanted to hear the "good news" of vulnerabilities being discovered on their platform.

We needed to provide the device to these "researchers" who would use special expertise and tools to poke and bang (electronically) the devices to try to break into them, break the authentication, force bad code onto them, and essentially validate (or not) the security measures that had been designed in. We learned a lot through this process. We learned that although the devices proved resilient and not corrupted, and no eavesdropping of data communication could be made to occur; there were recommendations to further harden the device, with a goal of preventing even hints of methodology from being discovered.

Looking all the way back to when we first learned of a reported telematics device hack in 2014, we now felt much better about the answer we could prepare in response. By peeling back the onion skin a little and looking into device security, it made us all the more sensitive to review the broader cybersecurity of the "back end" of our telematics platform, the servers, and outward-facing applications on the Internet.

We had been engaging a reputable third-party consultant for penetration testing of our "back-end" and server applications the prior year. We would get the report, see the new recommendations, implement, and retest. What I learned was new information can be learned from two penetration tests of the same system by different "researchers" under a similar statement of work (SOW). Such is the nature of this type of work. It is important to have multiple and different eyes and approaches used in penetration testing to more completely expose any possible vulnerabilities. Cybersecurity is a journey, not a destination.

There are at least a couple of models that IoT platforms can follow to develop, manage and secure information systems. One is the NIST Risk Management Framework

(NIST 800-37) "Guide for Applying the Risk Management Framework to Federal Information Systems." This is a requirement as the title suggests for implementation of IT within Federal Government systems.

Another is the ISO 27001:2013 - Information security management systems requirements. Incidentally, the NIST RMF noted above as cross reference to ISO 27001 controls. At a high level, these IT Security management programs are similar. It is at the detailed level where differences in how to validate controls are in place where differences emerge.

We recommend others adopt one of these IT system security management tools as a means to ensure all required system aspects are considered and a continuous improvement process is put in place.

At Geotab, we have developed a TOMS (Technical and Organizational Data Security Measures Statement) that includes our overall security program and embraces the NIST RMF. We also publish an annual letter from our CSO (Chief Security Officer) of key cybersecurity events, tests, and exploits annually.

Some of the key aspects of the TOMS document are

- Employee policies
- Segregation of duties
- Access control (for each identified function and area)
- OBD device security
- Systems monitoring
- Penetration testing/vulnerability scans
- Audits
- Incidents
- Data availability and backups
- Data retention, correction, and deletion
- Business continuity

Our goal in this chapter was to provide an introduction to the evolution of the vehicle from a classic car of the 1970s to the nonconnected yet ECU-sensor-based vehicle to finally the connected vehicle, and to consider how our rules, regulations, and even borders of nation states have been developed in the physical world, yet now apply in a cyber world as well. We need to now consider the far-reaching ramifications of the cyber world, in the modern cyber-physical vehicle. We discussed the concepts of security applied to the new cyber-physical vehicle and how in this new world... the vehicle is not a discrete machine, rather an integrated IoT component of a broader IT system. That, just like any IT system, requires a dedicated IT Security program that includes a heavy dose of good computer hygiene. We then provided references to the current standards, guidelines, white papers, and blogs that provide the technical details for engineers and other technical personnel who are responsible to design, build, validate, and maintain telematics platforms in the current reality of the connected commercial vehicle.

About the Author

Glenn began working with Geotab as a consultant in March 2014 and joined as an employee in January 2016. Glenn has over 25 years' experience, which includes proactive safety design analysis, product safety/company risk mitigation techniques, and product safety litigation defense. He is currently working with universities, industry associations, and authorities to proactively advance telematics product safety and cybersecurity at Geotab and more broadly within the industry. Glenn has a BSc in mechanical engineering from the University of Saskatchewan and a MEng from McGill University. Glenn was also Director EHS and Product Safety for Husky Injection Molding systems (global). He was chair of the ANSI SPI committee for 7 years, harmonizing EU and North American HIMM standards (horizontal injection molding machines).

The Promise of Michigan: Secure Mobility

Karl Heimer

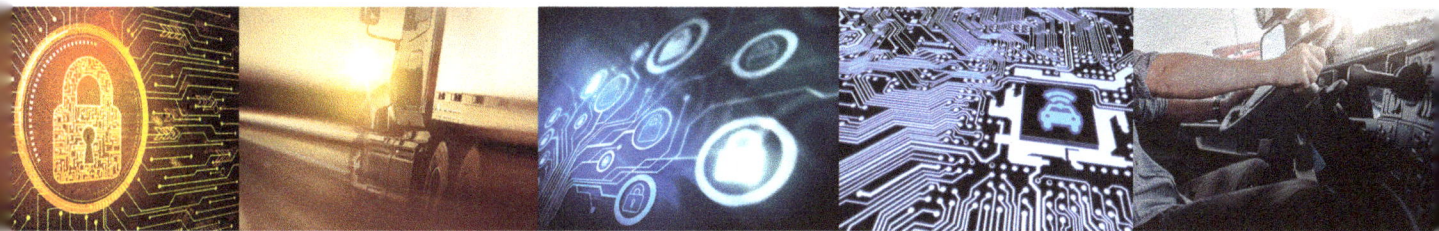

9.1 Governor's Foreword for "The Promise of Michigan"

By Governor Rick Snyder

Michigan revolutionized how the world moved more than 100 years ago. Today, we're at it again, transforming the auto industry to the mobility industry and setting the world on a new course. This new and exciting industry will be shaped by technology designed to make us more aware and safer as we're driving. By recognizing the potential of technology and aligning our state's policies, we will remain the leader the rest of the world sees as its biggest competition.

From MCity to the Planet M philosophy, from the American Center for Mobility to landmark legislation that allows the operation of autonomous vehicles on our roadways, Michigan has supported numerous efforts that will help expand the mobility industry here in the state.

But as vehicles become more connected and intelligent, we must remain steadfast in defending this technology from cybersecurity threats. Cybersecurity is an essential building block to making connected transportation systems safe for drivers, passengers, and all of

FIGURE 9.1

society. I am proud that Michigan is at the forefront of developing ways to protect this technology and the people who rely on it.

Michigan is home to world-class universities and state-of-the-art testing sites. We are a leader in research and development. We have the skills, talent, and resources the mobility industry needs, and we share an unwavering commitment to grow this industry faster than anywhere else in the world.

Michigan is second-to-none when it comes to automotive and technological innovation, and we remain committed to making sure that stays true for generations to come.

Rick Snyder is the 48th governor of the State of Michigan.

9.2 **Introduction**

Michigan has so many faces and aspects; when you "think" of Michigan it might be to imagine a quite cove, just right for catching fish. When you think of Michigan, it might be the outstanding brew-pub scene. When you think of Michigan, it might be the natural splendor of the northern shores or the wildness of the "U-P" (Upper Peninsula). If you are a history buff, then when you think of Michigan, you might be stirred by the incredible call to arms of Ford's bomber plant at Willow Run which produced so many bombers and helped win World War II, or if you are a sports fan, Michigan has great hockey, football, baseball, and basketball teams. If you are a musician, you might tap your feet to Motown music when you think of Michigan. And EVERYBODY thinks of cars and automotive engineering when they think of Michigan. What nobody thinks of, yet, is cybersecurity, but you will and for good reason: Michigan is surging in the cyber domain and will add it as a core and critical component of the state's priorities and expertise.

This strong state interest in cybersecurity comes from two sources. Governor Rick Snyder brings a strong business background from the information technology space, understands its value, and has been aggressively developing computer and cyber capabilities within Michigan. But perhaps an even stronger driver has been the natural evolution of the transportation sector. While automation has been broadly growing throughout the transportation industry, it was the concept of self-driving and perhaps "hacking" automobiles that enervated the cyber industry in Michigan. This interest played on several stages. Michigan-based military organizations partnered with DARPA (a Department of Defense research and development organization) to create "hack-proof" vehicles and used Michigan-based engineers and technologies to develop some of these systems and to integrate cybersecurity products and capabilities from all over the world. The automakers also took keen interest in creating self-driving cars, protecting fleets of vehicles, and continuously improving the cybersecurity protections in their automotive products. Lastly, several government and academic studies about how "connected" cars would interact with each other and with a "connected" infrastructure" of the city of the future were conducted in Michigan at universities and other institutions. Michigan quickly became the epicenter of connected and autonomous vehicle (CAV) research and design. And, of course, cybersecurity is a critical component of self-driving cars because if you can't be secure, then all the rest of the self-driving vehicle concept is moot.

And the state's cyber focus isn't just all about cars. While automotive is Michigan's iconic industry, Michigan is also a freight and transit "main line" for international commerce. The United States and Canada are among the world's largest bilateral trading partners, with the cross-border movement of goods between them exceeding $1.8 billion per day in 2014. Michigan's Ambassador Bridge is the busiest commercial border crossing in the nation, with almost 2.5 million trucks crossing in 2014 (as reported by the Public Border Operators Association). Michigan's Blue Water Bridge in Port Huron is the second-busiest on the northern border, with nearly 1.6 million trucks. Road based transportation equipment is the leading way of carrying freight product crossing Michigan's border with Canada. Goods from this trade are transported into and through Michigan to the rest of the country, making freight and trucking protection of central importance to Michigan. This commerce corridor and the importance of freight efficiency also makes Michigan an ideal location to test the potential of truck platooning—which is a modification of the self-driving vehicle and which is right in the middle of Michigan's ongoing research into CAVs.

This focus on cyber hasn't taken too long to be recognized. *Wired* magazine quoted an industry study concluding that Michigan was one of only five states that were prepared for cyber incidents and activities (the others being California, Maryland, New York, and New Jersey). Also, Govtech (govtech.com/computing/digital-states-2016.html) awarded Michigan an "A grade"—one of only five states to earn this mark (along with Utah, Virginia, Ohio, and Missouri) and the only state to earn an "A" in four consecutive years. But this is just the beginning. Michigan has created "Planet-M" to develop and/or attract mobility solutions and capability for the transportation sector, foster research, and create informed policy regarding CAVs. Planet-M debuted in January 2017 at the North American International Auto Show in Detroit.

Transitioning a state (or an industry) from largely mechanical engineering to software development and cybersecurity isn't haphazard (at least, not if successful), and Michigan has a distinct strategy for vaulting into the first rank of cyber-capable states.

9.3 **The Cyber Strategy**

Michigan's high-level cyber strategy with respect to vehicles and transit systems is a "three-legged stool" approach to ensure that the technical, the legal, and the capacity requirements for an advanced and cyber-secure environment are available, high quality, and reinforcing.

Michigan has created a framework of laws and policies that encourage safe research, development, and technology use within a specific and achievable overarching state vision while encouraging cooperation among the state and all its resident organizations from industry to federal to academic to state government. Michigan also provides consistent and reinforcing direction and resources to projects and initiatives and is fearless in accepting new challenges and capturing the research for the state.

Talent development remains a key to Michigan's technology future, and it is engaged in several programs to create, attract, and retain cyber talent for its industries. It is no accident that much of the connected and autonomous research is performed within Michigan. Talent is a central component of both technical (properly educated for the vehicular cyber domain) and "capacity" (having sufficiently many of a well-trained workforce).

Michigan purposely pursues an objective of engagement with other state partners as well as private and federal organizations within the state. The relationship between federal organizations such as United States Army Tank Automotive Research, Development, and Engineering Center (TARDEC) and Michigan's other transportation concerns is exemplary, and organizations such as Michigan Department of Transportation (MDOT) frequently conduct collaborative exercises with TARDEC. Also, ongoing outreach helps develop state capabilities (e.g., the American Center for Mobility at Willow Run) which can serve both federal needs as well as state and private industry needs, and by sharing intent and development plans, the state has ensured it can deliver the best environment, with the least overlap and at the lowest reasonable cost—all helping to make Michigan the best resourced and most efficient transportation (including cyber) center for R&D as well as verification and validation.

Organizations such as UMTRI and the American Center for Mobility (ACM) focus on research and create technical centers of excellence. MDOT is actively building out an extensive "smart highway"—the largest in the nation—and when paired with

ACM's test environments, gives Michigan the highest capacity to measure research and practical/operational environments and provide a responsive and cyber-secure transportation infrastructure.

9.4 Laws and Policies

On December 9, 2016, Governor Snyder signed four laws into existence that solidify Michigan's leadership position as a research, test, and deployment site for CAVs. These laws help show that Michigan should figure prominently in any CAV effort. Some key provisions of these new laws are

1. Driverless cars shall be permitted on all public roads.
2. Self-driving ride-sharing cars (e.g., self-driving taxis, and on-demand shared vehicles) can be operated if the vehicles are supplied or controlled by the OEM (automobile manufacturer).
3. "Platoons" of automated trucks can travel together at set speeds.
4. The public can buy and sell autonomous vehicles after the vehicles have passed testing and certification.
5. The Michigan Council on Future Mobility shall be created to recommend policies and regulate vehicle networks.
6. The automated driving system (if used) will be considered the driver and operator for the purposes of Michigan motor vehicle laws. This would lift texting bans on all vehicle occupants while in automated driving mode.
7. Manufacturers will be protected from liability where the automated system has been modified, and repair shops will be protected if the repairs are made according to manufacturer specifications.

The Michigan Council on Future Mobility advisory committee was formed by early spring 2017. Michigan now has the most forward-looking and development- and test-friendly laws in the country with respect to CAVs. These laws and the subsequent policy development firmly place Michigan as the "go-to" place to research and validate self-driving technologies.

9.5 Capability Development

We focus on three examples of developing or extending cyber capabilities within Michigan. First we look at the first large-scale truck "platooning" exercise on an active US roadway. Next we showcase a new national-level asset for CAV test and evaluation called the American Center for Mobility. Finally, we look at another Michigan-unique capability—a cyber National Guard-like capability which leverages and enables known cyber-experts to assist state authorities in times of cyber-crisis and operating under and coordinated by state authorities. This capability is called the Michigan Cyber Civilian Corps (or MiC3).

9.5.1 TARDEC-MDOT I-69 Platooning Exercise

On a bright sunny day in April 2016 at a rest area on Interstate 69 near Capac, Michigan there was an inordinate amount of activity. State Police vehicles were out in force as were four odd-looking semi-tractor trailers in camouflage paint, lots of news vehicles, and two large tents. Hundreds of people were milling about the parking lot, waiting—waiting for the first live road test/demonstration of truck platooning on a major American highway. After speeches and media questions, the four trucks drove east then returned west to stage again at Capac and review the drive. Each truck was equipped with a special set of sensors, communications equipment, and software; software used to manage the task of driving for the three rear convoy vehicles, driving on an active roadway with normal people, not test study drivers, and sharing the interstate with them. Each truck also had a driver, just in case. The results were brilliant, and the future of transportation leapt forward. The day in which convoys of trucks could deliver products with a single lead driver moved much closer (and promises to offer substantial price reductions due to lower transportation costs). On the military side, the day when the Army can use computers to drive "soft target vehicles" such as logistics support trucks instead of exposing soldiers to attack on convoys passed a new milestone. It is noteworthy that this Michigan-based test of a platooning convoy operation took place in the same month, April 2016, when several European truck manufacturers held their own autonomous driving demonstration (also successful).

Let's look at the particulars of the I-69 Platooning Exercise.

The MDOT partnered with the TARDEC to develop and execute this truck platooning/semiautonomous vehicle test. This partnership serves as a reminder of how well Michigan's state and federal organizations work together and demonstrates the strategic priority of unity of vision and purpose. Each of these organizations, one federal and one state, brought critical skills and equipment the other needed to fulfill the exercise, and each benefited from the partnership. This kind of strategic partnering is the norm in Michigan and is one great reason why the state has been able to rapidly develop meaningful cyber capabilities in the transportation space. MDOT and TARDEC set out to evaluate what was pure science fiction just a couple years earlier—trucks that would follow each other on operational roadways with human drivers whizzing around them—and to validate the safety of the system to everyone on the highway. TARDEC brought secure coding, secure communications, and excellent test and measurement skills. MDOT brought the largest connected and intelligent highway system in the Western Hemisphere (which will have more than 350 miles of connected roadways by 2019) and involvement of other needed state assets such as the Michigan State Police and the MDOT Operations Centers that monitor and help manage traffic.

The initial tests also used something called the DSRC radio (short for "dedicated short-range communications" which is a radio using a variant of the same Wi-Fi signal you use in your home or office—specifically 802.11p). DSRC made an excellent choice because it is the same radio system the US Department of Transportation (US-DOT) is considering mandating for vehicle-to-vehicle (V2V) communications, the signals cars will send among themselves to help drivers and maintain safety on future highways.

Michigan had already conducted DSRC research for several years and is home to much of the evaluation studies for V2V regarding both cars and buses. Army trucks would be expected to use DSRC (at a minimum) as soon as US-DOT issues the requirement. DSRC is also used for vehicle-to-infrastructure (or V2I), and Michigan's 350 miles of smart highway is already instrumented for V2I.

DSRC radio equipment allows rapid two-way communication between devices by using a radio frequency of 5.9 GHz dedicated by the Federal Government for intelligent transportation systems. This form of communication allows safety-critical information to be exchanged such as the location of the leading vehicle's bumper or the location of the edge of the road, between devices at a rate much faster than other communication technologies.

FIGURE 9.2

© SAE International

TARDEC has been developing autonomous vehicle capability as part of its automotive R&D portfolio for more than two decades. TARDEC's autonomy programs extend operational requirements compared to automaker's autonomy programs because the military vehicles must operate in severely degraded or even nonexistent road environments. TARDEC's vehicles also have advanced software to help ensure the vehicles aren't compromised or have faulty driving commands given to them while platooning.

The test itself used four semiautonomous TARDEC vehicles traveling as a platoon in which the lead vehicle was driven by a human and the trailing three vehicles autonomously followed it within carefully proscribed safety guidelines. The trailing three vehicles all had professional drivers in them ready to take the wheel, but none had to, since the platooning exercise worked perfectly. The vehicles drove a 40-mile loop from the Interstate-69 Capac Road rest area eastbound to Lake Pleasant Road, then they traveled westbound to return to Capac Road. Onboard units recorded the DSRC V2V communications from all the vehicles; this data described information such as speed, acceleration, deceleration, positioning information, and braking information. The onboard units also recorded the V2I communications to show the road was providing appropriate information to the autonomous vehicles; this information included bridge height data, curve speed warnings, local speed limits, lane closure and maintenance information, and even rest area advisories. The V2I communications allowed MDOT to remotely monitor the location and behavior (speed, etc.) of the TARDEC trucks during the entire exercise and will allow MDOT to actively manage Michigan's highways for future freight and military autonomous traffic. During the test, MDOT used two specially outfitted vehicles—one as an "emergency vehicle" and a second as a disabled vehicle in the roadway that the platoon would have to adjust to. The platoon properly reacted to the "disabled" vehicle and properly adjusted for the "emergency vehicle" DSRC broadcast messages. Both the V2V and the V2I communications worked perfectly; some of the information displayed in the onboard DSRC equipment is shown below.

FIGURE 9.3

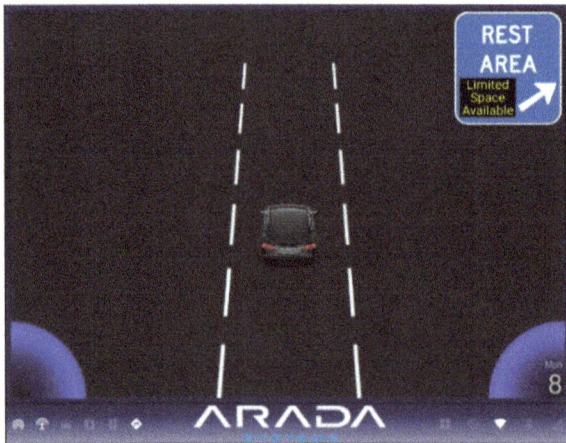

© SAE International

FIGURE 9.4

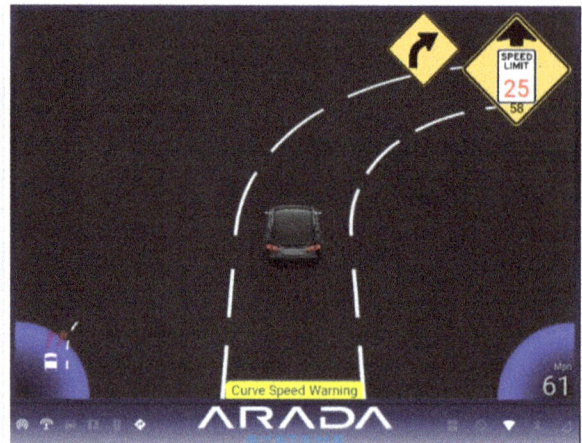

© SAE International

FIGURE 9.5

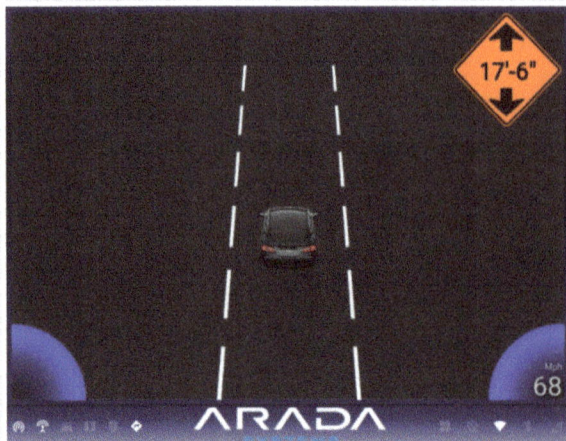

© SAE International

FIGURE 9.6

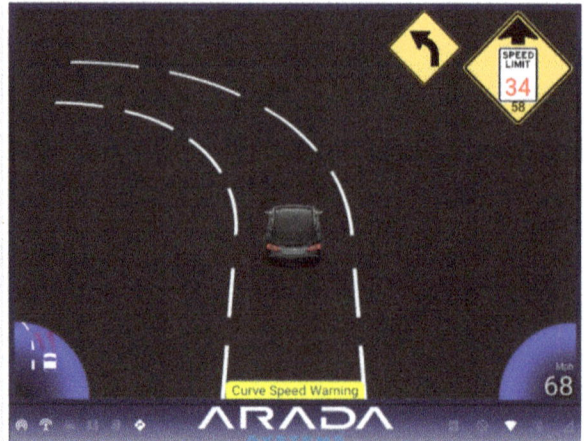

© SAE International

FIGURE 9.7

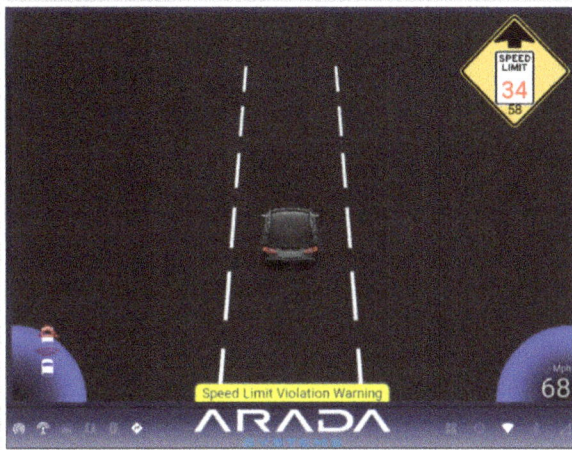

FIGURE 9.8

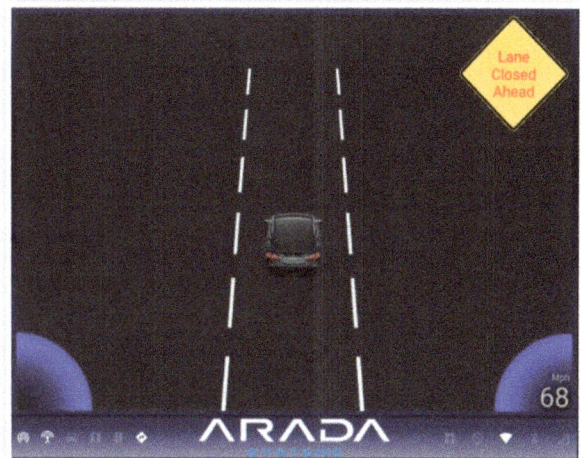

The Interstate-69 Corridor Platooning Exercise tested the functionality of the DSRC radios installed in TARDEC's trucks and their ability to communicate with the MDOT radios used for the smart highways. This was TARDEC's first test of autonomous vehicles on a public roadway, and it was also the first test of MDOT's smart highway infrastructure DSRC radios. Both were validated by this highly successful test. A testing byproduct was proving that the military and state systems networked correctly and provided appropriate awareness and situational visibility to both parties. Building off this test, MDOT is poised to conduct future interoperability tests with commercial vehicles. MDOT is also preparing for additional tests with TARDEC to explore additional capabilities in 2017.

MDOT views cybersecurity as a crucial element in ensuring reliable and safe highways in this new age of CAVs and intelligent transportation system highways. To address the cybersecurity requirements, MDOT is part of Michigan's new mobility initiative, Planet-M, which includes in its priority stack a mission to develop, implement, and research on-vehicle and V2I cyber mitigations and protections. MDOT is also partnering with the new ACM and its cyber-mobility initiatives.

The MDOT continues its tradition of being a cutting-edge state transportation department and a leader among transportation agencies around the country. The department is preparing for the paradigm-shifting technologies of CAVs that have the potential to provide the citizens of Michigan with safer and more efficiently managed roads. MDOT's smart roadways have already created a safer and more efficient transportation network to serve as the basis for future highway upgrades and modification throughout Michigan, ensuring the effective movement of goods and people around the state, the nation, and the world.

9.5.2 American Center for Mobility

The American Center for Mobility (ACM) at Willow Run is a nonprofit testing, innovation, and product development facility as well as a proving ground for future mobility. Designed to enable safe testing, validation, and self-certification of CAV mobility technology, to accelerate the development of voluntary standards and to provide educational opportunities, ACM incorporates a purpose-built test track environment directly integrated with active highway corridors and a combined corporate/academic technology park campus. This facilitates effective government and industry collaboration and information sharing, as well as a focus on educating and developing the mobility workforce for the future.

Creating ACM in Michigan shows the state's commitment to CAV research, validation, and development. ACM, combined with Michigan's 350 miles of smart, connected highways (portions to be completed by 2019) create the largest instrumented connected vehicle test environment in the Western hemisphere and create unique opportunities to conduct "in situ" studies on controlled and public roads in urban, suburban, highway, and rural environments. Perhaps more importantly, ACM is a multimodal test facility with an active railhead, an active runway/airport, and roadways for commercial and passenger vehicle tests. In late Phase I/early Phase II, ACM will create cybersecurity training, analytics, and test centers in cooperation with private industry, state, federal, and international research organizations.

FIGURE 9.9

© SAE International

The ACM's groundbreaking ceremony was held on November 21, 2016, at the historic Willow Run site, made famous both as the home of "Rosie the Riveter" and as a B-24 production site during World War II. ACM has an initial seed funding from Michigan Economic Development Corporation (MEDC) and other Michigan stakeholders of more than $80 million and is ahead of schedule for initial operating capacity in 2018. Current projections show the initial highway loop operational in late 2017. In January 2017, US-DOT announced that ACM was one of 10 winning facilities across the nation for a competitive designation as a federal vehicle proving ground under the federal Automated Vehicle Proving Grounds program.

Through successive land purchases, ACM's total area is at 500 acres as of May 2017 and has many different road surfaces and infrastructure components both existing and planned. Two of the first roadways that will be operational at ACM are the "Highway Loop" and the "Public Road CAV Corridor on US-12."

The Highway Loop at the Willow Run facility is a 2.5 mile continuous highway speed loop to test real-life scenarios, including straights, curves (standardized, fixed, and changing radius), elevation changes, varying lane configurations and widths, merging and exiting, bridges, underpasses, and tunnels. The testing speeds along this loop will vary between 50 and 75 mph "posted speeds," or higher in both clockwise and counter-clockwise directions. It will feature various two-, three-, and four-lane configurations, all with representative geometry, surfaces, markings, guardrails, signage, lighting, etc. An additional and adjacent 1.3 mile straight/gently curving roadway segment will complement the loop to simulate rural 55 mph two-way traffic. These will be capable of testing variable, critical, and failed-condition scenarios in a safe manner, including light and heavy commercial vehicles. The Highway Loop will be operational by December 2017.

The Public Road CAV Corridor on US-12 is planned to be a joint ACM and MDOT construction to deploy a public road corridor on US-12 Michigan Avenue with traffic signals and pedestrian and bicycle crossings, equipped with DSRC/Cellular devices and data acquisition equipment, including all features necessary for testing and developing connected and automated vehicles and applications. Testers and users linked with the ACM facility will be able to transfer data directly to and from the ACM communications systems, as well as to and from MDOT's data system and V2V and V2I applications. Future plans could then be to join US12 to I275 to I94 for a public road zone with ACM as a central data hub.

FIGURE 9.10

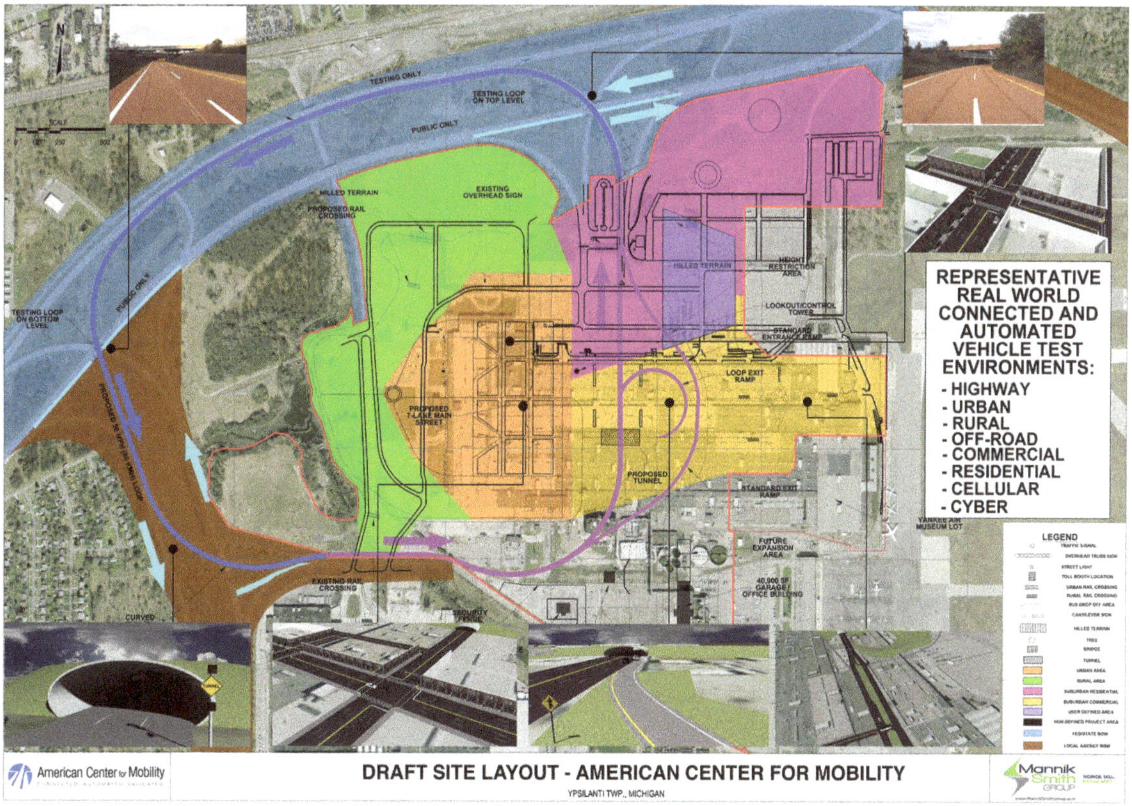

DRAFT SITE LAYOUT - AMERICAN CENTER FOR MOBILITY
YPSILANTI TWP., MICHIGAN

Planned intermodal research (including a manufacturing robotics lab) based at ACM are designed to accelerate development of technology through applied research, testing, validation, verification, and voluntary standards efforts. Targeted technologies should be at manufacturing readiness levels (MRLs) of 4–7 (which means able to produce the technology in a lab all the way through general use production for the technology). Activities/study areas include

- Adaption and reconfiguration of an existing on-site warehouse for truck to warehouse material conveyance
- Intermodal material-handling equipment
- Development of robotic vehicle test platform for material conveyance
- Automated refueling/recharging
- Highway loop intelligent transportation system deployment for truck platooning
- Automated truck maneuvering to loading dock
- Railhead integration for rail to truck material conveyance—which would have robots automatically move freight from trains to trucks (and vice versa)

9.5.3 Michigan Civilian Cyber Corps

Michigan Civilian Cyber Corps (MiC3) is a newly created, "unique-in-the-nation" cyber defense capability. This is an all-volunteer force of highly qualified cyber defenders who can be called to service to support Michigan State Police and/or the Michigan National Guard in times of crisis. Governor Snyder announced the concept and creation of the MiC3 during the 2013 North American International Cyber Summit (link: http://www.govtech.com/security/Michigan-Launches-Volunteer-Cybersecurity-Corps.html).

MiC3 volunteers must pass multiple technical and ethical certifications and then be vetted by state experts and government leaders. This process ensures the best possible defense force is available in times of need. The MiC3 also conducts periodic drills—just as the National Guard would do—to keep fresh and to exercise a team approach to cyber defense. The MiC3 is partnered with DTMB (Department of Technology, Management, and Budget), the Merit Network, and the Volunteer Registry System of the State Department and has teams assigned to each of Michigan's 10 "Prosperity Regions," thereby providing umbrella protection across the entire state geography and well suited to defending geographic and regional assets such as roadways and the smart transportation system and vehicles which use them.

In July 2016, DTMB partnered with the SANS Institute and certifying MiC3 members with the GCIH credentials (ANSI-certified) which satisfied DOD Directive 8570 requirements for the cybersecurity workforce within the military. As a result, Michigan now has a cadre of civilians who satisfy military standards for cybersecurity. This is unusual even for state National Guard teams, and definitely a first for volunteer civilian teams.

Also in 2016, Governor Snyder's 21st Century Commission suggested expanding the MiC3 to 200 members (link: http://www.miinfrastructurecommission.com/

21st-century-infrastructure-commission-report). Interested parties can apply online at https://www.micybercorps.org.

With respect to freight operations through Michigan—as a properly constituted and authorized body of expert cyber defenders with reporting and command chains to law enforcement and state governance organizations—the MiC3 offers unmatched protections for through-state transit; it offers immediate, highly trained, and practiced teams of experts who routinely practice defending infrastructure, thwarting cyberattack, and providing analysis. Freight through Michigan will be safer, and the chance of technical errors or hazardous fleet behavior reduced while operating under this protective umbrella. The MiC3 is a glimpse of what proactive governments, regions, and infrastructures must aspire to in the connected vehicle age and will help assure the safe delivery and transit of freight while within Michigan.

9.6 Michigan-Based Education and Training

Michigan seeks to develop, attract, and retain critical cyber talent—especially in the automotive and transportation domains. Developing a cyber-friendly ecosystem and cultural environment is as important as providing good jobs and educational opportunities; however, we will start with training and education for the talent development in this section which focuses on some outstanding training opportunities unique to Michigan.

The Michigan Cyber Range (MCR) is a unique-in-the-nation public-private partnership that provides realistic testing and training capabilities (in an unclassified environment) for academia, government, and industry. MCR acts as a collaborative platform which hosts a full-scale training program incorporating individual, small group, and collective training. This combination allows for processes, procedures, and integration plans to be implemented and tested in a controlled, isolated environment.

MCR delivers a superior training environment because

- MCR uses the Merit Network's high-speed optical backbone to provide accessibility to participants. This high-speed and high-capacity backbone enables MCR to deliver high-quality access to researchers, engineers, and technicians around the world to partner with Michigan researchers and projects.

- MCR highlights hands-on, experiential learning. This kind of educational engagement (hands-on with a real-world feel) is better able to determine and showcase student mastery than other kinds of training.

- MCR offers a long-term, persistent environment for practicing and experimenting in a controlled environment. It also makes extensive use of "virtual" technology allowing users to create and execute test cases that yield reproducible results. This virtual technology lets MCR create any infrastructure and any set of machines, giving exceptionally flexible and responsive training environments at a low cost.

"Alphaville" is a virtual city that MCR created and uses for training and testing. Alphaville is able to run up to 30 copies, or "instances," of itself simultaneously allowing for many parallel tests and to train larger user communities in a realistic setting. Using simultaneous "instances" also allows both competitions and tests against a control environment or control sample.

Alphaville is equipped to train and test vehicular systems and specialists in an environment that presents the same interaction and challenges as the real world. Since this testing is done virtually, different scenarios can be quickly implemented, tested, and archived before investing time and money in building physical test platforms. New or specially configured systems and subsystems can be exchanged to allow observation of different combinations of components.

While most testing focuses on interaction between onboard systems, Alphaville also allows designers and engineers to experiment with the vehicle's interaction with systems outside their control. These "off-board" systems include elements of smart cities, consumer electronics, and other types of vehicle. Scenarios concerning information exchange, privacy, and man-in-the-middle attacks are brought to light by having the vehicle interact with these off-board entities and can have positive influence on policy and procedure work.

MCR is working with ACM to apply Alphaville as a system for testing and evaluating "virtual ECUs"—or, put another way, using Alphaville to test extremely well-detailed software models of planned in-vehicle computers to determine how they interact with each other and with the smart city infrastructure during different cyberattack scenarios. These examinations could be performed while the in-vehicle computer was still in the design and specification phase and can save both time and money in the development process and lead to more secure products.

Walsh College began offering its first course in a planned vehicle cybersecurity curriculum in spring 2017. Walsh partnered with Detroit-area suppliers to develop a practicum-based, heavily lab/hands-on oriented environment for learning vehicular cybersecurity. This is one of the first, if not the first, general vehicular cybersecurity college programs in the United States at the undergraduate level. Walsh is contemplating adding a graduate curriculum.

The SAE CyberAuto Challenge is a Michigan-based annual event which helps to train and inspire the next generation of cyber-automotive engineers and to act as a "matchmaker" between the best available cyber student talent and the OEMs, suppliers, and CAV service providers in the industry. It is heavily hands-on in nature, competitive to attend, and has earned both industry and government accolades. It is cited as one of the three events used to understand cybersecurity issues by the Alliance of Automobile Manufacturers (AAM) and Global Automakers (link: https://www.globalautomakers.org/media/letter/global-alliance-response-to-senator-markey-on-cyber-security-and-privacy-concerns) and specifically cited in NHTSA's October 2016 "Cybersecurity Best Practices for Modern Vehicles" (link: https://www.nhtsa.gov/staticfiles/nvs/pdf/812333_CybersecurityForModernVehicles.pdf) as one of three "by name" events that serve as exemplars of modern vehicle cybersecurity training programs. The CyberAuto Challenge is held in Greater Detroit each summer and is open to high school STEM students and both undergraduate and graduate college students. Since the work is practicum based, many levels of knowledge and practice can benefit from the training. Contact SAE Training for more information (link: www.sae.org/cyberauto).

The US Army TARDEC & Commercial Vehicles Cyber Challenge sponsored by MEDC and the Michigan Defense Center as part of the Protect and Grow Initiative held its inaugural event in June 2017 in Greater Detroit. College students teamed with industry, academia, government (State and Federal), and hackers to assess common commercial systems as well as military hardware. Two days of guided vehicle and systems assessments (hacking) followed two days of training by world-renowned cybersecurity experts. The companies participating were provided with a debriefing of any findings, and then all data was destroyed and all participants committed to an NDA (nondisclosure agreement) regarding the particulars of the event. This event not only helped develop and showcase future talent for the workforce; it also helped foster a common understanding and commitment in the participating companies and government agencies and helped develop an engineering community of interest regarding vehicular cybersecurity for trucks and heavy equipment. Again, Michigan is leading the way in developing talent and in proving itself to be a forward-thinking state which creates and conducts interesting cyber programs—and strengthens the cyber community. Contact the Michigan Defense Center for more information (contact: defensecenter@michigan.org).

9.7 **Conclusion**

The transportation freight industry is facing a looming crisis as routes become increasingly congested and the number of people entering the field is declining. It is expected that there will be a shortage of truck drivers as the need for on-demand and just-in-time delivery increases, driven by consumer demand for products delivered to their door within days or even hours. In the case of this test, the platoon of multiple trucks was led by a vehicle with a human in command of the controls.

Michigan organizations (MEDC, MDOT, and Michigan Defense Center) are actively partnering with industry, academia, and other government organizations in testing a wide range of connected and autonomous technology at various sites around the state. Some additional partnerships include

- MCity: An automated vehicle testing center at the University of Michigan's Transportation Research Institute developed in partnership with the University of Michigan (U of M), MDOT, and Michigan's automotive industry.

- Ann Arbor Connected Vehicle Test Environment: Started as a pilot project to test the effectiveness of connected vehicle technology in a real-world environment around the city of Ann Arbor, it is now being transitioned from a test deployment to an operational deployment. The project is also being transitioned away from government funding to more sustainable long-term funding.

- Southeast Michigan Connected Vehicle Environment: A project consortium to create a Connected Vehicle Environment, including MDOT, General Motors, Ford, and the University of Michigan. The project will be developed primarily along I-96/I-696 and I-94 and will also include part of US-23 and the Connected Vehicle Pilot site in Ann Arbor. MDOT has identified 400-470 locations where roadside units could be located. The project includes numerous applications, both V2V and V2I, such as emergency electronic brake lights, forward collision warnings, left turn assist, work zone warnings, signal phase and timing, and border wait time applications. Although a recent $20 million pilot site application submitted to US-DOT to further support the development of the project was unsuccessful, development of the Connected Vehicle Environment will continue.

Through these and other resources and programs, Michigan fully embraces connected vehicle technology and, following Governor Snyder's leadership, has created unique-in-the-nation and even unique-in-the-world technical and support capabilities to safely and securely develop, test, and provide assured deployment of CAV on the world's highways. Michigan offers test, evaluation, and self-certification proving grounds to both industry and governments and has unequaled capability in ensuring cyber protections to freight fleets operating in Michigan and expert guidance in developing secure products. Michigan—open for business and leading the connected vehicle challenge!

About the Author

Karl Heimer is the Sr. Technical Advisor to Michigan's Auto Office and the Michigan Defense Center. He is also the principle consultant at Heimer & Associates, LLC; founder of the SAE CyberAuto Challenge; co-founder of the CyberTruck Challenge, a member of the editorial board of SAE's new quarterly *Transportation Cybersecurity and Privacy: An SAE International Journal*; a member of the Information Technology/Cybersecurity Advisory Board of Walsh College; sits on SEMA's Vehicle Electronics Task Force as a cybersecurity member; is a member of the Government Fleet Manager Steering Committee; and is a long-time cybersecurity practitioner.

Karl has been in the cybersecurity domain since the 1990s performing on red teams, running evaluation and development activities, turning basic research and ideas into practical solutions and leading organizations developing products and providing analysis, services, and consultation to industry and government customers.

CHAPTER 10

How the Truck Turned Your Television Off and Stole Your Money: Cybersecurity Threats from Grid-Connected Commercial Vehicles

Lee Slezak and Christopher Michelbacher

The increasing connectivity of commercial vehicles through multiple communication channels has created a significant cybersecurity threat to and from these connected vehicles and the movement of people and goods, as detailed in other chapters of this book. However, these communication pathways are not the only avenues for disruptive cybersecurity threats. Technological advances in electric drivetrains and energy storage are increasingly being integrated into medium- and heavy-duty commercial vehicles, from local delivery trucks to transit buses to over-the-road long-haul tractor trailers, as an effective method of lowering fleet operating costs, complying with regulations in certain states and jurisdictions, and reducing emissions for a cleaner employee working environment. As more and more commercial vehicles are equipped with battery charge depleting drivetrains that must be recharged by connecting to the electric grid, they are improving national security by reducing petroleum consumption. Unfortunately, they are creating a cybersecurity threat not just to the commercial vehicles but to other grid-connected vehicles and the entire electric grid.

Information and data are shared between the grid-connected electric vehicle and the charging station, and in the case of "smart" chargers between the charging station and the network operator and/or the local electric utility when the vehicle is connected to a charger to recharge the vehicle's battery. The information that is shared in the case of smart chargers and vehicles can include specifications on the charge rate for the vehicle's on-board charging system, which regulates the flow of direct current (DC) electricity to the vehicle's battery pack, the state of charge and total capacity of the vehicle's battery, what time the operator would like the charging to start, at what electricity price point the vehicle owner would like the charge to begin, and when the vehicle owner would like the charge complete. The use of this information and how it is shared between the vehicle and the charging station is addressed in multiple standards developed by the SAE such as Power Quality Requirements for Plug-In Electric Vehicle Chargers (J28894) and Plug-In Electric Vehicle (PEV) Interoperability with Electric Vehicle Supply Equipment (EVSE). However, standards that address the safeguarding of this information and ensuring that this information is not manipulated have not been issued. By misusing this information through code manipulation either on board the vehicle or in the charger, it would be possible to negatively impact the charging and operation of the vehicle. One example could be preventing the vehicle from ever charging, which would render the vehicle inoperable once the battery is depleted. Another, more serious example, would be preventing the charging of the vehicle from stopping, which could result in the battery and vehicle catching fire in the case of serious overcharging of the battery. The charging of the vehicle could also be started at a much higher charge rate than the vehicle's on-board charging system was designed for, which could damage this equipment and require its replacement before the vehicle could be recharged or operated depending on the extent of the damage to the internal electrical architecture. Furthermore, as vehicle electric ranges increase, primarily through larger on-board high-voltage traction batteries along with advances in the amount of energy a single battery can store (energy density), the need to charge at higher power to reduce overall charge time becomes a necessity. To facilitate quicker charge times, the vehicle charger, or EVSE, converts the alternating current (AC) electricity to DC electricity which it then passes directly to the vehicle instead of relying on the on-board vehicle charge controller to make the conversion. So why is having the EVSE convert grid power to DC power a cybersecurity vulnerability? It turns out, the charging station is no longer responsible for just converting AC power to DC power but is now acting as the vehicle charge controller. This was largely done to save cost along with reduce power electronics complexity and volume on board the vehicle. However, this engineering decision has placed a larger burden for security on the communications pathways between the vehicle charger and the vehicle. Higher charge powers can result in larger consequences to the on-board vehicle electrical architecture should communications be maliciously compromised. These consequences could range from significantly damaging or destroying the vehicle's electrical systems so that the vehicle is inoperable until significant and costly repairs are made to causing system overloads to the point that systems and the vehicle could catch fire.

There are more than 16,000 public EVSE charging stations, with more than 43,000 charging connections/outlets in the United States alone. The overwhelming majority of these stations are operated by one of the charging networks that exist today. Depending on one's geographical location, several charging networks may be available and should be thought of as members-only institutions where credentials of some kind must be verified prior to recharging your vehicle. Although these networks are independent, work is well underway to make these networks seamlessly interoperable. Achieving true interoperability that will allow for grid-connected vehicles to charge at any publicly available charging station nationwide without having a network card for each network requires communications and data sharing between the charging station, the host network, and the network the vehicle owner is signed up with. This communications pathway opens yet another avenue for malicious code to be transmitted from a compromised grid-connected vehicle to thousands of charging stations nationwide where the code can then potentially infect the tens of thousands of grid-connected vehicles on the road today, not just commercial vehicles but any grid-connected vehicle charging at one of the infected charging stations.

Spreading malicious code to tens of thousands of grid-connected vehicles presents a huge security threat that could be used to simultaneously disrupt the operations of commercial vehicles delivering products to stores and materials to factories, and those providing services, such as transportation, as well as privately owned vehicles used by thousands of people every day for commuting and other purposes. An example of the threat this poses: a potential security breach and infestation of malicious code to significant numbers of grid-connected vehicles could be used to cause all infected vehicles to simultaneously cease operation, whether being operated on road or parked and recharging, at a designated time or when signaled from a remote source that could emanate from anywhere in the world. As large as this threat is to the safety of drivers and the functionality of the nation's transportation sector, it pales in comparison to the threat posed to the electric grid.

Since grid-connected vehicles are recharged from charging stations, whether public access, residential, or at a fleet facility, they are linked to the electric grid. It is possible to utilize a grid-connected vehicle to access the EVSE charging station and from there to access the back office of the network operating the charging station. These networks are connected directly to the various electric utilities providing electricity to the charging stations across the network. This presents a path to the back office of the electric utilities that can be exploited to create localized, regional, and national disruptions to the electric grid that is the lifeblood of society today.

As the number of grid-connected vehicles, both commercial and noncommercial, dramatically increases, which it is doing, electric utilities must be more closely linked to the charging of these vehicles. Without knowing when and where vehicles will be charging and at what charge rate and for how long, the electric utilities will be faced with negative impacts, such as circuit and equipment overloading, on their distribution system and system load and harmonization issues impacting overall operations. To minimize these negative effects the utilities must be in closer communications with the grid-connected vehicles in their service territory. This enhanced communications will allow for the utilities to balance demands and load and prevent operational issues. This becomes more critical as vehicles come to market with higher and higher charging powers to reduce charge time. Today's typical battery electric vehicle and plug-in hybrid electric vehicle for commercial applications typically charge at 6.6 to 7.2 kW for AC level 2 charging and approximately 50 kW if they are DC Fast Charged. Within the next few years, chargers and vehicles will enter the market with the ability to charge at up to 400 kW. Utilities will have to have rapid, precise, and secure communications with the vehicles and charging networks to avoid grid imbalances associated with sporadic, high-power, short-duration loads. Many electric utilities, vehicle manufacturers, regulators, and vehicle owners are now beginning to explore the use of controlled charging and other management approaches to utilize grid-connected vehicles to provide grid services, such as peak shaving, frequency regulation, and load management, to name a few. To implement these vehicle grid services will require an even greater level of communications, data sharing, signaling, and controls between the vehicle, the charging station, the network operator (if the charger is a public station on a network), and the electric utility, thus creating a greater cybersecurity vulnerability. This greater integration of grid-connected vehicle charging into the everyday operations of the electric grid nationwide amplifies the impacts of cyber-induced disruptions. Simultaneously stopping all vehicles that are charging or starting all connected vehicles waiting to charge could create a massive load imbalance that could bring entire legs of the distribution network down, if not entire regions of the grid. This not only impacts the grid-connected vehicles, which would not receive a charge for continued operations, but would negatively impact every person relying on the affected portion of the grid. Not only did the compromised vehicles just turn off your TV, but it turned off every device using grid electricity in your house, all the other houses on the same transformer, probably every building on the same distribution feeder, and possibly every building, streetlight, and traffic signal in your utilities service area and potentially beyond. The impact could make the Northeastern United States power outage in 2003 look like just a flicker of a light.

From the grid side, commonly seen events such as voltage sags can trigger different responses from different PEVs which may exacerbate and/or prolong and propagate the grid event (**Figure 10.1**). For example, three different vehicle types are charging at AC level 2 when a gird-induced voltage sag is encountered. Vehicles A and B begin to draw more current in an attempt to maintain a constant charge power while Vehicle C sees the abnormality and stops charging for 5 s. All vehicle responses are less than ideal for the electric grid as Vehicles A and B would increase the load on the grid and Vehicle C would shed load back onto the grid then begin to draw power only 5 s later, creating imbalances in all conditions.

Now imagine there are thousands of PEVs on a utility's network and a similar event is triggered either through an abnormality in the normal grid operations or through an intentionally induced service disruption. The response brought by the vehicles can have a large enough impact on the electric grid that the original event is longer in duration and with greater magnitude. Given enough vehicles, the event can propagate to other nodes and regions of the utilities network that were previously insulated from the original event. This situation can be further magnified with the onset of higher power charging, such as 50 kW capable DC Fast Charge systems and

FIGURE 10.1 PEV grid impact demonstration—vehicle response to a voltage sag to 100 V RMS for 12 cycles (0.2 s). Source: Don Scoffield et al., Idaho National Laboratory, EV Charging/Electric Grid Interaction (Level 2 and DC Fast Charging), July 2017.

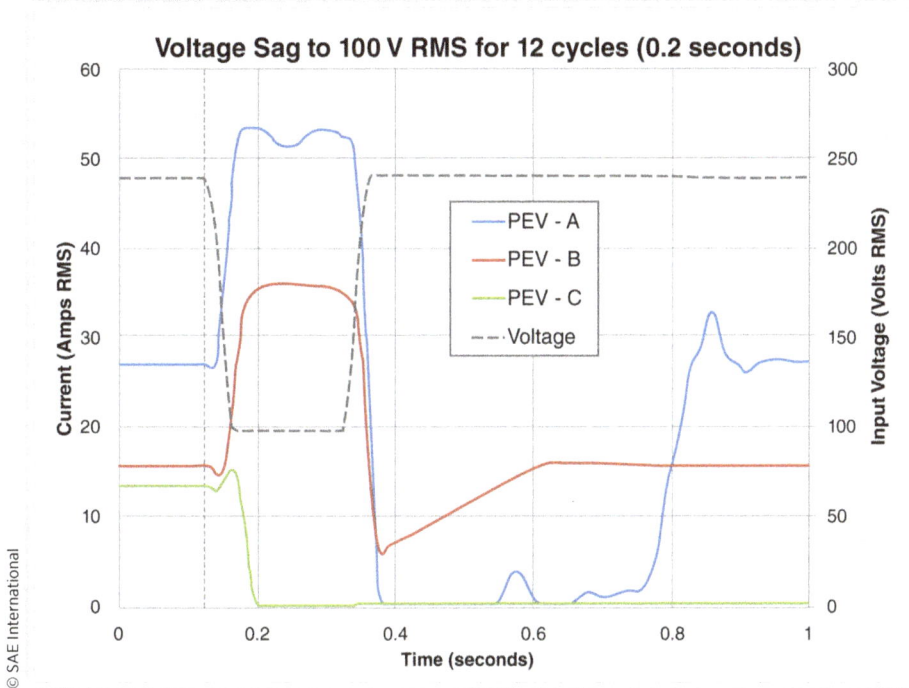

Voltage Sag to 100 V RMS for 12 cycles (0.2 seconds)

© SAE International

eventually 400 kW DC Fast Chargers. These EVSE already demand much larger loads from the electric grid than their AC level 2 counterparts, loads which are sporadic and shorter in duration. The utility may not have the reserve power in the case of even higher power draw from the EVSE, the excess dispatchable load to absorb a sudden increase in available baseload in the case of the EVSE going off-line, or the response time needed in order to cope with and/or recover from the event which could possibly result in a grid outage. As discussed in other sections, grid-connected vehicles offer new and additional communication pathways for which cyber and physical events such as voltage sags or other similarly common grid events can be exploited.

A new method of recharging grid-connected vehicles is beginning to enter the market, which transfers power inductively, i.e., without a physical connection, versus the plug that is used to connect the vehicle to the charging station in today's conductive chargers. The Department of Energy (DOE) previously funded two research and development projects, both with automotive manufacturer team members, to develop high efficiency inductive chargers with charge power to 20 kW. Another project currently being conducted is developing a high-power inductive charger with bidirectional capabilities for medium-duty commercial delivery trucks. The bidirectional power flow can provide electricity from the vehicle's battery to the building the inductive charger is connected to or to a microgrid at the facility. This wireless power transfer, or wireless charging, offers benefits over conductive charging through a cable such as ease of operation, automatic charging with no actions needed from the driver, and a lower likelihood of charging problems associated with cold ambient temperatures and inclement weather. With these benefits comes an increased risk from cybersecurity threats. These threats arise from the fact that the vehicle must utilize a wireless transfer of information and data from the vehicle to the wireless charging station. Unlike conductive chargers today where data transfer only begins when the vehicle is connected with a cable to the charger after the vehicle is <u>parked</u>, wireless charging will require the vehicle to begin communicating with the wireless charger to authorize and initiate the charger <u>before the vehicle parks over the charger</u>. This means that the vehicle's drivetrain will be fully operational and the vehicle will be in motion when this data is transferred. To prevent control of the vehicle's operation from being taken away from the driver for nefarious purposes will require cybersecure isolation of the charging system on board the vehicle from the drivetrain and other computer systems on board the vehicle. Although wireless charging is limited to a number of demonstrations with light-, medium-, and heavy-duty vehicles around the world today, most major automotive manufacturers are actively pursuing this technology, and several have announced plans to introduce the technology on their EVs by 2020. In addition, the introduction of fully autonomous vehicles with electrified drivetrains is likely to be accompanied by a wide-scale installation of wireless charging facilities or hands-free conductive stations, since an autonomous vehicle can't plug itself into a charger. This will eventually place thousands and potentially millions of vehicles on the road with no human driver that are at serious risk from cyber threats unless appropriate precautions are taken now.

In addition to all of the vehicle-specific information that is shared between the vehicles and the charger it is connecting to, whether conductive or inductive, all public access chargers and chargers connected to charging networks share another set of data. This is the financial data for the vehicle owner or the fleet operator that is paying the bills for the recharging of the grid-connected vehicle. The existence of this detailed financial information, which can include names, account numbers, credit card numbers, expiration dates, and security codes for the payee, but also can be linked to financial information for the charging network operators, creates a target for nefarious individuals and yet another cybersecurity threat for grid-connected vehicles. Not only can grid-connected vehicles be used to disrupt the transportation of people and goods throughout the country and to create massive disruptions to the electric grid, but they can be used for financial theft and manipulations. The whole system would need to be secure end to end.

One final cybersecurity threat faced by electric commercial vehicles is the manipulation of mode of operation zones or Geo-Fences. Geo-Fences offer a method for municipalities, working with electrified truck and bus operators and manufacturers of these vehicles, to limit how vehicles operate when in certain geographic locations. Although battery electric vehicles will always operate in electric-only mode, plug-in hybrid electric and other advanced hybrid electric vehicles are being demonstrated with the capability to operate in an electric-only mode when within a set geographic area, such as a city's center or any other designated Geo-Fence area, and in a hybrid electric mode utilizing an internal combustion engine when outside these areas. The battery pack size on these vehicles tends to be much smaller than those on a battery electric vehicle, and the range in electric-only mode for the vehicle is therefore limited, usually to around twenty or so miles. If someone were to hack into the system for the Geo-Fence via GPS manipulation or other means and manipulate the size of the Geo-Fence to significantly expand the area of electric-only operations, the operation of these Geo-Fenced commercial hybrid vehicles could be severely impacted. A large enough Geo-Fence could mean that vehicles entering the area and automatically switching to electric-only operation would not have sufficient range to complete their trip within or through the Geo-Fence. These vehicles would run out of on-board electricity in the battery pack and would be rendered inoperable until they were able to be recharged. Since the vehicles do not control the Geo-Fence and only respond to its signal, they would be powerless to override the manipulated Geo-Fence. This could create significant issues with the transportation sector and traffic inside the Geo-Fence and prevent commercial vehicles from operating as needed.

DOE to the rescue!! The DOE has been working with utilities and industry to address grid-related cybersecurity threats for years. In 2010 the Department began working with manufacturers of charging station equipment on cybersecurity issues. A thorough evaluation of the latest offerings from several manufacturers was conducted utilizing the appropriate resources and experience gained over the years by the Department's National Laboratories working in the cybersecurity space. These initial evaluations highlighted significant shortcomings in the ability of "cyber secure" charging stations to prevent and protect against cyber threats. As a result of these findings the Department launched several initiatives. The first of these is to develop a cybersecure diagnostic security module. This low-cost device can be seamlessly integrated into a grid-connected vehicle and into charging stations to help ensure that both the vehicle and charging equipment have not been tampered with. In the event that tampering is detected, the device that has been tampered with is immediately isolated and no data transfer or charging takes place. This module can also be integrated into a home or building energy management system to provide protection for equipment on the grid side of the charging station and to notify the building operations manager that a vehicle attempting to charge or a charger connected to the building has been compromised and has been isolated. In addition to this development initiative, the Department has been coordinating with the Department of Homeland Security, the Department of Transportation, branches of the Department of Defense, and other Federal agencies and commissions to develop a coordinated effort to address grid-connected vehicle cybersecurity issues. This group is working with vehicle manufacturers, charging equipment manufacturers, utilities, and universities to ensure that cybersecurity threats are being appropriately addressed across the transportation sector. Finally, the Department and key personnel from within the Department's National Laboratory network are working closely with the various standards setting organizations in the United States and abroad to develop and verify the effectiveness of standards that address cybersecurity threats for grid-connected vehicles.

About the Authors

Lee manages the Grid and Infrastructure Research and Development Activities for Vehicle Electrification within the Department of Energy's Vehicle Technologies Office. His responsibilities include developing and implementing plans for conducting R&D on high-power fast charging, including vehicle and grid impacts; wireless charging for dynamic charging; technologies and strategies to minimize grid impacts of charging; cybersecurity for vehicle charging; and supporting development of codes and standards for charging technologies. Previously, Lee managed the DOE's Vehicle Systems Program and the $400 million American Recovery and Reinvestment Act program for Transportation Electrification. Lee has been at the Department of Energy for over 23 years and previously managed the Federal Fleet AFV Program and the National Alternative Fuels Data Center. Lee has worked on advanced technology vehicle initiatives for the past 26 years.

Mr. Christopher Michelbacher joined Idaho National Laboratory (INL) in early 2009 as part of the Energy Storage and Transportation Systems (ES & TS) Department and is currently on a Management and Operating (M&O) assignment within the Vehicle Technologies Office (VTO) at the Department of Energy (DOE) headquarters in Washington, D.C. His roles include development and oversight of programs in the following research areas: (1) vehicle-centric cybersecurity (connected and autonomous vehicles, Smart EVSE), (2) grid modernization (Smart EVSE, vehicle to vehicle, vehicle to grid), (3) energy storage systems testing and research (lithium-ion advanced transportation battery systems research and benchmarking), and (4) electric vehicle fast charging (350 kW). Prior to his assignment at the DOE, Mr. Michelbacher supported the ES &TS Department in their state-of-the-art Battery Test Center (BTC) as a research engineer. Later he progressed to vehicle systems research followed by a return to battery research as a BTC Principal Investigator (PI). As PI, Christopher was responsible for all operations, analysis, and reporting of energy storage device testing within the BTC.

CHAPTER 11

CALSTART's Cyber Mission: HTUF REDUX

Michael Ippoliti

CALSTART originated the Hybrid Truck Users Forum (HTUF) with the assistance of the U.S. Army National Automotive Center (NAC) and Tank Automotive Research Development and Engineering Center (TARDEC), nearly two decades ago. The NAC is the Army's outreach arm to the commercial transportation industry, and it is charged with both understanding the capabilities of the commercial vehicle industry and working to increase the capabilities of the commercial industry to build advanced vehicles and technologies that can support emerging Army and military needs. The military was exploring hybrid trucks as one of the key future technologies for reducing fuel use and providing field electrical power generation [1]. Fuel use is a critical issue—military commanders were extremely concerned about the cost, both in dollars and in risk to personnel, of transporting fuel to deployed forces. A delivered gallon of fuel can cost hundreds of dollars, and the risks to fuel convoys are extreme.

The military had a real need, but as Bill Van Amburg, Senior Vice President of CALSTART, said "in 2001, no major North American truck maker was publicly discussing, much less supporting, the hybridization of medium- and heavy-duty vehicles. Despite the emerging success of light-duty passenger car hybrids, most notably the Toyota Prius, the prime focus of North American truck and engine makers was on how to achieve the significant emissions reductions required under the U.S. Environmental Protection Agency's 2007/2010 engine standards" [2].

Not only were truck OEM engineering resources focused elsewhere, but the business case for hybridization was unclear, as was customer demand. No one knew what real-world benefits would be gained by hybridization, what duty cycles or vocations were best suited for these new designs, or if end customers/purchasers cared enough to pay for the fuel economy improvements.

By 2007, the picture was very different. Every major truck maker had a hybrid moving into the market or under development, and the benefits were being proven in multiple field trials. Hybrids had become indicators of technology leadership, with truck makers touting their latest models. Many things changed during those 6 years, including higher and more volatile fuel costs plus growing Corporate Social Responsibility programs with a focus on climate change and carbon reduction. Other factors were present as well, but all "…were insufficient conditions alone to accelerate the pace of hybrid development and the move into production. What was required was focused demand from users, targeted first applications upon which manufacturers could focus, and a process to bring together users and manufacturers in a risk-reducing partnership" [3].

The HTUF was created to fill this market-technology gap. The military was both a large user (a truck fleet larger than any commercial enterprise) and a buyer of commercial equipment—trucks are core to mission achievement. Importantly, the NAC had already been working on hybrids and advanced truck drivetrain technology. CALSTART could leverage industry contacts and expertise, along with the credibility and insights from the Army, to form a new approach, "…a nimble, fast-track process to speed hybrids to market by focusing attention on and aggregating the needs of the key players who could change an industry's direction: the users themselves" [3].

HTUF was a great success. At one point, there were working groups active in six different applications and need areas, including long-haul Class 8. The Utility working group comprised more than 40 fleet users representing roughly 20 percent of the purchasing power in what is approximately a ten-thousand-truck-per-year segment. During the significant engineering demand experienced by truck OEMs to meet 2010 engine emission requirements, the speed of simultaneous hybrid development was remarkable.

Now, in 2017, a new area of technology development has appeared, with strikingly similar challenges. Hybridization helped enable battery developments, electric auxiliaries, and power electronics development that formed a foundation for rapid advances in electric and fuel cell cars and trucks. In the same way, cybersecurity is both a prerequisite and a foundation for platooning, autonomous driving, V2V and V2X systems, and other connected vehicle platforms.

CALSTART's nonprofit mission is to reduce emissions and fuel consumption in transportation. In 2017, the organization is celebrating 25 years of working on that mission. By operating as a membership-based entity, CALSTART has brought together diverse stakeholders in the clean transportation industry, ranging from small start-ups to the Big 3 automakers and the top truck OEMs, totaling over 160 members currently. Working across fuels, trucks, buses, and light-duty (cars), CALSTART continues to impact policy and product development strategies to accelerate the transition to clean transportation, to help reduce pollution and petroleum use, and to create economic growth and good jobs.

For CALSTART, the appeal in highly automated commercial vehicles (HACVs) is that many of the technologies, (such as platooning, where multiple trucks travel together with vehicle separation computer controlled and optimized) will have real impacts on fuel efficiency. Increased fuel efficiency will reduce emissions even if the fuel being saved is electricity. Cybersecurity is therefore something to be addressed in support of the CALSTART mission, because these HACV technologies must be cybersecure for them to be accepted. In the same way, the military is again an ideal partner. Due to their clear interests and needs around cybersecurity, they have already done a great deal of work and bring credibility and technical expertise that is unique and valuable. Just as their need for tactical capabilities like silent watch drove the development of electrification for trucks, their need for secure military systems can drive the development of secure commercial systems. In fact, it is often the same hybrid technology suppliers who are part of the HACV revolution, and they need to be including cybersecurity in their product development.

Cybersecurity today is analogous to hybridization 20 years ago in additional ways. Awareness is low, and the potential magnitude of the benefits—or threats in the case of cybersecurity—is not well understood by fleets or the public. Cybersecurity is not being treated as critical to future developments except by a few leading stakeholders. Even more essential, cybersecurity is no longer just a back-office computer issue. Trucks themselves are now an attack surface that leads to the back office. All data and operations for a company, not just the single load or the single truck, are at risk.

Logically then, the approach for a user group focused on truck cybersecurity will be much the same as for HTUF. There is a large need to educate and socialize the issue, uncover user needs that are not yet expressed, and find solutions that will enable other technologies dependent upon cybersecurity. Educational webinars, user interviews, and coordination with the teams formed by the National Motor Freight Traffic Association (NMFTA) and American Trucking Associations' Technology and Maintenance Council (ATA TMC), along with the Automotive Industry Information Sharing and Analysis Center (Auto-ISAC) will be critical steps in the process. CALSTART and TARDEC will be working to define the research that must happen, addressing immediate user needs. CALSTART will also lead in bringing the commercial bus segment into the cybersecurity arena. Transit buses are ahead of trucks in adoption of zero-emission technologies, with both fuel cell and battery electric buses in revenue service across several agencies nationwide. The need for cybersecurity in transit bus operation is equal to that of goods movement—both involve large vehicles with multiple communications systems, carrying valuable cargo on public roadways, logging millions of miles every day.

The structure for this working group has already been defined, through the work in cybersecurity across other fields. TARDEC is proposing the formation of an Information Sharing and Analysis Organization (ISAO). The foundation for these organizations was set by Presidential Executive Order 13691 (Obama) and is fully outlined in the documents created by the ISAO Standards Organization [4, 5]. "The intent of the TARDEC-supported Cybersecurity Truck ISAO is to be in front of the technology and developing countermeasures to cyber threats. The Truck ISAO will lead the definition of user needs, rather than just reacting to situations or threats that are uncovered, as with an ISAC" [6].

As with HTUF, the initial work will be collecting information, researching the status of the technology and the industry, and engaging with first-mover stakeholders. Unlike the Auto-ISAC which limits government agency involvement, TARDEC and other government agencies will be full contributors to this early effort. Unlike standard market research input to OEMs and technology developers, these users will not be "cherry-picked" or have their input filtered by other agendas. The goal will be developing a strategic framework for how the ISAO will operate and what priorities it will follow. Finding this user-driven framework is critical. Once the framework is in place, the team will set the working model for the ISAO, which can be used to build a 1-year plan and a 5-year plan based on the current needs and desires of the participants.

In this way, the HTUF model will again accelerate the advancement of critical technology for the trucking industry, the military, and our nation.

References

1. Van Amburg, B., "Effective Heavy-Duty Hybrid Market Development: The HTUF Commercial Military Model," Weststart-CALSTART, 2007, pp. 2-3, www.calstart.org/libraries.

2. Van Amburg, B., "Effective Heavy-Duty Hybrid Market Development: The HTUF Commercial Military Model," Weststart-CALSTART, 2007, p. 1, www.calstart.org/libraries.

3. Van Amburg, B., "Effective Heavy-Duty Hybrid Market Development: The HTUF Commercial Military Model," Weststart-CALSTART, 2007, p. 2, www.calstart.org/libraries.

4. ISAO SO (Information Sharing and Analysis Organization Standards Organization), "ISAO 100-1 Introduction to Information Sharing and Analysis Organizations (ISAOs) v1.01," October 14, 2016.

5. ISAO SO (Information Sharing and Analysis Organization Standards Organization), "ISAO 100-2 Guidelines for Establishing an Information Sharing and Analysis Organization (ISAO) v1.01," October 14, 2016.

6. U.S. Army TARDEC, "CALSTART HTUF Statement of Work Draft," Paragraph 4.2.2, April 2017, unpublished.

About the Author

Michael Ippoliti is highly knowledgeable in connected and automated technology. He was until January 2018 the director of clean transportation solutions at CALSTART, the clean transportation technologies consortia. He is now the Autonomous and Connected Vehicle Program Manager for HDR Engineering, based in Long Beach. While at CALSTART Mike led work with ports, fleets, and technology developers to improve the performance of their transportation operations and advance clean transportation technology. Mike focused on heavy-duty zero-emission vehicles as well as on connected/automated vehicles and cybersecurity under CALSTART's High-Efficiency Truck Users Forum (HTUF) program with the U.S. Army/TARDEC.

Prior to joining CALSTART, Mike was Research Director for Telematics and Automotive at ABI Research. He has worked as a consultant in new product development and intellectual property research. Michael brings many years of automotive OEM experience, with roles in market research and product planning at Volvo Cars of North America and in the Volvo Monitoring and Concept Center, where he led projects investigating the future of mobility and automobiles. He has a Bachelor of Science degree from Tufts University in mechanical engineering and ergonomics with advanced work in engineering design, and an MBA from the Tepper School of Business at Carnegie Mellon University.

CHAPTER 12

Characterizing Cyber Systems

Jennifer Guild

12.1 Introduction

This chapter was written to provide a level playing field for all to understand how to assess a cyber system. That said, some might find the material to start simply enough, then become deeply technical. As such, someone new to the cybersecurity field may find this to be a challenging chapter.

Most people automatically interpret the strength of the security capabilities of a system based upon their knowledge, experience, cultural background, and the association of that system to its functionality. As technology has progressed, so has our understanding of systems. A system is no longer just the desktop computer, but mobile phones, wearable devices, newer classes of automobiles, recreational vehicles, and aircraft.

A vulnerability assessment is the methodical examination of a system to determine the adequacy of security measures, identify security deficiencies, collect evidence from which the effectiveness of said security measures can be predicted, and confirm the adequacy of such measures against a single instantiation or model of a system [1].

12.2 **Assessment Models**

A number of aspects are considered when assessing a system, each of which must be modeled. As knowledge of the system increases, the content of these models will go from generalized to specific. As an assessment is rarely a single, continuous event, these individual models are iteratively addressed so as to represent each impression of the system's capabilities and the operational environment and correlate the models to the evidence and to representatively correlate or map the completed models to the empirical evidence of the assessment. Examples of operational environments are:

Vehicle

- Air (plane, drone, etc.)
- Land (car, semitruck, etc.)
- Submersible (deep-submergence vehicle (DSV), etc.)
- Afloat (boats, Jet Skis, ships, etc.)
- Space (satellites, etc.)

Stationary/land based

- Computer (server or workstation)
- Mobile (laptop, tablet, phone, watch, etc.)
- Network operations center

These environments are broad and generalized. A specific operational environment is a situational instance or state which reflects a physical characterization of the operational environments (such as parked vehicle vs. a vehicle in motion). However, it is not necessary to model the system in all of its possible states. Each state will be individually modeled for the aspects considered. These individual models will comprise the overall assessment model. The models will be represented using set theory.

The following sections contain examples that provide realistic modeling, but are not actual representations of the systems.

12.2.1 **Flaw Models**

The Committee on National Security Systems (CNSS) is representative of many U.S. government organizations and is responsible for the protection of national security systems (NSS) across the Intelligence Community, Department of Defense, and the Civil Agencies [2]. CNSS Instruction (CNSSI) 4009 is a U.S. government standard that defines a flaw as an error of commission, omission, or oversight in a system that may allow protection mechanisms to be bypassed. At the start of an assessment, the flaws of the system may or may not be known. As the assessment is conducted, and flaws are identified, the model is updated.

Flaws (F) exist in both the technical system and in all environments of operations and are defined as follows:

F represents the set of all flaws.

$F_T \subseteq F$ is the set of technical flaws.

$F_O \subseteq F$ is the set of operational environment flaws.

It is assumed that all flaws can be designated with an origin that is either technical, operational, or may be designated uncategorized (unknown technical or environmental flaws and will be represented as F_U), and therefore

$$F = F_T \cup F_O \cup F_U$$

The following is an example of the possible groups of flaws in a vehicle ($F_{VehicleBaseline}$), such as a car, where F_{T1} and F_{O1} are the technical and operational flaws for system s1:

- System of systems including system interactions (F_{SoS}) $\subseteq F_{T1}$
- Tire pressure monitoring system (TPMS) including wireless communication of TPMS measuring the pressures of tire and reporting those pressures (F_{TPMS}) $\subseteq F_{T1}$
- Ethernet including connecting a not-hardened/unpatched laptop to maintenance port ($F_{Ethernet}$) $\subseteq F_{T1}$
- CAN bus (neural network of a vehicle) including connectivity to radio frequency (RF) communications (F_{CANBus}) $\subseteq F_{T1}$
- RF communications including telematics systems ($F_{RFComms}$) $\subseteq F_{T1}$
- 802.11 connectivity including being a hotspot ($F_{802.11}$) $\subseteq F_{T1}$
- Connectivity including communicating with its manufacturer ($F_{Internet}$) $\subseteq F_{O1}$

The vehicle is a complex system of systems, which may have only been interconnected once implemented within the vehicle. Connectivity is an operational flaw in the vehicle because of its consistent connectivity to the manufacturer.

To be clear, during the assessment of a system, technical flaws must be addressed, because the technical flaws are always part of the system, including when it is instantiated into an environment. In addition, there is a need to assess the specific operational environment flaws, categorized appropriately, at the time the system is instantiated. Such flaws may expose previously unknown technical flaws, as well as operational environment flaws. Therefore, the set of flaws examined for system s1 will simply be

$$F_{s1} = F_{O1} \cup F_{T1}$$

However, the above formula-indicated flaws are static or unchanging. Therefore, the model requires expansion by allowing these sets of flaws to be defined dynamically.

The first computer was aptly named the Turing state machine. Computers alter states every time a binary decision is completed. So, computers and networks exist in fluidity, each constantly changing. However, the current common practice is to assess a single state, the state at which the system exists at the time of assessment.

In the 1950s, the Rand Corporation estimated nuclear explosions by modeling numerous "states" of the explosion. Similarly, the U.S. Centers for Disease Control (CDC) forecasts the infection rate of a contagious disease by modeling the state or spread of infection at specific time increments, such as 24 hours, 48 hours, 1 week, 1 month, 3 months, etc. [3, 4]. By applying these similar concepts to a system, it is possible to model the fluidity of systems' states without modeling every single state.

To model the dynamic nature of the system, the sets of flaws for a specific system are defined when it is in a particular state. In other words, the operational states that affect the environment of the system. The flaws for a specific state will be those possible flaws given the value of the state variables. The notation $(F_{T1})_n$ represents the set of technical flaws for system s1 in state n. Similar notations will be used for the other sets of flaws. A sample of possible operational states will be modeled for the vehicle.

Car powered off, parked in the owner's driveway $(F_{S1})_{ParkedOff}$

- RF communications including remote start allowing wireless connectivity $(F_{RFComms}) \subseteq (F_{T1})_{ParkedOff}$
- RF communications including wireless unlock $(F_{RFComms}) \subseteq (F_{O1})_{ParkedOff}$

RF communications has two aspects. The technical aspect is the actual communication systems themselves (hardware, firmware, and software). The operational environment RF communication flaws allow access to the vehicle.

Car in motion, being driven by owner $(F_{S1})_{OwnerDriving}$

- Baseline $(F_{VehicleBaseline}) \subseteq (F_{T1})_{OwnerDriving}$
- RF communications including possible geolocation $(F_{RFComms}) \subseteq (F_{O1})_{OwnerDriving}$
- Satellite communications including possible geolocation based upon satellite communications $(F_{SatComms}) \subseteq (F_{O1})_{OwnerDriving}$

In this case, the satellite communication flaws are those specific communications the vehicle has with satellites. The methodology of the geolocation is also different.

The preceding examples of the vehicle provide a glimpse into possible states, both technically and operationally. If $(F_{T1})_n$ and $(F_{O1})_n$ represent the set of technical and operational environment flaws for system s1 in state n, then the set of all flaws for system s1 in state n is

$$\left(F_{S1}\right)_n = \left(F_{T1}\right)_n \cup \left(F_{O1}\right)_n$$

An assessment of system s1 will then consist of an assessment of the system with respect to flaws in all possible states of the system. If $Ev_F(F_{s1})$ represents the assessment of the flaws in system s1 and i represents one of the possible k states, then

$$Ev_F\left(F_{s1}\right) = \bigcup_{i=1\ldots k} Ev_F\left(\left(F_{s1}\right)_i\right)$$

12.2.2 Countermeasure Models

In the National Institute of Standards and Technologies (NIST) Special Publication (SP) 800-53, a countermeasure is defined as any actions, devices, procedures, techniques, or other measures that reduce the vulnerability of an information system. During the initial assessment, countermeasures will be generically mapped to a flaw area (such as $F_{802.11}$). This allows for the creation of a high-level representation of the system, which will be updated throughout the assessment with the final countermeasure model mapped directly to a flaw model.

The idea is to identify groups of countermeasures that would apply to groups of flaws and then to correlate one or more countermeasures to one or more flaws as the evidence is acquired. The set of technical and operational countermeasures examined for system s1 will simply be

$$M_{s1} = M_{O1} \cup M_{T1}$$

Just as flaws exist in a fluid state, so do countermeasures. This is made more so by the fact that countermeasures are not just on the same system, but exist in more than one layer and in more than one component. Just as with flaws, to model the dynamic nature of the system, the sets of countermeasures for a specific system are defined when it is in a particular state.

The countermeasures for a specific state will be those countermeasures that are possible given the value of the state variables. The notations $(M_{T1})_n$ and $(M_{O1})_n$ will be used to represent the set of countermeasures for system s1 in state n, then a single state of countermeasures for system s1, including the null set, is

$$\left(M_{S1}\right)_n = \left(M_{T1}\right)_n \cup \left(M_{O1}\right)_n$$

An assessment of system s1 will then consist of an assessment of the system with respect to flaws and countermeasures in all possible states of the system. If Ev(s1) represents the assessment of the flaws and countermeasures of the system s1 and i represents one of the possible k states, then

$$Ev\left(s1\right) = U_{i=1\ldots k} \; Ev_F\left(\left(F_{s1}\right)_i\right) \cup U_{i=1\ldots k} \; Ev_M\left(\left(M_{s1}\right)_i\right)$$

It is not possible to map all of the permutations of flaws and countermeasures [5]. Adversaries are constantly attempting to avoid or overcome countermeasures, adding to the fluidity of countermeasures and their associated flaws.

12.2.3 Vulnerability Models

Vulnerabilities (V) are those flaws that are not completely mitigated by countermeasures. Since vulnerabilities are a subset of flaws, vulnerabilities are defined as follows:

V represents the set of all vulnerabilities.

$V_T = \{f \mid f \in F_T \text{ and } M_C(f) = \varnothing\}$ is the set of technical vulnerabilities.

$V_O = \{f \mid f \in F_O \text{ and } M_C(f) = \varnothing\}$ is the set of operational environment vulnerabilities.

Vulnerabilities have fluidity from the constant changing states of both flaws and countermeasures. To model the dynamic nature of the system, the sets of vulnerabilities are defined for a specific system when it is in a particular state. Considering a car (s1), this time powered off, parked at the mall $(V_{s1})_{ParkedMall}$:

- RF communications including remote start allowing wireless connectivity $(F_{RFComms}) \subseteq (F_{O1})_{ParkedMall}$
- RF communications including wireless unlock $(F_{RFComms}) \subseteq (F_{O1})_{ParkedMall}$
- Physical security including wireless unlock $(V_{PS}) \subseteq (V_{O1})_{ParkedMall}$
- Keyless start $(V_{KS}) \subseteq (V_{T1})_{ParkedMall}$

In this case, if the adversary can unlock the car, the adversary can drive the car to another location.

The vulnerabilities for a specific state will be those flaws and countermeasures that are possible given the value of the state variables. The notations $(V_{T1})_n$ and $(V_{O1})_n$ represent the set of vulnerabilities for system s1 in state n, including the null set:

$$\left(V_{S1}\right)_n = \left(V_{T1}\right)_n \cup \left(V_{O1}\right)_n$$

An assessment of system s1 will then consist of an assessment of the system with respect to flaws and countermeasures in all possible states of the system. If Ev(s1) represents the assessment of the flaws and countermeasures of the system s1 and i represents the possible k number of states, then

$$Ev\left(s1\right) = \left(\bigcup_{i=1\ldots k} Ev_F\left(\left(F_{s1}\right)_i\right)\right) \cup \left(\bigcup_{i=1\ldots k} Ev_M\left(\left(M_{s1}\right)_i\right)\right)$$

The operational environments have unique sets of flaws and countermeasures, some technical and some specific to the environment. A DoD research project, ARPANET, conducted by the Advanced Research Projects Agency, was the progenitor to the Internet [6, 7]. Unfortunately, the designers of that original packet-switched network, circa 1968, did not account for the threat of someone intentionally behaving maliciously [7]. This was for two reasons, one because it was a network with extremely few users, each of which were known to the others, and more importantly, it did not occur to the designers that someone would behave badly.

12.2.4 Threat Models

A threat assessment, per CNSSI 4009, is a process of formally evaluating the degree of threat to an information system and describing the nature of the threat. Threats are from a variety of sources, such as adversaries, disgruntled insiders, and natural disasters, and there are many definitions of cyber threats. In this paper, a threat (TR) is a tuple consisting of some combination of one or more threat source (TS) and that threat source's one or more capabilities (TC), with one or more threat source's motivation(s) (TSM) for exploiting a vulnerability and is defined as follows:

$$TR = TS^+ \times TC^+ \times TSM^+$$

Unlike flaws and vulnerabilities, threat sources are defined based upon the intent to exploit:

TS represents the set of all threat sources.

$TS_{Intentional} \subseteq TS$ is the set of threats intentionally exploiting a vulnerability.

$TS_{Accidental} \subseteq TS$ is the set of threats where a situation or method may accidentally trigger a vulnerability [8].

The accidental threat source includes such things as nature (TS_{Nature}) such as an earthquake or hurricane, as well as the accidental human ($TS_{AccidentalHuman}$), which is when a person performing a task that triggers a vulnerability unintentionally.

The major categories of intentional human threat sources could be expressed as state sponsored ($TS_{StateSpon}$), insider ($TS_{Insider}$), terrorist ($TS_{Terrorist}$), hacker (TS_{Hacker}), and organized crime ($TS_{OrgCrime}$). The threat source of state sponsored not only indicates adversaries in the direct employ of a nation state, but also those that act on behalf of the nation state out of some motivation. The set of a threat source's intentional exploitation of a vulnerability contains the set of accidental threat sources and is represented by

$$TS_{Intentional} = TS_{StateSpon} \cup TS_{Insider} \cup TS_{Terrorist} \cup TS_{Hacker} \cup TS_{OrgCrime} \cup TS_{Accidental}$$

Some would consider that state-sponsored threat source to have the greatest capability. However, the word capability is not sufficiently concise. The NIST Special Publication 800-30 R1 defines threat source capabilities as [8]

TC represents the set of all threat source capabilities.

$TC_{LevelOfExpertise} \subseteq TC$ is the set of a threat source's levels of expertise.

$TC_{Resources} \subseteq TC$ is the set of the number of resources of a threat source.

$TC_{Success} \subseteq TC$ is the set of a threat source's levels of success.

A threat source has one or more capabilities. A single capability will have one or more resources with each resource having varying levels of expertise and subsequently that capability will have varying degrees of success:

$$TC = \left(TC_{LevelOfExpertise}\right)^{+} \times \left(TC_{Resources}\right)^{+} \times \left(TC_{Success}\right)^{+}$$

A more concise breakdown of capabilities is [8]:

$TC_{LevelOfExpertise}$ = {VerySophisticated, Sophisticated, Moderate, Limited, VeryLimited}

$TC_{Resources}$ = {Unlimited, Significant, Moderate, Limited, VeryLimited}

$TC_{Success}$ = {MultipleContinuousCoordinated, MultipleCoordinated, Mulitple, Limited, VeryLimited}

Therefore, a state-sponsored threat source may have differing levels of expertise, resources, and success based upon that nation's focus on cyber warfare. However, the technical capabilities don't capture the motivation. It is often not modeled because there may be no motivation (nature), it may change for each category of threat, and within each threat source based upon situations. Some examples of motivations include financial ($TSM_{Financial}$), national security ($TSM_{NationalSecurity}$), power (TSM_{Power}), information gathering (TSM_{Intel}), and forcible change (TSM_{Change}).

$$TSM = TSM_{NationalSecurity} \times TSM_{Power} \times TSM_{Intel} \times TSM_{Financial} \times TSM_{Change}$$

Obviously, a state-sponsored threat source may include all of the above motivations, whereas a terrorist threat source motivation may include only

$$TSM_{Terrorist} = TSM_{Financial} \cup TSM_{Change}$$

Threat sources, capabilities, and motivations are fluid and change with the situational instance, and at times the technical system. Therefore, the threat states are a combination of all possible threat sources, threat source capabilities, and threat source motivations:

$$TR_{States} = TS \times TC \times TSM$$

The threat sources, capabilities, and motivations will also change based upon the operational environment category and so sets of threats for a specific system are defined when it is in a particular state. The state-sponsored entity of Russia ($TS_{StateSponR}$) may have financial, national security, information gathering, power, and forcible change motivations to exploit a vulnerability within a commercial vehicle in Ukraine ($V_{CVUkraine}$):

$$TR_{Staten} = \left(TS_{StateSponR} \times \left(TSM_{NationalSecurity} \cup TSM_{Power} \cup TSM_{Intel} \cup TSM_{Financial} \cup TSM_{Change}\right) \times \left(TC_{LevelOfExpertise}\right)_{VerySophisticated}\right) \cup V_{CVUkraine, State\,n}$$

The threats for a specific state will be those that are possible given the value of the state variables. The notation $(TR_{s1})_n$ represents the set of threats for system s1 in state n:

$$\left(TR_{S1}\right)_n = \left(\left(TS_{T1}\right)_n \times \left(TC_{T1}\right)_n \times \left(TSM_{T1}\right)_n\right) \cup \left(\left(TS_{O1}\right)_n \times \left(TC_{O1}\right)_n \times \left(TSM_{O1}\right)_n\right)$$

An assessment of system s1 will then consist of an assessment of the system with respect to vulnerabilities, and threats in all possible states of the system. Just as in the prior sections, if Ev(s1) represents the assessment of the vulnerabilities and threats of the system s1 and i represents the possible k number of states, then

$$Ev(s1) = \left(\bigcup_{i=1\ldots k} Ev_V\left(\left(V_{s1}\right)_i\right)\right) \cup \left(\bigcup_{i=1\ldots k} Ev_{TR}\left(\left(TR_{s1}\right)_i\right)\right)$$

Motivation implies some amount of probability[1] of attack occurrence, as does a situational instance, such as a U.S. aircraft in an adversary's airspace during a time of kinetic war. For every threat, there will be associated probabilities. By including the probabilities, which are extremely subjective, correlations may be expressed between threats, vulnerabilities, and attack vectors.

12.2.5 Probability Models

There are three probabilities regarding exploitations to consider during an assessment. There is the probability that a threat source will attack (PA). Then there is the probability of the success of the attack (PS). Finally, there is the probability of certainty (PC) of the knowledge. The overall probability (P) that an attack will occur with some levels of covertness and success would then be defined as a tuple:

$$P = PA \times PS \times PC$$

To provide the greatest repeatability, all three probabilities must be included in calculations. For conciseness, each probability will be individually identified. The probabilities are based upon those in the NIST Special Publication 800-30 R1 for likelihoods:

PA = {AlmostCertain, HighlyLikely, SomewhatLikely, Unlikely, HighlyUnlikely}

PS = {AlmostCertain, HighlyLikely, SomewhatLikely, Unlikely, HighlyUnlikely}

However, the categories of likelihood do not correlate to the probability of certainty of knowledge. By substituting the word "certain" for the word "likelihood," the vocabulary is correct and the values are maintained:

PC = {AlmostCertain, HighlyCertain, SomewhatCertain,
      Uncertain, HighlyUncertain}

---

[1] NIST Special Publication 800-30 refers to this as likelihood.

As with everything else, the probabilities exist for both the technical system and operational environments:

P represents the set of all probabilities.

$P_T \subseteq P$ is the set of probabilities for technical capabilities of a system.

$P_O \subseteq P$ is the set of probabilities for an operational environment.

The probability of attack, probability of attack success, and probability of knowledge certainty are fluid and change with the state. Therefore, the probability states are a combination of all probabilities of attack, probabilities of attack success, and probabilities of knowledge certainty:

$$P_{States} = PA \times PS \times PC$$

The operational environment also alters the probabilities. To model the dynamic nature of the system, sets of probabilities for a specific system are defined when it is in a particular state. As an example, the state-sponsored entity of Russia has motivations to exploit a vulnerability within a commercial vehicle in Ukraine, and there is an almost certain probability that Russia will attack, with a highly likely probability of success, based upon highly certain probability of knowledge:

$$P_{State\,n} = TR_{Russia,State\,n} \cup V_{UAVinUkraine,State\,n} \cup \left(PA_{AlmostCertain} \times PS_{HighlyLikely} \times PC_{HighlyCertain} \right)$$

The probabilities for a specific state will be those that are possible given the value of the state variables. The notation $(P_{S1})_n$ represents the set of probabilities for system s1 in state n:

$$\left(P_{S1}\right)_n = \left(\left(PA_{T1}\right)_n \times \left(PS_{T1}\right)_n \times \left(PC_{T1}\right)_n\right) \cup \left(\left(PA_{O1}\right)_n \times \left(PS_{O1}\right)_n \times \left(PC_{O1}\right)_n\right)$$

An assessment of system s1 will then consist of an assessment of the system with respect to vulnerabilities, threats, and probabilities in all possible states of the system. If Ev(s1) represents the assessment of the vulnerabilities, threats, and probabilities of the system s1 and i represents the possible k number of states, then

$$Ev\left(s1\right) = \cup_{i=1\ldots k} Ev_V\left(\left(V_{s1}\right)_i\right) \cup Ev_{TR}\left(\left(TR_{s1}\right)_i\right) \cup Ev_P\left(\left(P_{s1}\right)_i\right)$$

No matter the probabilities, without a physical mechanism/vector through or by which the exploit may be conducted against a vulnerability, the exploit won't succeed.

12.2.6 Attack Vector Models

An attack vector (AV) is a physical (analog or digital) mechanism or vector through which an exploit by a threat source that may be conducted against a vulnerability. Examples of categories of attack vectors are cyber (AV_{Cyber}), kinetic ($AV_{Kinetic}$), radio frequency (AV_{RF}), supply chain ($AV_{SupplyChain}$), and unknown ($AV_{!Known}$).

$$AV = AV_{Cyber} \cup AV_{Kinetic} \cup AV_{RF} \cup AV_{SupplyChain} \cup AV_{!Known}$$

The obvious attack vector for most systems is cyber (AV_{Cyber}). Wireless communications such as 802.1× ($AV_{802.1\times}$) and cellular broadband (AV_{Cell}) are well-known radio frequency (AV_{RF}) attack vectors. Attack vectors are defined as follows:

AV represents the set of all attack vectors.

$AV_T \subseteq AV$ is the set of attack vectors for a technical system.

$AV_O \subseteq AV$ is the set of attack vectors for an operational environment.

It is important to note that the physical connection between the threat source and the vulnerability may not necessarily be persistent, and must exist only once for the compromise to occur. Attack vectors are categorized into those threat sources with persistent connection ($AV_{Persistent}$) and all others ($AV_{!Persistent}$):

$$AV = AV_{Persistent} \cup AV_{!Persistent}$$

Attack vectors are not static, but are fluid and change with the state and, at times, the technical system. Therefore, the attack vector states are a combination of all possible attack vectors:

$$AV_{States} = AV_{Cyber} \cup AV_{Kinetic} \cup AV_{RF} \cup AV_{SupplyChain} \cup AV_{!Known}$$

Just as with the other aspects, to model the dynamic nature of the system, the sets of attack vectors for a specific system are defined when it is in a particular state. To continue the prior example, the state-sponsored entity of Russia has motivations to exploit a vulnerability within a commercial vehicle in Ukraine using RF communications or supply chain attack vectors, and there is an almost certain probability that Russia will attack with a highly likely probability of success based upon highly certain probability of knowledge:

$$AV_{State\,n} = TR_{Russia,State\,n} \cup V_{UAVinUkraine,State\,n} \cup \left(PA_{AlmostCertain} \times PS_{HighlyLikely} \times PC_{HighlyCertain} \right)$$
$$\cup AV_{RF} \cup AV_{SupplyChain}$$

If $(AV_{T1})_n$ and $(AV_{O1})_n$ represent the set of attack vectors for system s1, in state n

$$\left(AV_{S1} \right)_n = \left(AV_{O1} \right)_n \cup \left(AV_{T1} \right)_n$$

An assessment of system s1 will then consist of an assessment of the system with respect to vulnerabilities, threats, probabilities, and attack vectors in all possible states of the system. If Ev(s1) represents the assessment of the vulnerabilities, threats, probabilities, and attack vectors of the system s1 and i represents the possible k number of states, then

$$Ev\left(s1\right) = \cup_{i=1\ldots k} Ev_V\left(\left(V_{s1}\right)_i \right) \cup Ev_{TR}\left(\left(TR_{s1}\right)_i \right) \cup Ev_P\left(\left(P_{s1}\right)_i \right) \cup Ev_{AV}\left(\left(AV_{s1}\right)_i \right)$$

As shown above, not all threat sources would employ every attack vector. Just as not every attack vector would be available to every threat source.

12.2.7 Impact Models

An impact (I) is the variable result of a threat exercising an attack vector on a vulnerability. Normally, impacts are defined in terms of the magnitude of harm. The NIST

Special Publication 800-30 R1 categorizes harm as damage to operations (I_{OPS}), assets (I_{Assets}), organizations (I_{Org}), and the nation (I_{Nation}). Impacts will be expanded to include loss of human life (I_{Life}), impact to allies (I_{Allies}), and global impact (I_{Global}):

$$I = I_{OPS} \cup I_{Assets} \cup I_{Org} \cup I_{Nation} \cup I_{Life} \cup I_{Allies} \cup I_{Global}$$

To provide clarity, each of the above categories will be further identified to include whether the impact is technical (I_T) or operational (I_O).

I represents the set of all impacts.

$I_T \subseteq I$ is the set of impacts for a technical system.

$I_O \subseteq I$ is the set of impacts for an operational environment.

Impacts are a direct consequence of a vulnerability being exploited through an attack vector with a certain level of probability of attack and success by a threat source with specific capabilities and motivation. Impacts are just as fluid and changing at those aspects on which they depend. The dynamic nature of the system will be modeled by defining the impacts for a specific system when it is in a particular state.

The Russian attacks on the Georgian websites [9] provide an example of actual impacts. Those impacts were of a highly likely threat ($PA_{HighlyLikely}$) of a state sponsored ($TS_{StateSponR}$) (($TC_{LevelOfExpertise}$)$_{VerySophisticated}$) successful attack ($PS_{HighlyLikely}$) via the network ($AV_{Network}$) attack vector to an unsecured server (V_{LU}) motivated to forcibly change (TSM_{Change}) the behavior of the Georgian government:

$$I_{Stater} = \left(TS_{StateSponR} \times TSM_{Change} \times \left(TC_{LevelOfExpertise} \right)_{VerySophisticated} \right)$$
$$\cup \left(PA_{HighlyLikely} \times PS_{HighlyLikely} \times PC_{HighlyCertain} \right) \cup V_{LU} \cup AV_{Network}$$

The preceding example provides a possible state for impacts for that state of a specific system in an operational environment. If $(I_{T1})_n$ and $(I_{O1})_n$ represent the set of impacts for system s1, in state n, then

$$\left(I_{S1} \right)_n = \left(I_{O1} \right)_n \cup \left(I_{T1} \right)_n$$

An assessment of system s1 will then consist of an assessment of the system with respect to threats, probabilities, attack surfaces, and impacts in all possible states of the system. Just as in the prior sections, if Ev(s1) represents the assessment of the threats, probabilities, attack surfaces, and impacts of the system s1 and i represents the possible k number of states, then

$$Ev(s1) = \cup_{i=1...k} Ev_V \left(\left(V_{s1} \right)_i \right) \cup Ev_{TR} \left(\left(TR_{s1} \right)_i \right) \cup Ev_P \left(\left(P_{s1} \right)_i \right) \cup Ev_{AV} \left(\left(AV_{s1} \right)_i \right) \cup Ev_I \left(\left(I_{s1} \right)_i \right)$$

12.2.8 Risk Models

Risk (R) is defined to be the probability of threat source(s) with the capabilities of exercising attack vector(s) to exploit vulnerability for specific motivation(s), the probabilities of success of those attack(s), the certainty of the knowledge, and the resulting impact(s). To provide the greatest clarity, the risk model will be defined in stages.

First, the set of threats, which consists of a threat source, its associated capabilities, and its associated motivations, is represented.

$$TR = \left(\{TS\} \times TC^+ \times TSM^+ \right)^+$$

For every threat, there is a probability that the threat source will attack, which is represented by Threat Attack set (TA):

$$TA = \left(TR^+ \times \{PA\} \right)^+$$

For every vulnerability, there are one or more attack vectors, whose pairing creates the set Attack Vector Vulnerability (AVV), as represented by

$$AVV = \left(AV^+ \times \{V\} \right)^+$$

The probability of a successful attack can only occur if the threat has attack vector(s) to a vulnerability that it can exploit, which is represented by the set Attack Possibilities (AP):

$$AP = \left(TA^+ \times \left(\{PS\} \times \left(AVV \right)^+ \right)^+ \right)^+$$

The information just modeled all depends upon the certainty of knowledge of this information. However, that certainty level may vary depending upon whether the information is in regard to a threat, the probability of attack, the attack vector, or the vulnerability. As such, the above models become

TR = ({TS} × TC⁺ × TSM⁺)⁺ × {PC})⁺

TA = (TR⁺ × ({PA} × {PC}))⁺

AVV = ((AV⁺ × {PC})⁺ × ({V} × {PC}))⁺

Again, risk is the probability of threat source(s) with the capabilities of exercising attack vector(s) to exploit vulnerability for specific motivation(s), the probabilities of success of the attack(s), the certainty of the knowledge, and the resulting impact(s).

$$R = \left(\left(TA^+ \times \left(\{PS\} \times AP^+ \right)^+ \right)^+ \times I^+ \right)^+$$

System level risk is never a single value because there is never just one flaw and countermeasure equating to a single vulnerability, and no one threat or impact to consider. Risk is not static but fluid. As risk is a set of states, risk is refined to be the probability of a threat source with the capability of exercising an attack vector to exploit a vulnerability of a situational instance at an opportunity in time for a specific motivation, the probability of success of that attack, the certainty of the knowledge, and the resulting impact. Representatively,

$$R = \left(TA^+ \times \left(\{PS\} \times AVV^+ \right)^+ \right)_n \times I_n$$

If $(R_{T1})_n$ and $(R_{O1})_n$ represent the technical and operational risk for system s1, in state n, then

$$\left(R_{S1} \right)_n = \left(R_{O1} \right)_n \cup \left(R_{T1} \right)_n$$

An assessment of system s1 will provide the risk for system s1. If R(s1) represents the risk the threats, probabilities, attack surfaces, and impacts of the system s1 and i represents the possible k number of states, then

$$R\left(s1\right) = \cup_{i=1...k} \, Ev_V\left(\left(V_{s1}\right)_i\right) \cup Ev_{TR}\left(\left(TR_{s1}\right)_i\right) \cup Ev_P\left(\left(P_{s1}\right)_i\right) \cup Ev_{AV}\left(\left(AV_{s1}\right)_i\right) \cup Ev_I\left(\left(I_{s1}\right)_i\right)$$

The dynamic interaction between threats exploiting flaws and defensive entities implementing countermeasures for flaws is a constant battle. This fluidity requires the risk not to be a single instance in time decision, but regularly reassessed.

12.3 Assessment Methodology

The reason assessments are conducted on systems is to measure the confidence that the security of the system is implemented correctly and determine the risk that the system poses to the operational environment. It is very important to understand that the Model Methodology (MM) is meant to provide an objective characterization of the system.

The MM provides the mechanisms to map the evidence to mathematical models to represent assessment findings. The use of the models increases objectiveness, explicitness, repeatability, and knowledge of system robustness.

12.3.1 Stages

Within the MM, there are multiple stages with each stage correlating to the progression of the exposure to a system. Key stages include:

- Initial exposure
- System familiarization
- Assessment
- Data correlation

At each stage, the individual models are iterated to represent the impression of the system's capabilities and correlated to the evidence available at that stage. As knowledge of a system increases, the content of these models will go from generalized to specific as the assessment progresses.

The number of stages and the stage at which a model is created will vary wildly based upon the system functionality, and the point in the life cycle in which the system enters the MM, and the information available at that time.

12.3.1.1 Initial Exposure to a Cyber System: The initial exposure to a system identifies the function of the system, its complexity, and possible states. The first exposure to a system provides a very basic overview of the system and its requirements. The models may be formed and possible states considered.

Immediately upon being contacted, there will be initial thoughts about the system that must be represented. The following are some of the possible of flaws to initially consider for any system:

- Operating System $(F_{OS}) \subseteq F_{T1}$
- Internal microphone $(F_{Microphone}) \subseteq F_{O1}$
- RF communications $(F_{RFComms}) \subseteq F_{T1}$
- RF communications $(F_{RFComms}) \subseteq F_{O1}$

Therefore, the initial flaw models would be something similar to

$$F_T = F_{OS} \cup F_{RFComms}$$
$$F_O = F_{RFComms} \cup F_{Microphone}$$
$$F_{s1} = F_T \cup F_O$$

The following are some of the possible mitigations to consider:

- Operating System implements Discretionary Access Control $(M_{OS}) \subseteq M_T$
- Internal microphone not muted and disabled $(M_{MicrophoneNotMutedAndDisabled}) \subseteq M_O$
- RF communications are encrypted $(M_{RFComms}) \subseteq M_O$

Therefore, the initial countermeasure models would be something similar to

$$M_T = M_{OS}$$
$$M_O = M_{RFComms} \cup M_{MicrophoneNotMutedAndDisabled}$$
$$M_{s1} = M_T \cup M_O$$

The models are created and updated as necessary. An initial mapping of the actual system instantiation to the models will be completed. This is done by identifying the actual hardware, software, and firmware as it is associated to the flaw, countermeasures, etc., and then correlating the evidence used to map the system instantiation to the models. The following is a possible mapping of the system to the flaws:

- Operating System is unpatched Windows 10 as documented in section 4.1 of High Level Design Document (HLDD) $(F_{OSisUnpattchedWin10}) \subseteq F_{T1}$
- Internal microphone not muted and disabled as documented in section 4.2 of HLDD $(F_{MicrophoneNotMutedAndDisabled}) \subseteq F_O$
- Wireless not using WPA as documented in section 4.4 of HLDD $(F_{802.11NoWPA}) \subseteq F_T$

Therefore, the flaw models would be updated:

$$F_T = F_{OSisUnpattchedWin10} \cup F_{802.11NoWPA}$$
$$F_O = F_{Microphone}$$
$$F_{s1} = F_T \cup F_O$$

The following are some of the possible mitigations to consider:

- Operating System implements Discretionary Access Control as documented in section 4.1 of HLDD $(M_{OS\_DAC}) \subseteq M_T$
- RF communications encrypted as documented in section 4.4 of HLDD $(M_{RFCommsEncrypted}) \subseteq M_O$

Therefore, the countermeasure models would be updated:

$$M_T = M_{OS\_DAC} \cup M_{RFCommsEncrypted}$$
$$M_O = M_{RFComms}$$
$$M_{s1} = M_T \cup M_O$$

The veracity of identified vulnerabilities must be considered and may look like:

- Operating System is not patched as documented in section 4.1 of HLDD ($V_{OSNotPatched}$) $\subseteq V_T$
- Wireless not using WPA as documented in section 4.4 of HLDD ($V_{802.11NoWPA}$) $\subseteq V_O$

Therefore, the initial vulnerability models would be something similar to

$$V_T = V_{OSNotPatched}$$
$$V_O = V_{802.11}$$
$$V_{s1} = V_T \cup V_O$$

The key is to model the system as evidence is acquired. Everything noted is to be modeled, as those perceptions of the system are what are important. Certain points may jump out, such as a possible attack vector or probability of success of an attack, which though not founded in evidence, should be modeled because as the knowledge of the system is acquired, perceptions change, and these initial models will allow reference back to those ideas when the system was new.

The architectural review is the foundation for the assessment of the system. Many of the aspects of the system will be modeled, aspects of the system mapped to the model, architectural evidence correlated to the models, and initial risk of the assessment documented.

Architectural evidence is specific to a system. This evidence is usually some combination of diagrams, documents, and scans. As assessments may be for any combination of software, firmware, and hardware, the content of the evidence will vary widely. It is important to consider the functionality of the system when identifying possible states.

12.3.1.2 System Familiarization:

System familiarization is the basis for the complete mapping of the system to the models. The complete document review is the brunt of the mapping of the system to the models. If no threat assessment has been conducted, research into possible threats must be conducted. The possible threats to the system within its possible states should be identified.

The mapped models are integrated into the states. As an example, the following are some possible states of a vehicle:

- Vehicle parked, powered off, in owner's garage ($Vehicle_{ParkedOffGarage}$)
- Vehicle parked, powered on, in owner's driveway but unoccupied ($Vehicle_{ParkedOnUnoccupied}$)
- Vehicle in motion ($Vehicle_{InMotion}$)
- Vehicle parked, powered off, at the mall ($Vehicle_{ParkedOffMall}$)

Using the mappings above and others, an example integration of the flaws to the possible states:

$$F_{T, ParkedOffGarage} = F_{RFComms} \cup F_{802.11NoWPA}$$

$$F_{O, ParkedOffGarage} = F_{RFComms} \cup F_{802.11NoWPA}$$

$$F_{ParkedOffGarage} = F_{T, ParkedOffGarage} \cup F_{O, ParkedOffGarage}$$

$$F_{T, InMotion} = F_{RFComms} \cup F_{802.11NoWPA}$$

$$F_{O, InMotion} = F_{TPMS} \cup F_{802.11Interference} \cup F_{802.11NoWPA}$$

$$F_{InMotion} = F_{T, InMotion} \cup F_{O, InMotion}$$

12.3.1.3 Assessment:
The assessment stage is the official and final assessment of system. Tasks that were completed in previous stages should be reviewed as the first task in the final assessment to make sure the entire team is on the same page prior to starting the actual assessment.

Any previous testing conducted must be revisited. Testing conducted in this stage must be correlated to the models. The mapping of evidence to the system will include the models of these states.

At this point, attempts to exploit modeled flaws and vulnerabilities using modeled attack vectors, bypass modeled countermeasures, and exfiltrate data from an insider perspective will be conducted. In simple terms, the vast majority of evidence and the reality of system's behavior will be identified.

12.3.1.4 Data Correlation:
The data correlation stage is the official correlation of evidence to the models. This will be the most complete correlation of system evidence to the models to this point. This translation of evidence to the models is a key point. Evidence is anything that provides verification of a compromise, as well as anything that provides verification of the inability to compromise.

There may be no documented evidence to correlate threat sources, their capabilities, and probably not their motivations. As previously noted, the probability models provide the capability to model subjective perspectives with regard to probability of and success of an attack because there will probably be very little evidence to map to these models. The probability of the certainty of knowledge provides the capability to model the confidence in knowledge, for which there may or may not be evidence to map.

12.4 Conclusions

The individual models are iteratively developed, fulfilling the need to represent an initial impression of the system's capabilities, to represent the system's capabilities as it is assessed, and, finally, to representatively correlate or map the completed models to the empirical evidence of the assessment. The models, and the results they yield, provide an equal level of understanding for those that implement them, as well as those that interpret their results.

The MM provides an objective characterization of the system by providing the mechanisms to map the mathematical models to the evidence from the assessment findings. There are multiple stages within the MM, with each stage representing and refining the impression of the system. As with all assessment methodologies, the MM is not meant to be a checklist, and it is not meant to provide a formula to grade a system. In conclusion, the MM characterizes a system and its situational states using mathematical models and correlates assessment evidence to those models.

References

1. Committee on National Security Systems, "Committee on National Security Systems Instruction 4009: Committee on National Security Systems Glossary," April 26, 2010, Retrieved September 24, 2013 from US Department of Defense, Committee on National Security Systems: https://www.cnss.gov/CNSS/openDoc.cfm?shh5A9gWEjxsV+001OxD3g==.

2. Committee on National Security Systems, "FAQ: Committee on National Security Systems," May 2016, Retrieved May 2, 2017 from Committee on National Security Systems: https://www.cnss.gov/CNSS/about/faq.cfm.

3. Johnson, T., "Mathematical Modeling of Diseases: Susceptible-Infected-Recovered (SIR) Model," 2009, Retrieved July 23, 2013 from University of Minnesota Morris: www.morris.umn.edu/academic/math/Ma4901/.../Teri-Johnson-Final.pdf or https://www.researchgate.net/publication/242272678_Mathematical_Modeling_of_Diseases_Susceptible-Infected-Recovered_SIR_Model.

4. Belik, V., Geisel, T., and Brockmann, D., "Recurrent Host Mobility in Spatial Epidemics: Beyond Reaction-Diffusion," *European Physical Journal B (EPJ B)* 84 2011: 579-587, doi:10.1140/epjb/e2011-20485-2.

5. Stephenson, P.R. and Prueitt, P.S., "Towards a Theory of Cyber Attack Mechanics," N.D., Retrieved August 3, 2013 from OntologyStream: http://www.ontologystream.com/gFT/Towards%20a%20Theory%20of%20Cyber%20Attack%20Mechanics.PDF.

6. Waldrop, M., "DARPA and the Internet Revolution," 50 Years of Bridging the Gap, 78-85, N.D., Retrieved July 26, 2013 from US Defense Advanced Research Projects Agency: www.darpa.mil/WorkArea/DownloadAsset.aspx?id=2554.

7. Oppliger, R., *Internet and Intranet Security*, (Norwood, Artech House, 2001), 12, ISBN:1580531660. Retrieved July 12, 2016 from Artech House computer security series.

8. National Institute of Standards and Technology, Special Publication 800-30 R1: Guide for Conducting Risk Assessments. National Institute of Standards and Technology, September 2012, Retrieved August 11, 2013 from National Institute of Standards and Technology Computer Security Resource Center: http://csrc.nist.gov/publications/nistpubs/800-30-rev1/sp800_30_r1.pdf.

9. Markoff, J., "Georgia Takes a Beating in the Cyberwar With Russia," *The New York Times*, August 12, 2008, Retrieved September 24, 2013 from The New York Times: http://www.nytimes.com/2008/08/13/technology/13cyber.html?_r=0.

About the Author

Jennifer Guild, is a professional in the cybersecurity mission area, providing cybersecurity guidance implementation support, and assessment of cyber requirements throughout the Department of Defense.

Dr. Guild joined the Department of the Navy's (DoN's) Naval Sea Systems Command (NAVSEASYSCOM) Naval Undersea Warfare Center, Division Keyport, in October 2012, focusing her efforts on elevating the security posture of government equities to minimize risk of cyber threat by functioning in the capacity of the Authorizing Official Designated Representative. As a recognized expert with over 15 years of service, Dr. Guild has served in defense and independent organizations leveraging her knowledge and expertise in specialized security and the cyber mission area. Prior to her work at NUWC Keyport, Dr. Guild was the Sr. IA Engineer, SPAWAR Atlantic, where she provided SME level leadership for vulnerability assessments, including cross-domain solutions, and cyber engineering.

Dr. Guild graduated from California Lutheran University with a Bachelor of Science in Computer Science in 1997. In 2004, she earned her Master of Science in Computer Science with emphasis in Information Assurance from the United States Naval Postgraduate School. She earned her Doctor of Philosophy in Computer Science from the University of Idaho in August 2016.

Dr. Guild is a Certified Ethical Hacker (CEH), Certified Information Systems Security Professional (CISSP), Certified Secure Software Lifecycle Professional (CSSLP), and Information Systems Security Engineering Professional (ISSEP).

CHAPTER 13

"...No, We Should Be Prepared"

Joe Saunders and Lisa Silverman

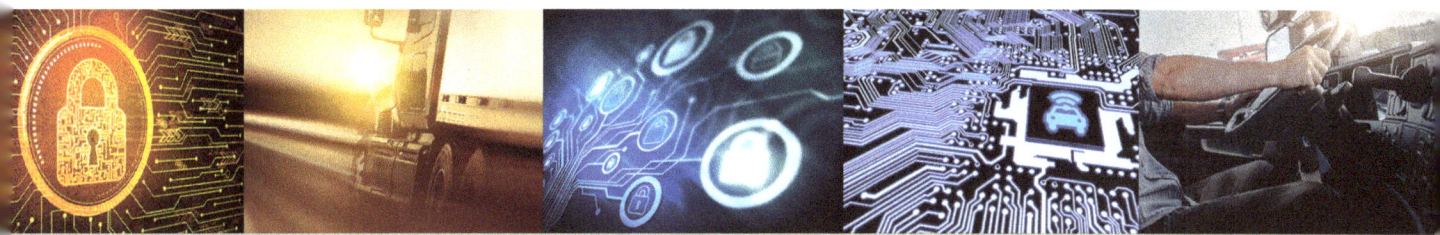

13.1 Introduction

It's a beautiful, clear morning in Columbus. You're on your second cup of office coffee and you've just reviewed the information available through your new telematics system on your truck fleet. Stan, your truck driver, has been on I-95 since 5 am and will deliver ahead of schedule. Bob, another one of your drivers, drove through the night and is napping at a truck stop before heading on to Amarillo. The rest of your 25 drivers are operating their rigs within speed limits. You're feeling pretty good about your decision to install the system before the electronic logging device (ELD) compliance date of December 2017. All of a sudden, you notice that seven of the trucks on the East Coast have slowed down to 10 mph. Next you get a panicked call from Phil in Denver that his horn has a mind of its own and his lights won't stop blinking. He's been pulled over by a state trooper. Then you get an alert that the brakes on Dave's truck in Pittsburgh aren't working, although they were just inspected. What's causing this nightmare scenario?!

In layman's terms, your fleet has been hacked. Someone has gained unauthorized access to your trucks by exploiting vulnerabilities in your vehicles' internal networks. Each vehicle is essentially a computer or series of computers on 16 wheels. Every internet-connected device or electronic control unit (ECU)—installed in the cab or under the hood—expands the target area (attack surface) for hackers. The list is long and includes engine sensors, tire pressure monitors, airbags, alarms, adaptive cruise control, Global Positioning System (GPS), Bluetooth, telematics, and accident avoidance systems that provide warnings and diagnostics. It's not hard to theorize several financial motivations for causing disruption: hijacking high-value cargo, interrupting competitors, and extorting fleet owners. Unfortunately, there is always the threat of terrorism as well.

By way of background, in March 2016, Jose Carlos Norte, a Spanish security researcher used scanning software, found thousands of publicly exposed telematics gateway units (TGUs). These are small radio-enabled devices attached to heavy-duty vehicles' networks to track their gas mileage, location, and other data. Because of the lack of security on these units, Norte was able to look up the location of 3,000 vehicles. Although he didn't go further, Norte could have sent commands over the vehicles' network, the Controller Area Network Bus (CAN Bus), to affect steering, brakes, or transmission.

Last year, security researchers at the University of Michigan actually tested attacks on a 2006 Class-8 semitrailer and a 2001 school bus. They demonstrated the ability to accelerate a truck in motion, disable a driver's ability to accelerate, disable a vehicle's brakes, and change the readout of the instrument panel. The experiments were performed by injecting messages into the on-board diagnostic system (OBDS) via the open standard SAE J1939. The research of Burakova, Millar, and Weimerskirch focused on direct physical access, but it's reasonable to assume that a remote attack will follow soon.

So what can you do to be prepared? In this chapter, we'll explain a bit about vehicle communications and the J1939 standard, the state of the threat, our ideas about how to mitigate the vulnerabilities, and thoughts about the next steps.

13.2 What Makes the Rolling Computers You Call a Fleet Vulnerable?

Almost all heavy-duty vehicles on the road today, including trucks, buses, tractor-trailers, and ambulances, use the SAE J1939 protocol for sending and receiving messages over their CAN. There exists little or very weak authentication on these networks, which means that a network will accept **any** message sent across it. In order to disrupt a car's network, you'd need to know its make and model to tailor the attack. However, with a truck, the openness of the J1939 standard provides easy access for attacks across a wide range and number of vehicles. So basically, any ECU—like a sensor—using J1939 standard can transmit a message on a truck's network. And since the technical details regarding the message format of the standard are published, a hacker can fairly easily get the information needed to attack critical components. This means that once the bad actor discovers how to accelerate a truck remotely, she/he could apply that knowledge to almost every truck in the United States, allowing for a "one-size-fits-all" attack. For a hacker, it is routine to develop an exploit to inject malicious code through software vulnerabilities. This technique is known as Return-Oriented Programming (ROP) and can allow a hacker to take control of a vehicle's operation.

So let's assume the perpetrator can transmit messages on the truck's J1939 bus. It's easiest to do if she/he has physical access to the OBD port and can get the right hardware. With a dongle and just a few minutes of access to a vehicle, a hacker could get continuous control of the bus to create mayhem. In fact, researchers at the University of California at San Diego were able to develop a full CAN network attack via a dongle back in August 2015. They remotely altered the dongle's firmware to send CAN commands to a Corvette—turning on its windshield wipers and disabling its brakes, once again demonstrating the danger of insecure, internet-connected gadgets.

To summarize, ECUs transform a truck into an Internet-connected device. Use of the SAE standard means an attack can spread widely to an entire fleet or more than one fleet.

Since the term telematics was coined by French authors in 1978, computer processors have continued to shrink and are embedded in more and more devices. Telecom and wireless networks have grown exponentially and are able to move data between points regardless of where the host computer is, thanks to the cloud. Whether your trucker is driving through Southeastern Oregon or downtown Chicago, GPS units (which utilize on-board computers) can access either GPRS (cell phone networks) or GPS (satellite networks) to send real-time data to your office. In addition, you can monitor fuel usage, speed, and location of vehicles. Telematics systems support your efforts to route efficiently, making the most of your drivers' time, fuel, and resources. In fact, you are likely aware that the Federal Motor Carrier Safety Administration (FMCSA) has mandated that electronic logging devices (ELDs) be implemented by carriers by December 2017 in order to accurately track, manage, and share records of duty status (RODS) data. This requirement, along with contributions to fleet safety, productivity, and savings improvements, has led to the widespread use of telematics. Shaye Anderson of Trucks.com predicts that the number of telematics units in North American trucks will grow to 8.1 million by 2018, slightly more than a 100% increase in 5 years [1]. So what's not to like?

Well, as wireless networks have become more robust, telematics have become more sophisticated and act as two-way communications channels. Commercial vehicles use wireless communication to continuously transmit and receive data, making them vulnerable to cyber attacks. As internet-connected functionality is added, so are there more ways to attack. It's not that telematics systems are inherently evil—it's just that a hacker only has to exploit a single weak point in the security of a wireless device in order to get access. In addition to the opportunity to control a truck or fleet, access translates to a trove of sensitive data on cargo, location, routes, and delivery times. Having that information then allows business disruption through destruction, corruption, or disclosure of data, as well as the potential for piracy, ransom, or physical harm. Trucks may well be easier to hack remotely than consumer vehicles because of the increased attack surface created by telematics.

13.3 **The State of the Threat**

In order to classify the level of threat posed to fleet managers and to the trucking industry by cyber vulnerabilities, it's useful to consider a risk model. The DREAD algorithm takes into account the average of five categories (from which the model gets its acronym): **D**amage Potential + **R**eproducibility + **E**xploitability + **A**ffected Users + **D**iscoverability. The threat is rated by answering the following questions and assigning values representing severity, with 10 signifying the highest risk and 0 being virtually no risk. Let's take a look at how this works in practice [2].

- *Damage Potential*: If a threat exploit occurs, how much damage will be caused?
  - 0 = Nothing.
  - 5 = Individual user data is compromised or affected.
  - 10 = Complete system or data destruction.
- *Reproducibility*: How easy is it to reproduce the threat exploit?
  - 0 = Very hard or impossible, even for administrators of the application.
  - 5 = One or two steps required, may need to be an authorized user.
  - 10 = Just a web browser and the address bar is sufficient, without authentication.
- *Exploitability*: What is needed to exploit this threat?
  - 0 = Advanced programming and networking knowledge, with custom or advanced attack tools.
  - 5 = Malware exists on the Internet, or an exploit is easily performed, using available attack tools.
  - 10 = Just a web browser.
- *Affected Users*: How many users will be affected?
  - 0 = None
  - 5 = Some users, but not all
  - 10 = All users
- *Discoverability*: How easy is it to discover this threat?
  - 0 = Very hard to impossible; requires source code or administrative access.
  - 5 = Can figure it out by guessing or by monitoring network traces.
  - 10 = The information is visible in the web browser address bar or in a form.

Total scores (after adding all components and dividing by 5) of 3 or below are considered low threat, scores of 4-7 are moderate threat, and scores of 8 and above are high threat.

To complete this subjective analysis, we can rate our cybersecurity threat using DREAD—assuming we are considering your fleet only.

- Damage Potential = 8
- Reproducibility = 8

- Exploitability = 8
- Affected Users = 8
- Discoverability = 10

Dividing the total of 42 by 5 yields 8.4, a high threat-but you already suspected this.

Now let's transition the consideration from the micro, a single fleet, to the macro, the trucking industry. During 2015, trucks moved $13.2 billion worth of freight, or 68% of the total domestically, providing ample credence to the saying "if you bought it, a truck brought it." And for you futurists, that value of freight amount is projected to grow to $24.4 billion in 2045 [3]. It is patently clear that trucking presents a fat target for hackers with objectives as diverse as seizing high-value cargo for monetary gain to completely disrupting commerce as a form of terrorism.

Since November 2014, cyber attacks have targeted companies in the entertainment (Sony), online (Dyn), gaming (Las Vegas Sands), restaurant (Chick-fil-A), retail (Staples), healthcare (Anthem), banking (Citibank), finance (Morgan Stanley), transportation (Uber), news and business (Forbes), academia (Penn State University), hotel (Trump Hotel Collection), communication (WhatsApp), infrastructure (Bowman Dam), technology (MacKeeper), government (U.S. Department of Homeland Security), network management (Verizon Enterprise Solutions), political organizations (Democratic National Committee), and payment (Oracle MICROS) industries [4, 5]. It's not a stretch to believe that trucking is in the crosshairs of sophisticated hackers.

A whitepaper published by the American Trucking Associations [6] demonstrates that the DREAD score for the equivalent of a "denial of service" attack (e.g., a complete halt to all truck traffic) would easily be a perfect 10. During the first 24 h following a truck stoppage:

- Delivery of medical supplies to the affected area will cease, causing hospitals to run out of materials such as syringes and catheters; radiopharmaceuticals will deteriorate and become unusable.
- Service stations will run out of fuel.
- Manufacturers using just-in-time processes will develop component shortages.
- U.S. mail and other package deliver will cease.

During the next 3 days:
- Food shortages will escalate in the face of hoarding and consumer panic.
- Automobile fuel availability will dwindle, leading to sky-rocketing prices and long lines at gas pumps
- Assembly lines will shut down, forcing plant closures.
- ATMs will run out of cash and banks will be unable to process transactions.
- Garbage will start piling up in urban and suburban areas.

You get the picture. The impact on communities everywhere and the overall economy would be staggering. While the scenario of a full stop to all trucking and its consequences represents an extreme example, keep in mind it was difficult to imagine the horror of 9/11 before that happened.

13.4 Recommendations to Prepare Fleet Managers

So, with the potential high threats lurking, how can you protect your fleet, reduce attack vectors, and prevent attacks from doing damage to your fleet? The quick answer is that cyber hardening a fleet requires layers of security and continuous management. Whether one develops an internal discipline to manage the cyber risks of a trucking fleet or works with a software vendor or solution provider, there are several measures one can take.

The primary steps one can take immediately include:

- Protect telematics platform.
- Monitor for malicious messages.
- Ask trucking manufacturers to protect ECUs.
- Ask trucking manufacturers to install intrusion detection system.
- Share exploits with the industry.
- Periodically conduct penetration tests against known methods of attack.

Each recommended action is summarized below.

13.4.1 Protecting Telematics Platform

A major attack vector of a connected trucking fleet is its telematics platform. Whether recently having upgraded the telematics platform for compliance reasons (fuel-efficiency or ELD-related initiatives) or to gain further efficiency in fleet management, the connected communication of a telematics system exposes trucks to remote attacks.

For a motivated hacker, gaining access to the fleet through the telematics system is straightforward. One easy-to-employ solution is to implement a Runtime Application Self-Protection (RASP). See Appendix A—Runtime Application Self-Protection Examples—for two examples that work specifically for trucking industry.

There are many approaches, but the ideal solution should starve malware or exploits the deterministic uniformity it needs. One method, called Binary Randomization, is to make each truck's telematics system functionally identical (so it does all the same things from truck to truck), but logically randomized. Through Binary Randomization, exploits developed are rendered inert because the binaries, memory locations, and access to DLLs are not accessible as expected when the exploit was first written after reverse engineering the telematics platform.

Another technique is analogous preventing a hacker from taking *The Hunt for Red October* and resorting the letters and words to create Shakespeare's *Hamlet*. This approach, called Control Flow Integrity, means hackers cannot use functions within the written code in ways that differ from the original sequence or order of events. In other words, hackers can't use legitimate functions within the software code to create new outcomes the original software writer never intended.

13.4.2 Monitor for Malicious "J1939" Messages

Another approach is to monitor the messages on the internal messaging trucking platforms that allow separate systems from communicating with each other. The CAN Bus is the messaging platform, and trucks have adopted multiple components of the SAE J1939 Standard to facilitate this interoperable form of communications. For a summary of the J1939, see Appendix B—J1939 Overview.

One approach is to embed monitoring software on each CAN Bus and thwart malicious activity in real time. Whether implemented as a dongle inserted into the diagnostics port or installed in the CAN Bus itself, these solutions tend to be highly effective in stopping an attack in its place for known malicious messages that can cause harm to the fleet or truck. See Appendix C—Preventing Malicious Messages on the CAN Bus. These systems can also employ machine learning algorithms and develop known combinations of messages that when used together could be catastrophic. This approach should be done with approval from manufactures in an aftermarket way or via implementation by the manufacturer.

13.4.3 Install Intrusion Detection System Across the Fleet

Similarly, Intrusion Detection Systems (IDS) can be implemented as gateways monitoring traffic between ECUs. These systems can offer tremendous reporting if configured to deliver findings to a telematics platform and/or sent to the command center for real-time data review. With that said, as has been proven over the years, 1:1 IDS solutions are often beat by hackers and necessitate having additional layers used in conjunction with them.

13.4.4 Protect Software on ECUs

Assuming hackers can thwart any detection or prevention system, a fleet should also consider adding other layers to its security posture. Accordingly, working with manufacturers to identify ECUs that should be further cyber hardened should be a part of any fleets consideration.

Employing similar RASP techniques—such as binary randomization or control flow integrity (again see Appendix A)—against malware and exploits that expect uniformity from truck to truck will reduce attack surfaces. By employing these solutions, fleets can ensure malware won't propagate across the truck's platforms, sensitive data won't be stolen, and driver safety is ensured.

Ultimately, as hackers demonstrate their ability to get into a truck, they will be discouraged once thwarted if their exploits are difficult to execute.

13.4.5 Share Exploits with the Industry

The U.S. Intelligence Community cooperates via a concept often referred to as the "five eyes." Close allies share intelligence because each single entity cannot detect all threats around the world. In a similar way, sharing cyber exploits within the industry means all fleets can benefit from shared knowledge of what bad actors are doing. Trucking fleet managers should also consider the broader automotive and transportation industry for sources of new developments, the latest exploits, and the latest trends in hacker techniques.

13.4.6 Periodically Conduct Penetration Tests

Fleet managers should also consider penetration testing ("Pen Test"). Pen Test is an authorized simulated attack on a computer system(s) to identify and enumerate vulnerabilities. The banking industry as well as other components of critical infrastructure regularly conducts Pen Tests as a periodic review of one's cybersecurity posture.

For truck fleets, knowing whether the telematics system is an exposed attack surface or whether one can maliciously send CAN Bus messages or install malware on an ECU is critical to understanding the threat landscape one faces. Often, Pen Tests are done in conjunction with a demonstration of protection as well.

For example, as part of an evaluation process of a new security solution, one could develop exploits used during a Pen Test and then—after protection is offered—demonstrate that the security solution blocks the attacks against it.

Using this approach, one can demonstrate progress towards protection through periodic Pen Tests over time as well as can use a Pen Test to test new threats or thwart known vulnerabilities.

13.5 Future Considerations to Advance Preparation Levels

With these initial recommendations, one can see that protecting a fleet from a cyber attack is not a one-time event. As platforms change and the threat landscape evolves, fleet managers should consider ongoing improvement to the cybersecurity posture.

Accordingly, for truck fleets, especially ones with high-value cargo or competitive markets, engaging a broader community may make sense. Here are three additional steps the proactive fleet should take going forward:

A. Future Truck Designs

Work with manufacturers on your cybersecurity requirements. Conduct tests in their labs as you consider new purchases and want to engage them on your cybersecurity needs. Manufacturers and their suppliers can be extremely helpful in the process as they compete for a fleet's continued business.

B. Engage Standards Bodies

Like J1939 maintained by SAE, standards offer interoperability and collective improvement. Whether a mandated standard, a technical de facto standard, or an industry-driven standard, expressing your requirements will enhance the protection of your fleet and its improvement.

C. Involve Government Agencies

Lastly, there is no doubt that various government agencies have an interest in helping industry maintain safety and security. Engage them as well.

References

1. https://www.trucks.com/2016/05/17/long-haul-trucking-connectivity-brings-hacking-risks/.

2. https://www.owasp.org/index.php/Threat_Risk_Modeling.

3. https://www.transportation.gov/briefing-room/dot-releases-30-year-freight-projections.

4. http://www.heritage.org/cybersecurity/report/cyber-attacks-us-companies-november-2014.

5. http://www.heritage.org/defense/report/cyber-attacks-us-companies-2016.

6. http://www.trucking.org/ATA%20Docs/What%20We%20Do/Image%20and%20 Outreach%20Programs/When%20Trucks%20Stop%20America%20Stops.pdf.

7. Wagner, D., Foster, J., Brewer, E., and Aiken, A., "A First Step towards Automated Detection of Buffer Overrun Vulnerabilities."

8. cwe.mitre.org/top25/.

9. http://www.sae.org/misc/pdfs/J1939.pdf.

10. http://www.simmasoftware.com/j1939-presentation.pdf.

13A.1 Appendix A: Runtime Application Self-Protection Examples

One of the oldest and most common vulnerabilities in software is buffer overflows:

- Wagner et al. estimated over 50% of all vulnerabilities to be due to buffer overflow [7].

- Buffer overflow is number 3 in the 2011 CWE/SANS "Top 25 most dangerous programming errors" [8].

A buffer overflow occurs when a program or process tries to store more data in a buffer (temporary data storage area) than it was intended to hold. In buffer overflow attacks, the extra data may contain codes designed to trigger specific actions, in effect sending new instructions to the attacked computer that could, for example, damage the user's files, change data, or disclose confidential information.

There are two techniques that help prevent such return-oriented programming errors (ROP):

   A. **Basic block randomization**—"every app binary is unique"

FIGURE 13A.1

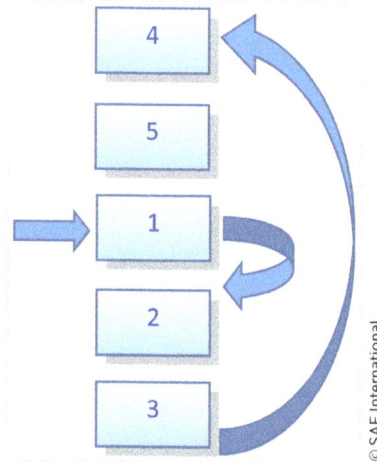

© SAE International

Randomization is a basic RASP technique to "inoculate" an app binary. Buffer overflows and memory corruption attacks are *effectively* rendered inert since the attacker would have to rebuild the attack for each and every binary, which would not be practical.

In each binary, blocks are reordered but the functionality is *identical*, with block execution from 1 to 5 preserved, as illustrated.

With binary randomization, every executable is *effectively* unique.

B. **Control Flow Integrity**—"guardrails on app execution"

FIGURE 13A.2

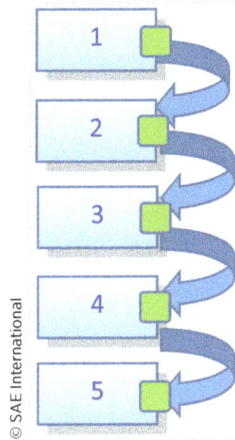

© SAE International

CFI is another RASP technique, putting execution "curbs" or "guardrails" around "jumps" and "returns".

As illustrated, the order of execution and functionality are preserved.

When an exploit is attempted, it is rendered inert.

13B.1 Appendix B: J1939 Overview

J1939 is a standard maintained by the Society of Automotive Engineers that defines how information is transferred across a network to allow ECUs (computers) to communicate within trucks and off-highway, agricultural, construction, and first responder equipment, as well as other vehicles. J1939 is defined using a collection of SAE documents (the top level document and 16 companion documents) based upon the layers of the Open System Interconnection (OSI) model for computer communication architecture [9]. J1939 can be thought of as a software specification that rides on top of the CAN [10]. And the CAN is a form of serial communication that transmits data with a binary model. The CAN uses four different types of frames for communicating information and the state of the bus:

- Data frame (to send data)
- Request frame (to request data)
- Error frame (to indicate an error)
- Overload frame (to insert a delay)

A CAN data frame uses the standard architecture of a header followed by data. The header is the identifier and will indicate whether the string is 11 or 29 bits.

13C.1 Appendix C: Preventing Malicious Messages on the CAN Bus

In July 2015, Jeep recalled 1.4 million vehicles because they were vulnerable to remote security hacking. White hat hackers used the design features of in-vehicle communications networks to interfere with safe and normal operations. Fleet managers, in government, commercial trucking, law enforcement, rental agencies, and taxi companies, with thousands of vehicles under management, also face stolen data, higher insurance costs, liability exposure, and driver safety issues.

Bad actors can take control and issue messages through the standard **CAN Bus**. The CAN Bus is designed to allow microcontrollers and devices to communicate without a host computer. The initial CAN Bus protocol, designed by BOSCH GmBH, was adopted by the automotive industry in 1986.

Vehicles today have as many as 30-70 software modules, and some have up to 100 microprocessors across **ECUs** that operate various subsystems, such as brakes, airbags, transmissions, power windows, mirror adjustments, and many other automotive components. With over 8.8 million vehicles under fleet management, there is an immediate need to lock down these vehicles to reduce risk exposure.

13C.1.1 The Problem

In a normal CAN Bus operation, an authentic signal, such as a signal from a Remote Control Door Lock Receiver module, is transmitted via the CAN Bus to all attached modules in the form of a combination of dominant and recessive bits (**Figure 13C.1**). The intended recipient of the signal, for example, the Body Control Module (**BCM**), executes the legitimate signal to unlock the vehicle's doors. The remaining modules disregard the signal. The effects of these signals can range from basic electronic actions to diagnostic messages with the potential to disable crucial vehicle systems, including the capability to apply the brakes.

13C.1.2 The Entry Point

If a remote attacker wanted to take control of vehicle operations for malicious reasons, the bad actor could exert control over the functionality of a vehicle by accessing the Vehicle

---

FIGURE 13C.1   CAN Bus example.

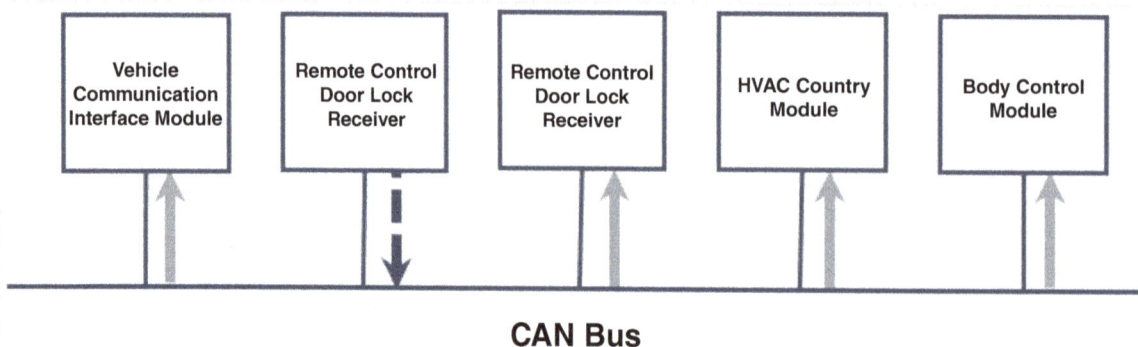

Vehicle Communication Interface Module — Remote Control Door Lock Receiver — Remote Control Door Lock Receiver — HVAC Country Module — Body Control Module

CAN Bus

© SAE International

Communication Interface Module (**VCIM**). VCIM provides the vehicle with a satellite connection used, among other functions, to allow for Global Positioning System (**GPS**) navigation or related services. By taking advantage of this telematics unit, an attacker may gain access to the internal CAN Bus system and prompt the VCIM to transmit unauthorized attacking messages.

Because the CAN Bus is designed as a closed system, signals transmitted by the attached modules are implicitly trusted and considered authentic. If an attacker is able to access one such module, he has the capacity to freely transmit any electronic instructions executable by the other modules across the entirety of the CAN Bus without hindrance or verification processes. As such, a majority of these types of attacks are in the form of diagnostic messages transmitted by the VCIM to the BCM, instructing the BCM to execute an action, such as activating windshield wipers, turning on the vehicle horn, or preventing application of the brakes.

13C.1.3 The Solution

On cars, by inserting a dongle that plugs into the on-board diagnostic (**OBD**)-**II** port, a driver or fleet manager can instantly thwart remote attacks (note, in alternate designs, this approach can be embedded in each CAN Bus). The solution prevents certain diagnostic commands from being transmitted across the CAN Bus. In can be analogized to a stateless firewall filter with the ability to block or allow individual packets (**Figure 13C.2**). Unlike other current gateway designs that require extensive integration or past traditional firewalls that are placed between two networks, this solution blocks packets internally, e.g., on a single CAN Bus, rather than acting in between networks.

Because of the CAN Bus architecture, the solution is capable of providing filtering for all of the devices connected to the bus without being placed between the misused module, such as the VCIM, and the targeted module, such as the BCM. This system feature gives the approach a one-to-many property, that is, the solution can protect all modules on the CAN Bus. This differs from conventional practices that are one-to-one solutions.

When the solution detects an illegitimate signal, it intentionally injects an error message of six dominant bits. All normally functioning modules on the CAN Bus will interpret this series of dominant bits as an error state causing all modules on the bus to drop the transmission (**Figure 13C.3**). In order to inject the errors, the dongle must analyze the transmitted packet and make a determination of its legitimacy at line speed.

FIGURE 13C.2 Remote attack access.

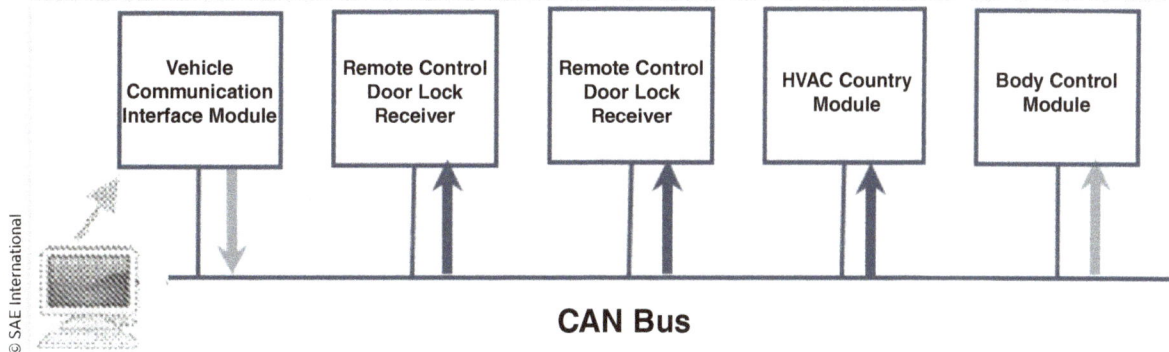

CAN Bus

FIGURE 13C.3   One-to-many protection via CAN Bus.

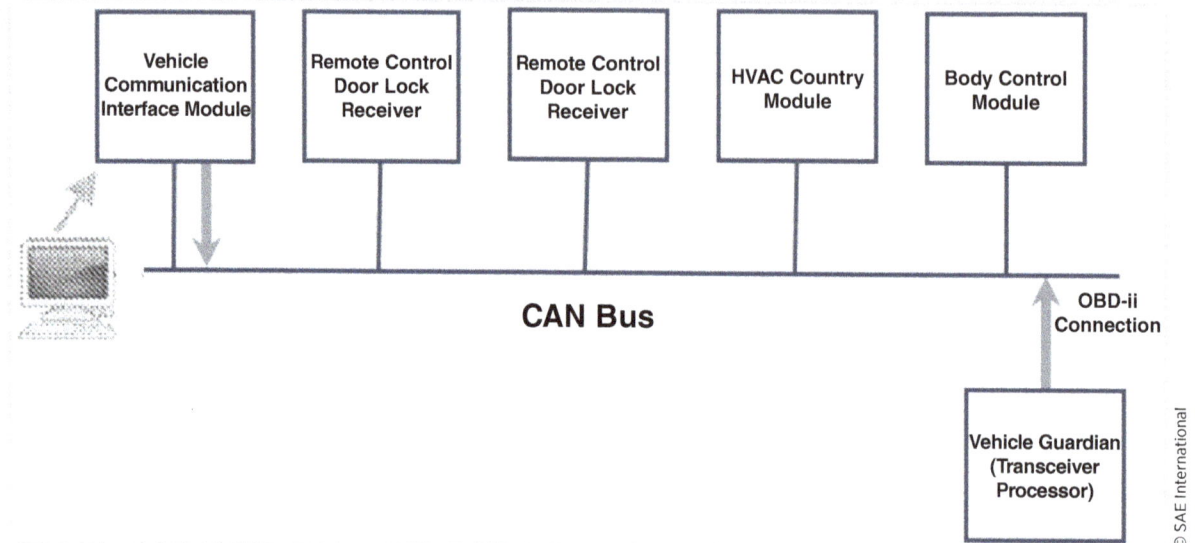

The solution may determine a message's legitimacy in several ways. First, if a diagnostic message originates from somewhere other than the diagnostic port on the OBD-II, the message is considered an illegitimate message. The dongle may be programmed with additional messages that have been learned to cause damage or interfere with safe vehicle driving operations. The solution may also be updated wirelessly or wired. During such an update, the solution would learn of new potentially dangerous messages that need to be filtered.

About the Authors

Joe Saunders is the CEO of RunSafe Security, the pioneer in cyber hardening embedded systems across automotive, military and national security and critical infrastructure sectors. Prior to founding RunSafe Security, Joe served as a director of Thomson Reuters Special Services where he led analytical solutions to help law enforcement agencies identify national security threats, including programs to target the theft of intellectual property. Joe previously was a member of the management team at TARGUSinfo, which was acquired by Neustar for $750M in 2011. Joe started his career as a consultant at PricewaterhouseCoopers, building quantitative models for financial institutions. Joe has a BS degree in mathematics from the University of Michigan, an MS in predictive analytics from Northwestern University, and an MBA from George Mason University. He is the founder of Children's Voice International, a 501(c) 3 nonprofit that restores the rights of abandoned, neglected, trafficked, and exiled children worldwide.

Lisa Silverman, VP of Marketing at RunSafe Security, is an accomplished executive with broad experience in utilizing new technology to expand product portfolios. Lisa's background includes product management, marketing, and strategic planning for both large enterprises and small businesses. Prior to RunSafe Security, Lisa served as director of product management for workplace safety at American Red Cross, where she introduced simulation learning for First Aid and CPR. Lisa drove the strategy, definition, and execution of new offerings as vice president of market development at Localeze, the local search division of TARGUSinfo. As director of product management for Total Care at AOL, she led all business initiatives surrounding the launch of a PC security offering. Previously, she directed new product management and partner marketing at Network Solutions and managed conferencing marketing MCI WorldCom. Lisa holds a BA in economics and mathematics from Brandeis University and an MS in management from MIT Sloan.

CHAPTER 14

Heavy Vehicle Cyber Security Bulletin

Issue Date: September 9, 2016

In 2013 there were approximately 10. 6 million heavy vehicles registered in the US. It is estimated that class 8 trucks, which constitute the heaviest of these vehicles, have a service life of 7-8 years, with approximately 150,000 new class 8 trucks added to the fleet each year.

Heavy vehicles - while having some obvious material differences - are substantially similar in network architecture to light vehicles. Therefore, there is no reason to believe that heavy duty vehicles are less vulnerable to cyber security threats than the average automobile. The difficult part of hacking vehicles is gaining access, ideally remote access. Indeed, while passenger vehicles are just now becoming "connected" through telematics systems such as OnStar, SYNC, Uconnect, etc., heavy vehicles have been more pervasively "connected" through satellite and cellular communications linking to telematics, fleet management, and engine management applications, for quite some time. Consequently, heavy vehicles currently have more avenues for remote access than light vehicles.

With hundreds and sometimes thousands of virtually identically configured vehicles, commercial truck fleets have a high level of electronic homogeneity that can enable an adversary to economically develop viable exploits that could attack large numbers of vehicles simultaneously.

Therefore, we are encouraging motor carriers to maintain awareness of potential issues and threats that can impact the safety and security of their vehicle fleets.

How can motor carriers help minimize heavy vehicle cyber security risks? There are several actions that can be taken to potentially mitigate the risks associated with the technological features of heavy vehicles.

Develop a CyberSecurity Program

Create an internal program to address cyber security with high-level organizational support, which includes people and policies, to assess and respond to cybersecurity issues. The nature and composition of the program and team will vary by the type and size of organization. A good starting point is the National Institute of Standards and Technology (NIST) Cybersecurity Framework (https:ljwww.nist.gov/cyberframework). Center for Internet Security (CIS) also offers a good starting point in their controls for effective cyber defense (https://www.cisecurity.org/critica1-controls.cfm).

Protect Your Networks

An easily targeted access point is office networks and those computers that are used to communicate with the vehicles. Attacks can include malicious websites, email attachments as well as access by a rogue contractor or disgruntled employee. Companies should employ basic network and computer security protocols.

Separate Networks	Segregate networks where computers have remote access to vehicle systems from those utilized for routine business functions (email, browsing the internet, working on office documents etc.).
Network Security	Protect your networks that communicate with vehicles with well configured firewalls, intrusion detection/prevention systems (IDS/IPS), as well as vulnerability management tools to help ensure your environment has the latest patches and is configured properly.
Lock Down Internet Access	Restrict internet access on all systems and computers that communicate with vehicles and consider removing internet browsers, PDF readers, and email clients etc. If outbound internet access is required, make sure to restrict internet access to a known set of safe destinations.
Change Default Passwords	Change the default passwords for all network and connected equipment from vendor supplied defaults.
Two Factor Authentication	Ensure a II systems that give remote access to vehicle communication and features are accessible only via two factor authentication, which prevents password sharing, phishing, and brute force password attacks.
Disaster Plans and Backups	Establish disaster recovery plans with backup processes and procedures which include offsite and "offline" backups, i.e. "air gapped." In the event of a incident such as fire, ransomeware, or malware the impacted system(s) could be restored.

Protect Your Vehicles

While short term solutions to vehicle computer designs are limited, there are a number of steps to take to reduce the risks.

Vendor Communication	Establish communication and notification avenues with manufacturers and third party product/service integrators to ensure that you are notified of any critical security issues or updates to your equipment and service.
Established Maintenance Plans	Establish documented maintenance plans for the vehicles which include requirements to ensure that the latest firmware and software patches/upgrades are applied to the vehicles systems within 30 days of release.
Customizations and Add-ons	Make sure that any modifications or additions to your vehicle such as third party tracking and telematics systems do not compromise the security of your vehicle or bridge networks which have been separated by the OEM.
Reduce Attack Surface	Disable and remove unused features that are not critical to the use and functionality of the vehicle, especially those that enable remote access.
Update Vehicle Pre-Trip Inspections	Add cybersecurity inspection points to the Pre-Trip inspection part for a driver to attain a CDL license. Drivers need to be aware of cyberthreats and looking for foreign devices mounted to accessible parts of the vehicle that can connect to the CAN bus is an easy way to prevent onboard access.

How can motor carriers prepare for a cyber security attack?

Given the increasing odds of a security breach, it is necessary to develop a plan to ensure you know how to recover and survive a breach or attack. A standard part of system security is an incident response plan. This plan outlines the process and procedures to follow in the event of an incident. It is highly recommended that all motor carriers immediately start working with heavy vehicle manufacturers and telematics providers, and associated third parties to develop a plan on how to recover.

Incident Response Plan

The following generic steps have been identified for responding to a major incident. Many steps can occur in parallel depending on the nature of the situation, e.g. multiple attack vectors or vulnerabilities, carriers, etc.

FIGURE 14.1

PREPARATION › IDENTIFICATION › ASSESSMENT › CONTAINMENT › ERADICATION › RECOVERY › FOLLOW-UP

© SAE International

Preparation	Create team
	Establish communication plan and crisis management structure
	Conduct exercises
Identification	Identify if attack has occurred or is ongoing
	Identify the impacted assets
Assessment	Assess the scope, impact and risk of the incident
	Investigate the cause and establish first course of action
	Collect forensics and critical data for next steps
	Create profile of affected units
Containment	Minimize and isolate the damage or risk
	Use profile to strategically contain affected units
	Implement contingency plans to maintain continuity of business
	Determine the root cause
Eradication	Conduct analysis on forensics data collected and assets
	Restore/ rebuild systems affected
Recovery	Implement irrevocable corrective actions
	Restore normal operations
Follow-up	Lessons learned collected and incident response plan is updated
	Identify other units with similar vulnerability and create remediation plans

Educate

Raise awareness within your company and the industry. Educate a II the different stakeholders as to the issues and potential impact regarding heavy vehicle cyber security. Additional information and resources are available through National Motor Freight Traffic Association, Inc. (customerservice@nmfta.org) and the American Trucking Association's (ATA) Technology & Maintenance Council (TMC), (tmc@trucking.org).

How can motor carriers help?

Incorporate Security in OEMs/Vendor Selection	Start including security evaluations as part of the product selection criteria. Ask questions to sellers. Does this product undergo adversarial security testing? Does it come with a field kit to or offer any other assistance in recovery in the event of a incident? If not, why not?
Get Involved	Participate in industry meetings and conferences on cybersecurity for transportation. Carrier participation is critical to the hardening of our nation's transportation infrastructure.

How do I report a cybersecurity incident?

The Internet Crime Complaint Center (IC3) is a reporting mechanism to submit information to the Federal Bureau of Investigation (FBI) concerning suspected

Internet-facilitated criminal activity and to develop effective alliances with law enforcement and industry partners. The information reported is analyzed and disseminated for investigative and intelligence purposes to law enforcement and for public awareness. The FBI's IC3 offers annual reports and multiple self-help documents for effective cyber defense. https://www.ic3.gov

Credits and Acknowledgements

This document was developed as a collaborative effort by NMFTA staff and a large number of other companies, associations, universities, and federal agencies. While we are not able individually recognize everyone involved, we are extremely appreciative of their assist a nee and continued involvement. Thank you for your support of the transportation community.

Disclaimers

The information contained in this document is subject to change without notice. The information contained in this document is presented in good faith, and is believed to be correct, but correctness and completeness is subject to the limitations of an expedited research and writing cycle. The information contained in this document is for information purposes only. NMFT A disclaims a II warranties, express or implied.

Trademarks

ClasslT, NM FC, SCAC, and NationaI Motor Freight Classification are a registered trademarks of the National Motor Freight Traffic Association, Inc. The names of actual companies and products mentioned herein may be the trademarks of their respective owners.

Law, Policy, Cybersecurity, and Data Privacy Issues

Simon Hartley

It's a small thermal exhaust port—
A precise hit will start a chain reaction which should destroy the station—Star Wars [1]

Executive Summary

This chapter reviews the existing law, policy, cybersecurity, and data privacy considerations around conventional, connected (CV), driver-assisted (ADAS), and fully automated vehicles (AV) [2]. It contends that—

(1) The risk of vehicle cyberattack is not just a problem for future model years, but a real risk to the public, consumer, and commercial fleet vehicles on the road today

(2) Potential harms are much greater than those of historical data breaches around smartphones, personal computers (PCs), and the cloud, ranging from driver distraction to Distributed Denial of Service (DDoS) and ransomware to property damage and bodily injury to death and debilitation of critical transport infrastructure [3].

(3) Mitigating cyber risk should be addressed by mandatory regulation, as far as is possible under the present administration, setting a common standard for engineering, legal, and cyberinsurance purposes across the industry, rather than relying on a patchwork of voluntary standards that are inconsistently applied to vehicles assembled just in time (JIT) from long supply chains.

Publication Note

This chapter is based on publicly available information up until April 8, 2017. The date is significant since changes occur daily in this fast-moving area.

15.1 Introduction

15.1.1 Physical Safety

Ralph Nader published the book "Unsafe at Any Speed: The Designed-In Dangers of the American Automobile" [4] just over a half-century ago in 1965. It accused the vehicle original equipment manufacturers (OEMs) of the time of focusing on new vehicle features and styling above safety for fear of alienating potential customers and losing competitive advantage. The media and political storm that it unleashed led to the introduction of *physical* vehicle safety systems including airbags, antilock brakes, and seatbelts, as well as the formation of the Department of Transportation (DoT)'s National Highway Traffic Safety Administration (NHTSA) to regulate the industry.

NHTSA's statutory mission is to "save lives, prevent injuries, and reduce economic costs due to road traffic crashes, through education, research, safety standards, and enforcement activity" [5]. A separate agency was created a generation later to focus on large trucks and buses, the Federal Motor Carrier Safety Administration (FMCSA) [6]. It follows a similar safety-related mission, balanced with efficiency around transportation of goods and public transport.

15.1.2 Accident Statistics and Human Error

In 2015, there were over a quarter billion vehicles (cars, trucks, and commercial vehicles) on U.S. roads [7]. NHTSA still reports over 35,000 U.S. fatalities [8] per year due to accidents, at a cost of $784 for every person in the United States and an economic impact of over $240 billion per year [9]. The most striking feature of today's accidents is that 94% are due to *human error* rather than vehicle failures, or other factors [10]. An old industry joke underlines the role of human error in accidents—"the most dangerous part of a vehicle is the nut behind the wheel!" [11]

That traffic deaths in the United States rose over 7% in 2015 compared to the previous year is even more surprising giving decades of *physical* safety improvements [12]. It is the largest single year increase in a half-century and expected to rise again once 2016 numbers are tallied [13]. The explanation is twofold, an increase in miles driven but also a rise in *distracted driving*, where drivers divert attention from driving to focus on some other activity. Distracted driving is becoming much more common as vehicles become more connected and automated, so in part software is contributing to the problem.

15.1.3 Vehicle Hardware Improvements

In the two centuries since the introduction of the first motor vehicles [14], there have been tremendous improvements in *hardware* from engines to transmissions, emissions, brakes, and suspension, leading to increased safety, performance, and efficiency. Improvements are no longer confined to top of the range models but feature across product ranges and types of vehicles. For example, Road & Track magazine recently reviewed [15] two 2016 sports cars, the $30,000 Mazda Miata and the $300,000 Ferrari 488 Spider, concluding that "virtually everything you could say about one of the cars can be said about the other." The differences were the price and Ferrari's brand.

The U.S. automotive sector has grown to represent 3% of GDP [16], with over 90% of households owning one or more vehicles [17] and over 70% of freight being moved by truck [18]. Given the sector's importance to our daily lives, transportation is considered a critical national infrastructure under Presidential Policy Directive 21 (PPD-21 [19]), much like the energy or telecommunications sectors.

15.1.4 Vehicles Become Data Centers on Wheels

Under the hood, software has been replacing hardware since the 1980s. Today, the carburetors and chokes of classic cars have long been replaced by electronic control units (ECUs), microprocessors that manage engines and emissions. The electrical and electronic share *by value* of vehicles is often over 40% [20]. By 2009, cars had become data centers on wheel, with the Institute of Electrical and Electronic Engineers (IEEE) reporting [21] that cars contained up to 100 million lines of code and up to 100 ECUs each.

Throttles and brakes are no longer tied to mechanical linkages but have been replaced by drive-by-wire controls just like modern airplanes, with software communicating over digital networks, the controller area network (CAN Bus) [22] in cars or the SAE J1939 [23] network in trucks. The existence of ECUs or vehicle architecture in general tends to reach the public's attention only where there are alleged issues, for example, in Volkswagen's emissions testing [24] or around the perceived safety of accelerator [25] and brake [26] pedals.

Underlining the importance of fuel efficiency and emissions regulations to cars, it was the Environmental Protection Agency (EPA) that mandated that all cars built since 1996 have to include the on-board diagnostic (OBD)-II port [27] for reporting purposes. A car's OBD-II port or truck's J1939 port can be leveraged for data capture and ECU updating by OEMs, captive repair shops, independent repair shops [28], or any interested third parties.

15.1.5 Rise of Connectivity, Automation, and Public Concerns

The most *publicly visible* improvements in vehicles have been *software related*, layering in new vehicle features such as Internet connectivity, infotainment systems, and advanced driver assistance systems (ADAS) [29]. ADAS are systems that help the driver in the driving process, for example, in blind spot monitoring, lane-keeping assist, or adaptive cruise control. ADAS already allow snow plows to keep passes and airports open even in the most challenging weather [30]. Where the vehicle is connected to the Internet, the door is potentially open for remote control.

Scotiabank estimated that 10% of cars shipped in 2014 included connectivity features, and Giesecke & Devrient, a German automotive security company, estimates that 75% of cars shipped in 2020 will include mobile connectivity [31]. Car and Driver magazine set this context for ADAS in today's vehicles—"commercials may have convinced you that every vehicle on the road is a computerized guardian angel, and there's no question that safety technology is advancing" but "just 8% offer *all* of the ADAS features possible" [32].

As the importance of software in vehicles has grown and vehicles become more and more like consumer devices, the Consumer Electronics Show (CES) in Las Vegas has begun taking over from the Detroit's North American International Auto Show for key industry announcements [33]. The year 2017's CES highlighted the prospect of ADAS extending *to fully automated or driverless vehicles* [34] and also the public's fears around vulnerabilities, with 59% of buyers *actively* concerned about the prospect of car hacking in one survey [35], and in another survey [36] 33% were *extremely* concerned that self-driving vehicles could be hacked to cause crashes.

As early as 2014, an episode of the popular TV show "Crime Scene Investigation (CSI) Las Vegas" featured a car-hacking storyline [37]. The 2017 installments of the block-buster "X-men" [38] and "Fast and Furious" [39] movie franchises showcase automated big rigs and vehicle hacking, respectively. The latest round of U.S. "WikiLeaks" media coverage lead with references to car hacking [40], although there was little of substance in the materials released to date [41].

15.1.6 Commercial Vehicle Fleets and Telematics

Twelve million vehicles, or almost 5% of the total on U.S. roads, are aggregated into fleets for business, government, utilities, police, taxi, and rental use [42]. Almost five million of these are connected to a telematics fleet management systems (FMS) [43]. The number of FMS-connected vehicles is slated to grow, with the truck and bus regulator FMCSA's mandatory electronic logging device (ELD) deadlines in late 2017 [44].

ELDs are intended to help create a safer work environment for drivers and to make it easier for fleets to accurately track, manage, and share records of duty status (RODS) data to the government. Another industry driver is auto insurance, with "70% of all auto insurance carriers expected to use telematics Usage-Based Insurance (UBI) by 2020" [45], as detailed in the UBI section.

Fully autonomous trucks are already in testing on public roads [46], and although truck software architectures are different than those of cars, they have been shown to share many of the same types of vulnerabilities. Full automation is likely to come first to the $700 billion trucking industry where fuel and time efficiency are key, drivers are in short supply, humans have an 11 h legal driving limit [47–49], and accidents are bad for business in every sense. SafetyFirst, a fleet driver safety company, notes that today "improved telematics technology will be integral for fleets in keeping an eye on driver safety" [50] However, IOActive, a cybersecurity company, equates fleet telematics with the "Holy Grail of automotive attacks" [51].

15.1.7 Gating Issue of Cyber Safety and Industry Tipping Point

Full vehicle automation (and the steps on the way to it) offers the promise of delivering safety as well as economic and other benefits. The *gating issue* to get to the benefits of connectivity and automation is the safety of software, especially its resilience to cyber threats.

In early 2017, we are at a tipping point, 50 years after Mr. Nader's "Unsafe at Any Speed" addressed hardware safety issues, but we now face software safety issues. Software's artificial intelligence (AI) and machine learning (ML) raise questions today that were first raised as pure science fiction 75 years ago [52].

2017 began with retaliation for Russian cyberattacks around the U.S. elections [53, 54], fears of infiltration of our nation's critical infrastructure [55–57], and Yahoo setting a dismal new record for data breach at over 1 billion accounts [58]. That breach came at a $350 million price tag [59] and had repercussions for company senior executives [60], and there are still over 40 lawsuits still pending [60].

2016 had seen the first vehicle automation-related fatality with an ex-Navy SEAL in the driver's seat of a Tesla killed while using "autopilot mode" [61, 62] but also the promising milestones with Google's AI beating a human world champion at the board game of "Go" [63] (seen as the ultimate test of AI technology), Google driving over 2 million fully automated miles [64], and Tesla clocking up over 100 million miles under its "autopilot" mode [65].

This chapter reviews the existing law, policy, cybersecurity, and data privacy considerations around conventional, connected (CV), driver-assisted (ADAS), and fully automated vehicles (AV) [2]. It contends that—(1) The risk of vehicle cyberattack is not just a problem for future model years but a real risk to the public, consumer, and commercial fleet vehicles on the road today; (2) potential harms are much greater than those of historical data breaches around smartphones, PCs, and the cloud, ranging from driver distraction to DDoS and ransomware, to property damage and bodily injury, and to death and debilitation of critical transport infrastructure [3]; and (3) that mitigating cyber risk should be addressed by mandatory regulation, as far as is possible under the present administration, setting a common standard for engineering, legal, and cyberinsurance purposes across the industry, rather than relying on a patchwork of voluntary standards that are inconsistently applied to vehicles assembled JIT from long supply chains.

15.2 The Promise of Software, Connectivity, and Automation

The increasing use of software and automation improves safety and also delivers important benefits in terms of efficiency, the environment, and the economy. The various categories of benefits tend to flow one in to another but share in common a *reliance* on software and communication *within* vehicles and *outside* to cloud data centers (directly or via third-party dongles or smartphones). Newer generations of vehicle are slated to add vehicle-to-vehicle (V2V) [66] and vehicle-to-infrastructure such a roads (V2I) [67] communications, collectively V2X (vehicle to everything).

15.2.1 Fuel Efficiency and Clean Air

Fuel has long been the largest cost for U.S. trucking fleets, accounting for almost 40% of the cost of ownership in 2013 [68]. Reducing fuel consumption through powertrain management reduces costs, leads to cleaner air, and improves the efficiency of logistics and the transport sector in general. Consumption can also be reduced by *directly* limiting engine idling and maximum speeds for fleet needs, with the majority of ECUs being reprogrammable by third parties [69]. Such modifications are legal after the 2015 exemptions to the Digital Millennium Copyright Act (DMCA) [70] much as rooting or jailbreaking phones. Telematics connectivity offers an *indirect* approach to fuel efficiency by logging driving style data such as idling, rapid acceleration, speeding, and hard braking to provide data to enforce fuel-efficient and safe fleet driving standards.

FMS combine wireless telematics systems with dynamic vehicle routing algorithms and vehicle-positioning systems to produce a telematics-enabled information system that can be employed by commercial fleet operators for real-time monitoring, control, and planning [71]. Future AI systems comparing current and previous journeys promise potential fuel savings up to 10% [72]. Future V2V communication enabling trucks to run in convoys or so-called platoons promise potential fuel savings from 5% to 10% [73], with slipstreaming reducing drag for trucks in the body of a convoy.

15.2.2 Routing and Parking Efficiency

Better routing reduces travel times and costs and leads to cleaner air. Safety is also improved through superior traffic and hazard detection. Heavy Duty Trucking magazine reported early in 2017 that "the industry wants to stay away from congestion and slow-downs to improve safety" [74] Similar improvements are available to consumers with better routing software such as Google's popular Waze app, which is beginning to be built directly into vehicles [75], and also with parking software, where today up to 34% of cars in downtown traffic are looking for parking [76].

15.2.3 Usage-Based Insurance (UBI)

Insurers monitor driver behavior directly via consumer dongles or fleet telematics systems, with improved safety correlating to lower premiums. Underwriters are better able to model risk with real usage data rather than relying on generic actuarial models per the National Association of Insurance Commissioners (NAIC) [45]. UBI data captured includes miles driven, time of day, Global Positioning System (GPS) location [77], speed, any rapid acceleration, hard braking, hard cornering, and also collision detection.

Some trucking telematics systems attempt to rank driver behavior to promote safety with so-called gamification [78], and even go so far as to trigger inward-facing video cameras in edge-case behavior. Future drowsiness detection systems may video drivers continuously [79]. The recent Cadillac CT6 car already boasts automatic 360° outside video surveillance capabilities [80] with off-line data storage in the trunk in case of vehicle tampering or an accident.

There is no reasonable expectation of privacy in *public spaces*, for example, in streets monitored by closed-circuit television (CCTV), patrolled by cruisers equipped with dash cams, and automated facial recognition (AFR) and license plate readers (LPR), although the degree of *internal* vehicle data captured raises privacy concerns discussed later.

15.2.4 Accident Investigation

In the case of collision, data are preserved in an event data recorder (EDR) [81] in anticipation of litigation, effectively constituting a vehicle "black box" record, much as for trains and planes. Tesla frequently leverages EDR data to allay public concerns over the functionality of its "auto pilot" technology [82]. Where automation is involved in an accident, who is liable—the driver, the manufacturer, the component supplier, or the software company?

15.2.5 Towards an Automated, Sharing, and Smart City Future

Companies such as Uber and Lyft have pioneered the ride-sharing marketplace and are beginning to reshape demand for private vehicles, recognizing that today's cars are parked 95% of the time [83]. The use of autonomous vehicles is potentially transformative for society. Ride sharing could transform cities by freeing 15% to 20% of the land presently used for parking spaces [84]. It could also potentially make transportation more affordable and accessible to millions of senior citizens and those with disabilities and eliminate the category of drunk drivers. Companies including Amazon [85] and Convoy [86] are building out similar offerings for trucks today.

15.3 Risk of Vehicle Cyberattack

The risk of vehicle cyberattack is not just a problem for future model years but a real risk to the public, consumer, and commercial fleet vehicles *on the road today*, with an industry-wide emphasis on building in security for *future model years* and less on independent testing and solutions that can mitigate risk across entire supply chains of suppliers.

15.3.1 Vehicle Attack Surfaces

The attack surface of a software environment is the sum of the vectors where an attacker can try to enter data or extract data from an environment. The attack surface of a modern vehicle is vast, which could include the ODB-II or J1939 ports accessing the CAN Bus or J1939 networks and reprogrammable ECUs, the USB port, Ethernet port [87], CD player [88] in older vehicles, charging port [89] on electric vehicles (EVs) or hybrid electric vehicles (HEVs), Bluetooth, cell, GPS [90], near-field communication (NFC) [91], radio data system (RDS), light detection and ranging (LiDAR), tire pressure monitoring system (TPMS), keyless entry system (KES), on-board Wi-Fi, third-party plug-in dongles, third-party smartphone environments, third-party telematics environments, over-the-air (OTA) updates, back-end cloud data centers, and future V2X communications. The Pentagon's latest and most advanced fighter jet the F35 has similar fly-by-wire technology to modern cars but contains 12-times less code, yet has its own struggles with cybersecurity issues [92].

15.3.2 A Brief History of Vehicle Hacks

In 2010, researchers first found and reported vehicle vulnerabilities [93], but the event went largely unnoticed by the *general public*, as did other related reports. It was commonly believed that automotive systems were little known to hackers, and individual vehicles were not an attractive financial target.

2014 saw the publication of the "Car Hacker's Handbook" [94] It was downloaded 300,000 times in the first four days [95, 96] tapping into an audience not of researchers but the community of mechanics and tuners [97]. The word "hacker" is used here in the sense of enthusiasts educating themselves in the inner workings of vehicles, risking only vehicle warranties and insurance coverage rather than as cyber criminals.

The publication echoed 1985s "Hacker's Handbook" [98] that introduced early home computer enthusiasts to dial-up communications with professional mini and mainframe computers 3 years before anyone would be prosecuted [99] under the then new Computer Fraud and Abuse Act (CFAA) [100]. It marked a symbolic breakdown of an older computer industry mindset that the relative obscurity of systems was itself a kind of security.

Over time, knowledge of even the most advanced computer systems democratizes to the commonplace. The team behind the car books recently used Kickstarter to successfully fund an inexpensive OBD-II dongle bundled with open-source software (OSS) and documentation to democratize car hacking to being "like rooting or jailbreaking a phone" [101].

In 2015, the first *widely public* demonstration of *remote* vehicle exploitation (via the infotainment system) was made by cybersecurity researchers Chris Valasek and Charlie Miller. They took control of a Jeep's dashboard functions, steering, transmission, and brakes from 10 miles away as the Jeep drove down the highway. It led to front page coverage around the world. The potential for that exploit to be repeated on any of up to 1.4 million Fiat Chrysler Automobile (FCA) vehicles lead to the *world's first cybersecurity safety recall* [102].

The research duo followed up a year later with a *local* vehicle exploit (via the USB and CAN Bus) showing non-connected vehicles to have vulnerabilities at highway speeds, answering naysayers that had sought to downplay the breadth of attack surfaces in legacy vehicles (and narrow industry focus only to the protection of future model years). The researchers described the work as "extremely difficult, time-consuming, and expensive," but significantly they had achieved their breakthroughs alone and mostly with self-funding [103]. They indicated that they were leaving the field to others, having widely disseminated their knowledge in reports, at industry events, and on the web.

Their breakthrough is analogous to that of Roger Bannister, an amateur runner and student. He was the first person to run a mile in under four minutes, generating front page publicity around the world [104]. Prior to that, it was thought to be a difficult, if not an impossible, barrier to break, much like hacking a car. In the next few years, his record was beaten several times by other amateur runners, and in today's team-based and well-funded athletics environment, it is regularly broken by high school students in routine meets. Since that first public vehicle hacking demonstration, numerous new vulnerabilities have been found, for example, in different manufacturer's vehicles including Tesla [105] and General Motors (GM) [106]; in commercial vehicles [107], by car thieves using laptops[108]; and most recently in smartphone apps used for keyless entry [109]. The many standards making and political reactions to these events are covered in the Law and Policy Section.

Unlike in running, hackers do not need to build up their own expertize or experience [110], instead they rely on tools created by a much smaller group of experts, to download exploits or run them as software as a service (SaaS) from the dark web [111]. Commercial fleets aggregate large groups of similar vehicle models making them relatively homogenous targets. The asymmetry of hacking economics means that defenders must defend *all* attack vectors, but an attacker needs to find just *one* vulnerability and can use it directly, or share to the others. Vulnerabilities do not necessarily fall within a single organization and its Gantt [112] chart planning processes.

15.3.3 Internet-of-Things (IoT) Hacks

October 2016 saw the world's largest DDoS attack, taking down the Internet for most of the U.S. Eastern seaboard [113]. The "Mirai" (Japanese for "the future") attack originated from basic IoT devices such as webcams, set-top boxes, and routers. Gartner Group estimates that there are over 8 billion such connected IoT devices in use in 2017, up a staggering 31% from 2016 [114]. The code for "Mirai" is publicly available for download on the web [115].

The scale of the attack underlined both the vulnerability of connected infrastructure to DDoS attack and the vulnerability of IoT devices themselves, where patching and updating may be infrequent, expensive, or unavailable, leading to potentially indefinite vulnerability windows.

AT&T notes that "in the past three years" it "has seen over a 3,000% increase in attackers scanning for vulnerabilities in IoT devices" [116] Underwriters Labs (UL), a global safety company, were asked, "Will we end up in another arms race here between hackers and automakers, like we've seen with personal computers?" Their answer was, "We're already in it. *People just don't recognize it*" [117].

15.3.4 The Issue of Legacy Vehicles, Updating and Recalls

The average vehicle age is over 11 years [7] and that of trucks over 14 years [118]. As illustrated by the Fiat Chrysler (FCA) recall, the issue of cybersecurity is not confined to future model years but instead falls squarely on the quarter billion vehicles *already on* U.S. roads, where updating typically requires a registered owner to be contacted to set up an in-person service appointment or, for some vehicles, that a memory stick be plugged into the USB port.

OTA updates familiar to PC and smart device users are offered on some newer vehicles by Tesla, Ford, and General Motors (GM) [119], but the majority of vehicles on the road are in a similar position to legacy IoT device manufacturers in lacking OTA update functionality.

Manufacturers are responsible for notifying and coordinating with NHTSA a *mandatory* safety recall campaign when a safety-related defect or noncompliance with federal safety standards is discovered, *whether the origin is hardware or software*. Today, that process misses about one in five vehicles on average each time [120], and overall one in four vehicles on U.S. roads have an outstanding safety recall [121]. A larger number of updates tied to software means more costs for vehicle manufacturers, potentially suppliers [122], fleets, and consumers.

The ability to update vehicles and the length of time of which updates are made available raise an interesting question. Might vehicle hardware at some point have a lifetime longer than that of software, especially in the case of trucks and commercial vehicles with their typically longer working lifetimes? In the consumer device world, once a produce reaches end of life (EOL), it may still function but any security vulnerabilities remain unpatched, raising questions of liability and insurability. In a recent study, 87% of Android-based smartphones were found to have security vulnerabilities due to a general lack of OTA provisioning and availability of the updates themselves [123]. Apple has largely avoided that issue with its iPhone devices, by providing regular OTA updates.

15.3.5 The Issue of End-to-End Hardening and Long Supply Chains

Cars and trucks are not built by vertically integrated manufacturers but instead assembled from parts drawn from global supply chains made up of thousands of subcontractors. Over 80% of the parts in a modern vehicle come from tier 1, 2, or 3 suppliers [124], and aftermarket parts are added post delivery such as the FMS telematics in trucks. Manufacturers own the statements of work (SOW) for suppliers and collaborate closely with tier 1 supporters both directly and through the new Auto-ISAC but have less strong relationships and visibility with the aftermarket, tier 2 or 3 suppliers as well as the wider white hat hacker community in industry and academia.

The CAN Bus [22] network architecture standardized communications between ECUs in vehicles from the mid-1980s. The two-wire design saved weight and easily allowed new ECUs and suppliers to be added. However, the architecture was never designed to be secure or to be connected to the Internet, a problem shared with other legacy critical embedded infrastructure in the military and energy sectors.

Each ECU broadcasts to all the others on the network, without having to identify itself. Essentially any illegitimate traffic on the network, whether introduced via outside connectivity or due to fake or malicious component in the supply chain, can compromise the entire vehicle.

In modern designs, there is a gateway or air gap between the outside world and the vehicle's network. In older designs, especially trucks and commercial vehicles, there is no such gap, making them particularly vulnerable to attack. To change the design would require coordination of all the affected suppliers over years, even if moving to a more modern design such as Ethernet [125], the standard in modern data centers.

Along with third-party ECU programming, dongles, telematics, connectivity, and integration of tablets, smartphones and wearables, the number of components to be hardened across an end-to-end attack surface is multiplied and then compounded when updates and upgrades are considerable. For example, in the original Jeep hack, supplier components from Harman's infotainment system, Blackberry's QNX operating system, and Sprint's network were compromised [126]. Even where each subcontractor thoroughly unit tests their component, there can be nonobvious interactions as components and subsystems are brought together that open vulnerabilities that can be exploited.

15.4 Potential Harms Due to Vehicle Cyberattack

Potential harms associated with vehicle cyberattack are much greater than those associated with historical data breaches around smartphones, PCs, and the cloud. The field of potential threat actors ranges from disgruntled auto industry insiders to hobbyists, to hacktivists, criminal gangs looking to steal vehicles or extort ransom with the threat of DDoS attacks, terrorists, and nation states and their proxies. Few victims of accidents, or fleets experiencing downtime, will be satisfied with a year of free credit reporting, the remedy most associated with historical data breaches.

15.4.1 Distracted Driving

A new 2017 American Automobile Association (AAA) study [127] reports that more than half of all drivers have texted while driving, run a red light, or exceeded the speed limit in just the last 30 days. Insurers increasingly blame distracted drivers as costs related to crashes outpace premium increases [128]. As noted in the introduction, 2015 was the largest single-year increase in accidents in the last half-century. Even a low-level prank vehicle attack, turning up the stereo, starting the wipers, or honking the horn could have serious real-world consequences, especially if coupled to existing distractions like talking on the phone or texting or if there are pets or children in the vehicle.

15.4.2 Distributed Denial of Service (DDoS) and Ransomware

As previously discussed in the section IoT Hacks, DDoS attacks against infrastructure are one of the simplest forms of attacks to mount since there are "increasingly powerful tools that anyone can download and use to trigger denials of service" [129]. When coupled with ransomware, they are potentially one of the most lucrative. Trend Micro, a cybersecurity company, reports ransomware attacks in general have increased 752% in 2016 compared to 2015 [130].

Commercial fleets are potentially the most vulnerable since they carry valuable or sensitive loads, run tight schedules, and make JIT deliveries. A ransomware attack could prove costly directly, in terms of reputation and future business. Amazon Web Services (AWS) recently suffered just a few hours of outage due to human error, which is estimated to have cost $150 million to the companies using the service [131].

The U.S. 911 emergency phone system was recently taken down for 12 h in October 2016 across a dozen states [132] in *the largest-ever cyberattack* on the U.S. emergency-response system. The attack, beginning as a prank by a student, exploited a bug in Apple's iPhone operating system.

15.4.3 Property Damage, Bodily Injury, and Death

In the past, software developers have not necessarily suffered when their products went wrong with enterprise licenses and end-user license agreement (EULA) contracts limiting their liability [133]. Historically, the share prices of companies hit with data breaches were relatively untouched [134] and remedies little more than providing free credit reporting. However, that is beginning to change with Yahoo setting new records for harms due to data breaches, as discussed in the Introduction.

Cyberattacks against vehicles have the potential to cause not just simple data breaches but cause property damage, bodily injury, and death, leaving manufacturers and likely their suppliers open to class action lawsuits, shareholder lawsuits, and regulatory action. Additionally, manufacturers and suppliers can face criminal charges as was the case with Takata, an automotive supplier, who recently agreed to plead guilty and pay $1 billion in penalties to settle criminal charges that it knowingly sold defective airbags to automakers, causing injuries and deaths [135].

15.4.4 Debilitation of Critical Transport Infrastructure

Where attacks extend to multiple vehicles, transport infrastructure itself may be impacted, blocking key road arteries, shutting down cities, and interrupting JIT distribution networks. Nation state actors and their proxies are highly motivated to prepare the cyber battlefield [136] given recent *cyber weapon* successes. Russia has already used cyberattacks as part of its campaigns against Estonia, Georgia, and the Ukraine [137].

In 2010, the United Stats and Israel launched "Stuxnet" [138] against Iran, the world's first publicly known cyber weapon. It was successful in degrading Iran's nuclear infrastructure despite their systems being disconnected or "air gapped" from the Internet, a claim sometimes made for existing vehicles, which ignores architectural issues in older vehicles and the presence of third-party connectivity. Stuxnet included four so-called zero-day exploits [139], for which there were no immediately available patches. 2012 saw Iran's response, with "Shamoon" [140] successfully deployed against Saudi Aramco's infrastructure, the world's most valuable company [141], as well as other targets within the United States.

15.4.5 **Data Privacy**

Data captured in vehicles includes miles driven, time of day, GPS location [77], speed, rapid acceleration, hard braking, hard cornering, and collision detection, potentially all audio and video within and around the vehicle. Intel estimates that the "average autonomous car will throw out four terabytes of data a day" [142]. Automation is intrinsically tied to data collection—without data, functionality is curtailed. For example, just to start an automated vehicle it must recognize the driver—are they an authorized adult, a new owner, a valet, a child, or potentially a thief?

For commercial vehicles, the sensitivity of vehicle data is self-evident for safety and commercial reasons, whether protecting delivery routes against "porch pirates" [143] a load of smart TVs from hijack, radioactive materials from terrorists [144], or just knowing the "10-20" location of a big rig in a yard, at a truck stop, a customer loading dock, or traversing a bridge or tunnel.

The Atlantic magazine sets the stage for vehicle data where "there's huge potential value to companies who mine individual data and use it for marketing and other services … For companies like Google and Uber, privacy issues are very important … That could kill a business, if you don't handle privacy properly" [145].

Justice Sotomayor in her concurrence in the 2012 Supreme Court case the United States v. Jones [146] warned that revealing *just GPS data* could disclose "trips to the psychiatrist, the plastic surgeon, the abortion clinic, the AIDS treatment center, the strip club, the criminal defense attorney, the by-the-hour motel, the union meeting, the mosque, synagogue or church, the gay bar and on and on." Data forensically extracted from smartphone already routinely figures in corporate litigation, personal injury suits, and divorce [147], and with vehicle data could potentially be a much more detailed and effective "smoking gun."

The book "Hate Crimes in Cyberspace" [148] underlines the risks where "[Cyber] stalkers are using very sophisticated technology … so they can track all of your interactions on the Internet … or installing GPS in your car so you will show up at the grocery store, at your local church, wherever and there is the stalker." There may be no need to install a new device on a vehicle, instead just leveraging tools *already part of the automotive ecosystem*. For example, Google's "Waze" routing app was found by researchers to allow third parties to follow anyone using it in real time [149].

Even anonymous sets of data from sensors within a vehicle and untied from Vehicle Identification Numbers (VIN) or typical Personally Identifiable Information (PII) [150] can be much more revealing than expected, amounting to digital fingerprints [151]. For example, "researchers can identify individual drivers with only 5 to 10 minutes' worth of data from the brake sensor in a vehicle." Identification of typists from keystroke dynamics and other so-called user behavior analytics (UBA) have been around for decades [152] in the traditional data center.

It is well known that consumer gadgets featuring Google Assistant, Samsung's Bixby, Apple's Siri, Amazon Alexa, and Microsoft's Cortana are listening but less well known that law enforcement has been legitimately able to collect audio and location data from connected cars for 15 years [153]. In 1979's Smith v. Maryland case [154], the Supreme Court held that a person has no "legitimate expectation of privacy in information voluntarily turned over to third parties." If there is an opening for law enforcement or intelligence services, there is a vulnerability to hackers also.

Vehicles are simply becoming more like other consumer devices where the enterprise license or EULA, opt-in and opt-out provisions, their availability, and communications around them are key. Recent consumer devices with high-profile privacy concerns include the "My Friend Cayla" doll [155] that encouraged children to have conversations with it, recording their responses and sharing them to the cloud for voice recognition and voice print analysis. Smart TV's have been shown to have the ability to not only share viewing choices to the cloud but all conversations heard around them [156], echoing the bidirectional functionality of "telescreens" in the novel "1984" [157]. Most recently, web browsing and app usage history can now be shared to third parties by telecoms following a recent reversal of an Federal Communication Commission (FCC) rule [158].

Legislators and standards organizations have recognized the vehicle issue, and one industry group has proposed a simple solution—extending the existing privacy models associated with Apple and Android smartphones to smart vehicles and the proposed SPY Car Study Act 2017 emphasizes data protection and clear communication with the public, as detailed in the Law and Policy Section.

15.5 Law and Policy

15.5.1 Brief Review of Government and Industry Reactions to Car Hacking

The last 2 years have seen a blizzard of government and industry reactions. Together, they indicate widely shared goals in improving vehicle cybersecurity and data privacy. The common theme to all these warnings, best practices, policies, and manifestos is that today *none of them are mandatory beyond the original physical safety provisions of NHTSA and FMCSA* as detailed in the Introduction.

15.5.1.1 Pre-2015—Proactive Research and Development (R&D): In 2012, the Department of Defense (DoD)'s Defense Advanced Research Projects Agency (DARPA) had anticipated many of these cybersecurity issues in launching its High-Assurance Cyber Military Systems (HACMS) [159] program, focused on hardening IoT systems found in military vehicles, drones, and medical devices. HACMS funded R&D with white hat cybersecurity researchers, including Chris Valasek and Charlie Miller.

Law enforcement vehicles are some of the most narrowly focused and connected model groups in the marketplace, which is why in October 2015 the Virginia State Police (VSP) began publicly testing the cybersecurity vulnerability of police cruisers to hacking and also the effectiveness of countermeasures [160]. For example, as part of testing, a cruiser's engine was disabled by car hacking. An aftermarket intrusion prevention system (IPS) developed from the HACMS research was plugged into the OBD-II port and was able to prevent the same attack [161].

15.5.1.2 2015—Senate Warnings, Auto Information Sharing and Analysis Center (ISAC): In February 2015, Senator Ed Markey (D-Mass.) issued a report

warning that the increasing technical complexity of vehicles was putting driver security and privacy at risk [162]. In July 2015, Senators Markey and Richard Blumenthal (D-Conn.) introduced the Senate bill "Security and Privacy in Your Car Act," the snappily named "SPY Car Act" (2015) [163] calling for the application of *mandatory* vehicle cybersecurity and data privacy standards along with penetrating (pen) testing. The bill was read twice and referred to the Senate Committee on Commerce, Science, and Transportation at the time. It was reintroduced in early 2017 (see section 2017).

A pen test is an *authorized* simulated attack on a computer system that looks for cybersecurity weaknesses, potentially gaining access to the system's features and data. Pen tests are regularly used in the commercial marketplace, for example, the industry-lead Payment Card Industry (PCI) Data Security Standard (DSS) *requires* pen testing on a regular schedule, and after system changes, and defines what constitutes a qualified pen testing organization [164].

In the same month, the auto industry created a closed industry group for "Automotive Information Sharing and Analysis Center" (Auto-ISAC) focused on a narrow group of manufacturers and tier 1 suppliers to "promote collaborative cybersecurity efforts" within that subset of industry stakeholders [165].

The idea of ISACs was first introduced in Presidential Decision Directive-63 (PDD-63) [166] two decades ago, where the government asked each critical infrastructure sector to establish sector-specific organizations to share information about threats and vulnerabilities. ISACs in theory raise the cyber maturity of industry sectors, and sharing can prevent threat actors simply moving from target to target with identical attacks. The Auto-ISAC established a secure platform for anonymous or attributed sharing, tracking, and analyzing intelligence about cyber threats and potential vulnerabilities around the connected vehicle.

15.5.1.3 **2016—FBI, DoT, NHTSA, FTC Warnings, and Multiple Standards:** In January 2016, the Society of Automotive Engineers (SAE) released a new *voluntary* standard SAE J3061, the "Cybersecurity Guidebook for Cyber-Physical Vehicle Systems" [167]. SAE, as the auto industry's own standards body, continues to collaborate with U.S. and international standards bodies on continued development [168].

In March 2016, the Federal Bureau of Investigation (FBI), DoT, and NHTSA put out a formal Joint Public Service Announcement warning that "Motor vehicles are increasingly vulnerable to remote exploits" [169]. In the same month, Tesla, General Motors (GM), and slightly later Fiat Chrysler Automobiles (FCA) set up public-facing vulnerability reporting "bug bounty" programs that are distinct from Auto-ISAC [170], to encourage the sharing of vehicle vulnerabilities. Consumer software heavyweights Google and Facebook had popularized this approach in the early 2010s, and by mid-2016 even the Pentagon had introduced a bug bounty program [171]. In theory, white hat hackers motivated by money or recognition share details of exploits to the companies.

Also in March 2016, the Government Accountability Office (GAO)'s report on vehicle cybersecurity found "vehicle attacks could potentially impact a large number of vehicles and allow an attacker to access targeted vehicles from anywhere in the world" [172]. The government's own fleet of vehicles is over 600,000 strong, and a recent public Request for Information (RFI) underlined the importance of cybersecurity hardening for telematics systems [173]. In April 2016, the organization Future of Automotive Security Technology Research (FASTR) was formed as an industry group including Intel and Uber, as a "neutral, inclusive nonprofit consortium that seeks to enable innovation in automotive security with a *vision* of self-healing vehicles" [174].

In July 2016, Auto-ISAC put out its own *voluntary* set of "Automotive Cybersecurity Best Practices" [175]. In September 2016, The National Motor Freight Traffic Association (NMFTA) issued cybersecurity *warnings* [176], underlining the important of cybersecurity to trucking. In the same month, NHTSA released new *voluntary* automated vehicle policies [177] (based on the earlier SAE work) and also new *voluntary* distracted driving guidelines [178].

In September 2016, the Federal Trade Commission (FTC) warned the public over the issue of data remanence from vehicle usage and smartphone syncing in rental cars—"unless you delete data before you return the car, other people may view it, including future renters and rental car employees or even hackers" [179]. The FTC has carried out a number of successful recent cybersecurity enforcement actions around "unfair or deceptive practices affecting consumers," including Wyndham Hotel Group *over sloppy data security practices* [180] and the ADT alarm company over its IoT alarm devices and their *vulnerability to hacking* [181].

In October 2016, NHTSA released new *voluntary* cybersecurity guidelines for vehicles [182]. In November 2016, cybersecurity experts *warned* Congress in IoT and DDoS hearings to "regulate cybersecurity or expect a disaster" [183]. In November 2016, the Drive Trust Alliance *proposed* that automated vehicles should offer a similar privacy choices to those developed for smartphones over the last decade—an individual or fleet driver's key should open *only* their data, data owners should have the right to securely erase their data, sell it, or transfer it to another vehicle [184].

In December 2016, the DoT *proposed* new rules for vehicle-to-vehicle communications (V2V), which have the potential to increase safety but also increase attack surfaces [185]. In the same month, the FCC released a *notice of inquiry* "seeking to accelerate the dialog around the critical importance of the early incorporation of cybersecurity protections in 5G networks, services and devices" [186]. 5G is important for automated vehicles given its speed and low latency. In the same month, the State of Michigan signed legislation to allow more autonomous vehicles onto state roads, with the goals of "quickening development of the technology and making roads safer" [187]. Michigan is one of around a dozen states along with Washington, D.C., that allow public road testing of autonomous vehicles [188].

15.5.1.4 Post 2017—New SPY Car Act and More Inclusive Auto-ISAC: January 2017 saw a new *bipartisan* House Bill, the "SPY Car *Study* Act (2017)" [189] introduced by Representatives Joe Wilson (R-SC) and Ted Lieu (D-Calif.) with the goal of devising cybersecurity standards for new vehicles and setting an execution *deadline*. In the same month, the "Developing and Growing the Internet of Things" or DIGIT Act (2017)" was reintroduced [190] seeking to *establish a working group*, including both federal and private-sector representatives, to identify the regulations or practices that might inhibit the growth of IoT. Also in the same month, the outgoing head of NHTSA expressed concern that the new administration would let auto safety

slide, explaining "For 50 years it's always been reactive. Unfortunately, reactive means you've got to wait until somebody gets hurt or killed before you're stepping in, the way the system is now. What you're talking about is premarket approval" [191].

In February 2017, Senator Gary Peters (D-MI) described [192] a "plan to develop regulatory flexibility for the auto industry in its drive to create self-driving vehicles, with an eye toward proposing legislation this year," and Mike Abelson of General Motors (GM) cautioned lawmakers about "over-regulating the new technology" [193]. In the same month, the FASTR think tank published its own cybersecurity "manifesto" [194].

In March 2017, the SPY Car Act (2017) was reintroduced by its original democratic sponsors, read twice, and once again referred to the Committee on Commerce, Science, and Transportation, as before. It includes the previous *mandatory* provisions, along with a strong emphasis on communication understandable to members of the public [195]. Most significantly, 2017 has seen the extension of Auto-ISAC to include commercial vehicles [196], and it recently indicated a potential openness to a wider representation of suppliers and vendors, including white hat hackers from industry and academia on a public "TU-Auto" cybersecurity panel [197].

15.5.1.5 **Innovation and Regulation:** In a recent interview, Airbnb's CEO, head of one of the largest so-called tech unicorn companies [198], talked about regulatory push back being part and parcel of innovation, from the invention of automated teller machines (ATMs) to video recorders to cars themselves, all of which were heavily opposed in the beginning [199]. Uber, at the head of that tech unicorn list, recently clashed with the authorities in San Francisco [200] and moved automated testing to Arizona temporarily [201]. Another automotive tech startup, Comma.ai, after clashing with regulators, announced it would give away its "OpenPilot" self-driving software as well as plans for compatible hardware for free [202].

The Economist magazine offered this insight into the tension of innovation and regulatory compliance—"firms that pioneer new technological trends do not always manage to stay on top. Think of Nokia and BlackBerry in smartphones, Kodak in digital cameras or Myspace in social networking. Much will depend on which firm *best handles the regulators*" [203].

15.5.2 Existing Cybersecurity and Data Privacy Standards

Of all industry sectors, the Federal government has the most existing cybersecurity and data privacy standards, including HSPD-12 [204], FIPS [205], FISMA [206], FedRAMP [207], and NIST 800-53 [208], all of which are *mandatory* for Federal agencies and their own long supply chains of subcontractors. Many of the examples of fake [209] or malicious [210, 211] components in the supply chain come from the U.S. government. The healthcare industry has *mandatory* standards such as HIPPA [212] and the credit card industry has the *mandatory* PCI DSS standard, which were discussed earlier. Forty-seven states and DC have data breach legislation in place [213].

Perhaps the most comprehensive and well accepted general cybersecurity standard is NIST's Cybersecurity Framework (CsF) [214], a *voluntary* framework to manage cybersecurity risk in critical infrastructure (PPD-21 [19]), which is widely used by all types of organizations. The CsF relies on a variety of existing international standards, guidelines, and practices to form its core and to enable critical infrastructure providers to achieve resilience [215] and benefit from economies of scale for adoption.

In February 2016, California's Attorney General (AG) reported that [216] the twenty "Center for Internet Security (CIS) Critical Security Controls" [217] are to be viewed as the "*minimum level of information security*" that all organizations handling personal data should use. The report posits *a bright-line rule*: "The failure to implement all the Controls that apply to an organization's environment constitutes a lack of reasonable security," i.e., they are mandatory. The last control CIS recommends is pen testing, much as for PCI DSS from the financial sector.

In December 2016, New York's Department of Financial Services (NYDFS) [218] revised the mandatory cybersecurity regulations it had proposed just over a year earlier following a long notice and comment. They too include pen testing.

15.5.3 A European Point of View

The European Commission is considering making many ADAS technologies mandatory for vehicles driving on European roads [219]. These include autonomous emergency braking, active lane-keeping assists, driver drowsiness and distraction monitoring, and seat belt reminders. A more controversial proposed provision is general speed-limiting, much as is already done in some U.S. trucking fleets (see section <u>Fuel Efficiency and Clean Air</u>).

In the United Kingdom, a recent Transport Bill [220], covers automated vehicles. Insurers will have to cover vehicles *both in traditional manual driving and autonomous driving modes*, side-stepping the need for injured parties to directly bring suit against manufacturers and suppliers. Another provision ties liability to failure to install updates, or for making unauthorized modifications.

15.6 Mitigating Risks and Balancing Interests

A dignitary was run over and killed on the opening day of the world's first railway 200 years ago [221] and the Wright Brother's plane killed a military aviator in its initial government trial 100 years ago [222], much as last year saw the first fatality associated with vehicle automation. Today, rail and air are some of the safest modes of transport … *and roads can be too.*

The key to drastically reducing road accidents and increasing vehicle safety is automation (and the steps towards it), but the *gating issue* is the safety of software, especially its resilience to cyber threats. As the <u>Law and Policy Section</u> made clear, there is no common cybersecurity standard for engineering, legal, and cyberinsurance purposes across the industry beyond older regulations for physical safety and safety recalls, only a patchwork of voluntary standards that are inconsistently applied across long supply chains.

This chapter contends that cyber risk should be addressed by mandatory regulation, *as far as is possible under the present administration*. President Trump has made clear that new regulation in general is not a priority [223, 224] and pending data privacy legislation has been rolled back [158] rather than embraced. Setting a common standard for engineering, legal, and cyberinsurance purposes across the industry is a balancing act of preserving public safety, industry competitiveness, and above all making sure that the right regulations are in place in a field where there is not yet universal agreement on which regulations should be adopted.

The auto industry has a unique opportunity to learn from the experiences of other sectors, R&D in the military, government, and law enforcement and proactively embrace government or self-regulation, or run the risk that a laissez-faire approach will create the conditions for an incident or incidents where major damages, loss of life, and interruption of services set back innovation for years, with public sentiment leading to harsher regulation than would otherwise have been the case.

15.6.1 Proposed Engineering Emphases

From the material covered in the <u>Risk of Vehicle Cyberattack</u> Section, there are three areas where broad principles can be adopted that drive effective risk mitigation for vehicles, building upon existing standards and initiatives and address "low-hanging fruit" while wider common standards are agreed.

15.6.1.1 (1) Systematically Running Pen Tests with Independent Testers:
The industry has made good progress in the last 2 years introducing cybersecurity best practices such as bug bounties and the Auto-ISAC program. However, a strong emphasis on independent penetration (pen) testing by external and attack-oriented "*red team*" testers is not there. It is this group that has the strongest expertise and motivation to find flaws in security. It is they who have had the most success in finding flaws to date and offer what may be the "biggest bang for buck" in terms of cybersecurity testing and hardening. Voluntary pen testing already figures in auto industry standards. Mandatory pen testing figures in PCI standards, the recent New York and California cybersecurity legislation, and has been proposed in both the 2015 and 2017 SPY Car Acts.

Pen tests are distinct from the internal, defense-oriented "*blue team*" testing associated with normal Software Development Life Cycles (SDLC). Pen testers have specialist skill sets and playbooks in hacking, with the goals of quickly finding exploitable vulnerabilities that traditional blue teams have missed. By analogy, the U.S. Navy uses special forces SEALs to test the security of its bases, ships, and airports, where their specialized skills and training allow them to breach what is intended to be impenetrable security developed and manned by regular warfighters. The thinking is that it is preferable for the SEALs to find a vulnerability than terrorists or agents of foreign powers [225].

The majority of auto manufacturers and suppliers have to focus on delivering new features as quickly as possible for commercial advantage, with internal testing geared to showing that code *does what it's supposed to do*. However, pen testing has a different focus, to see if that same software *can be made to do something else or be illegitimately controlled*. Where timelines are limited and there are no penalties or mandatory standards that need to be met, the temptation is to concentrate on blue team testing the last phase of product development, with team members subject to internal management pressures.

This system of incentives and disincentives is analogous to the situation Mr. Nader described back in 1965, where it was difficult for any single manufacturer or supplier to unilaterally introduce major rather than incremental hardware safety improvements for fear of being over-priced, or late to market, compared to a competitor simply following the status quo and able to market 100% compliance with existing government regulations and participation in industry bodies.

As the Office of Personnel Management (OPM) [226] and other high-profile government breaches underline, numerous and detailed *mandatory* standards alone are not sufficient. With the philosophy that a good offense informs a good defense, pen testers have a unique perspective into what it takes to resist a cyberattack [227]. The application of that knowledge takes the form of authorized pen testing by independent groups on an annual basis, or whenever there are significant updates or upgrades. The PCI standards already define who are qualified pen testers in the financial industry, although they may need to be redefined for the auto industry by SAE, NHTSA, and industry groups, with an inclusive Auto-ISAM ensuring collaboration.

15.6.1.2 (2) Over-the-Air (OTA) Updating for "Forgotten" Quarter Billion Vehicles:

There are a quarter billion vehicles on the road today. Working on cybersecurity standards only for new vehicles will *eventually* solve the problem, and it is well known that building in security is a much better approach than trying to retrofit it. However, it does not address the vulnerabilities affecting the vast majority of road users today, especially those most vulnerable in commercial fleets with their older vehicles and vulnerable architectures.

Apple iOS and Google Android both entered the smartphone market in 2008 [228, 229]. It took Apple until 2011 (partnered with Good Technology) and Android until 2014 (partnered with Samsung KNOX) to meet *mandatory* government cybersecurity standards just for *unclassified* devices [230, 231] despite their tight control of code bases an order of magnitude smaller than those found in vehicles [232]. A key element in achieving those standards was providing OTA updates, to help patch the various security flaws found mainly by industry and academic researchers.

As previously discussed, some OEMs have already implemented OTA vehicle updates and the DHS-sponsored Uptane initiative facilitates an expansion of that process. Since safety recalls are mandatory, holdouts will have a choice of forging on with the traditional process, despite its flaws, or moving to more cost effectively deal with updates with aftermarket solutions that can handle multiple updates without alienating consumer and fleet customers. Once again standards for solutions such as cell-enabled dongles need to be defined with SAE, NHTSA, and industry groups, with an inclusive Auto-ISAM ensuring collaboration. Updating is going to be most important for fleets where the risks are greatest and the Return on Investment (ROI) of cybersecurity is the greatest.

15.6.1.3 (3) Reduce Attack Surface Across Supply Chain, Mitigating Weak Links:

Given the architecture of vehicle networks, their vulnerabilities to any weak link from a supply chain covering the aftermarket, OEM, and tiers 1, 2, and

3 suppliers, it is clear that a transition from 100% unhardened components to 100% hardened components will a) take many years and b) that along that journey there will be a mix of hardened and unhardened components connected together.

It is proposed to emphasize approaches that leverage the experience of the DARPA, law enforcement, and other government R&D that can mitigate cybersecurity risk across the board, without imposing heavy burdens of costs, and time, or the need for expensive engineering resources such as retrofitting intrusion prevention and intrusion detection systems (IPS/IDS) and transformation of binaries using Run-time Application Self-Protection (RASP) algorithms that can be effective in mitigating both known and zero-day attacks.

15.6.2 Legal and Cyberinsurance

The move to ADAS systems over the last two decades, where computers rather than the driver are at times controlling the vehicle, have not yet moved accident liability and the need for insurance coverage from the driver to the product [233]. This is likely to change with potentially successful cyberattacks and over time as driving decisions move to being entirely the responsibility of the OEM. Google, Volvo, and Mercedes are assuming liability in order to accelerate the adoption of automation [234], by removing liability uncertainty and shielding both drivers and their own suppliers. Both OEM and underwriters are incentivized to focus on effective cybersecurity legislation that covers existing vehicles, end-to-end attack surface, and whole supply chains, rather than simply focusing only on new vehicles, OEM, and tier 1 suppliers.

15.7 Conclusions

This chapter reviewed the existing law, policy, cybersecurity, and data privacy considerations around conventional, connected, driver-assisted, and fully automated vehicles. It contends that—(1) the risk of vehicle cyberattack is not just a problem for future model years, (2) potential harms are much greater than those of historical data breaches, and (3) mitigating cyber risk should be addressed by mandatory regulation, as far as is possible under the present administration, setting a common standard for engineering, legal, and cyberinsurance purposes across the industry.

References

1. Star Wars, *A New Hope*, Episode IV, 1976, http://www.imsdb.com/scripts/Star-Wars-A-New-Hope.html

2. "Automated Driving—Levels of Driving Automation Defined in SAE Standard J3016," https://www.sae.org/misc/pdfs/automated_driving.pdf

3. "Transportation Systems Sector," https://www.dhs.gov/transportation-systems-sector

4. "50 Years Ago, 'Unsafe at Any Speed' Shook the Auto World," http://www.nytimes.com/2015/11/27/automobiles/50-years-ago-unsafe-at-any-speed-shook-the-auto-world.html?_r=0

5. National Highway Traffic Safety Administration (NHTSA), https://www.nhtsa.gov/about-nhtsa/nhtsas-core-values

6. Federal Motor Carrier Safety Administration (FMCSA), https://www.fmcsa.dot.gov/mission

7. "253 Million Cars and Trucks on US Roads; Average Age Is 11.4 Years," http://www.latimes.com/business/autos/la-fi-hy-ihs-automotive-average-age-car-20140609-story.html

8. "Fatalities in the United States," https://crashstats.nhtsa.dot.gov/api/public/publication/812349

9. "The Economic and Societal Impact of Motor Vehicle Crashes," https://crashstats.nhtsa.dot.gov/api/public/viewpublication/812013

10. "Critical Reasons for Crashes Investigated in the National Motor Vehicle Crash Causation Survey," https://crashstats.nhtsa.dot.gov/api/public/viewpublication/811059

11. Oxford English Dictionary (OED) contributor Barry Popik finds numerous references from the 1920s forward. http://www.barrypopik.com/index.php/new_york_city/entry/the_most_dangerous_part_of_an_automobile

12. "U.S. Traffic Deaths Rose a Whopping 7.2% in 2015," https://www.wsj.com/articles/u-s-traffic-deaths-rose-a-whopping-7-2-in-2015-1472506314

13. "Estimated 2016 Road Deaths Climb 6%," www.automotive-fleet.com/news/story/2017/02/2016-estimated-road-deaths-climb-6.aspx?utm_campaign=enews-friday-20170221&utm_source=Email&utm_medium=Enewsletter&omid=1009602784

14. "1801—Richard Trevithick Introduces His 'Puffing Devil,'" www.history.com/this-day-in-history/richard-trevithick-introduces-his-puffing-devil

15. "Ferrari 488 Spider vs. Mazda Miata Grand Touring: Open-Air Market," http://www.roadandtrack.com/car-culture/a29602/ferrari-488-spider-vs-mazda-miata/

16. "Center for Automotive Research 'Assessment of Tax Revenue' and 'Contribution of the Automotive Industry' Published January 2015," https://autoalliance.org/in-your-state/

17. "Vehicle Ownership and Availability," https://www.rita.dot.gov/bts/sites/rita.dot.gov.bts/files/publications/transportation_statistics_annual_report/2003/html/chapter_02/vehicle_ownership_and_availability.html

18. "Reports, Trends & Statistics," www.trucking.org/News_and_Information_Reports_Industry_Data.aspx

19. "Presidential Policy Directive—Critical Infrastructure Security and Resilience," https://obamawhitehouse.archives.gov/the-press-office/2013/02/12/presidential-policy-directive-critical-infrastructure-security-and-resil

20. "Automotive Industry—Innovation Driven by Electronics," http://embedded-computing.com/articles/automotive-industry-innovation-driven-electronics/

21. "This Car Runs on Code," spectrum.ieee.org/transportation/systems/this-car-runs-on-code

22. "Control Area Network (CAN Bus)," www.popularmechanics.com/cars/how-to/a7386/how-it-works-the-computer-inside-your-car/

23. SAE J1939, http://store.sae.org/j1939/contents/

24. "Why Volkswagen's Emissions Scandal Has No End," fortune.com/2017/01/11/volkswagen-emissions-scandal-2/

25. "The 2009 Toyota Accelerator Scandal That Wasn't What It Seemed," https://www.manufacturing.net/blog/2016/08/2009-toyota-accelerator-scandal-wasnt-what-it-seemed

26. "How to Deal with Unintended Acceleration," http://www.caranddriver.com/features/how-to-deal-with-unintended-acceleration

27. "OBD-II Background," www.obdii.com/background.html

28. "'Right to Repair' Battle Won by Truck Owners, Independent Shops," www.overdriveonline. com/right-to-repair-battle-won-by-truck-owners-independent-shops/

29. "Tesla Autopilot Crash Investigation Yields Autonomous Tech Insights," http://www. automotive-fleet.com/news/story/2017/01/tesla-autopilot-crash-investigation-yields-autonomous-tech-insights.aspx

30. "Driver-Assist Technology for Snowplows and Specialty Vehicles," www.its.umn.edu/ research/featuredstudies/driverassistive/plow/

31. "G&D Expects 75% of All Cars to be Shipped with Integrated Connectivity by 2020," www. iot-now.com/2016/02/16/42573-giesecke-devrient-expects-75-of-all-cars-to-be-shipped-with-integrated-connectivity-by-2020/

32. "Safety for Some: Are Electronic Driver Assists Really for Everyone?" blog.caranddriver. com/safety-for-some-are-electronic-driver-assists-really-for-everyone/

33. "Fiat Chrysler Chooses CES over Detroit Auto Show for Media Reveal," www.crainsdetroit. com/article/20161208/NEWS/161209855/fiat-chrysler-chooses-ces-over-detroit-auto-show-for-media-reveal

34. "Driverless Cars Finally Steer near Showrooms," www.autonews.com/article/20170103/ oem06/170109971/driverless-cars-finally-steer-near-showrooms?utm_source= feedburner&utm_medium=feed&utm_campaign=feed%3a+autonews%2fbreakingnews+ (automotive+news+breaking+news+feed)

35. "Harman Demonstrates First Ever Live Car Hack at CES," www.autocar.co.uk/car-news/ motor-shows-ces/harman-demonstrates-first-ever-live-car-hack-ces

36. "Americans Worry about Vehicle Hacking," www.automotive-fleet.com/news/story/ 2017/02/americans-worry-about-vehicle-hacking.aspx?utm_campaign=enews-friday-20170228&utm_source=Email&utm_medium=Enewsletter&omid=1009602784

37. "Crime Scene Investigation (CSI) Boston Brakes," 2014, http://www.imdb.com/title/ tt3421364/

38. "Logan," 2017, http://www.imdb.com/title/tt3315342/

39. "The Fate of the Furious," 2017, http://www.imdb.com/title/tt4630562/

40. "WikiLeaks Posts Thousands of Purported CIA Cyberhacking Documents," https:// www.wsj.com/articles/wikileaks-posts-thousands-of-purported-cia-cyberhacking-documents-1488905823

41. "Security Experts Shrug Off CIA Memo," www.autonews.com/ article/20170313/MOBILITY/303139955/security-experts-shrug-off-cia-memo?utm_source=feedburner&utm_medium=feed&utm_ campaign=Feed%3A+autonews%2FBreakingNews+ (Automotive+News+Breaking+News+Feed)

42. "US Automobile and Truck Fleets by Use," https://www.rita.dot.gov/bts/sites/rita. dot.gov.bts/files/publications/national_transportation_statistics/html/table_01_14. html_mfd

43. "The Installed Base of Fleet Management Systems in the Americas Will Exceed 13 Million Units by 2019," www.fieldtechnologiesonline.com/doc/the-installed-management-americas-million-units-by-0001

44. "ELD Implementation Timeline," https://www.fmcsa.dot.gov/hours-service/elds/ implementation-timeline

45. "Usage-Based Insurance and Telematics," www.naic.org/cipr_topics/topic_usage_based_ insurance.htm

46. "Autonomous Tech Could Make Driving Semi-Trucks Even Less Fun," https://www.wired. com/2016/04/autonomous-tech-make-driving-semi-trucks-even-less-fun/

47. "Robots Could Replace 1.7 Million American Truckers in the Next Decade," www.latimes.com/projects/la-fi-automated-trucks-labor-20160924/

48. "Self-Driving-Truck Startups Race to Take on Uber," https://www.wsj.com/articles/self-driving-truck-startups-race-to-take-on-uber-1488466802

49. "Hours of Service Limitations," www.truckingtruth.com/cdl-training-program/page93

50. "State-of-the-Art Fleet Technology: Safety," www.automotive-fleet.com/article/story/2017/02/state-of-the-art-fleet-technology-safety.aspx?utm_campaign=enews-friday-20170214&utm_source=Email&utm_medium=Enewsletter&omid=1009602784

51. "A Survey of Remote Automotive Attack Surfaces," www.ioactive.com/pdfs/IOActive_Remote_Attack_Surfaces.pdf

52. "Runaround" short story by Isaac Asimov (1942) detailing "The Three Laws of Robotics."

53. "Executive Summary of Grizzly Steppe Findings," https://www.dhs.gov/news/2016/12/30/executive-summary-grizzly-steppe-findings-homeland-security-assistant-secretary

54. "Obama Administration Prepares Sanctions, Retaliation for Russian Election Meddling," www.cnn.com/2016/12/28/politics/russian-election-meddling-sanctions-hacking/

55. "Russian Government Hackers Do Not Appear to Have Targeted Vermont Utility, Say People Close to Investigation," https://www.washingtonpost.com/world/national-security/russian-government-hackers-do-not-appear-to-have-targeted-vermont-utility-say-people-close-to-investigation/2017/01/02/70c25956-d12c-11e6-945a-76f69a399dd5_story.html

56. "Inside the Cunning, Unprecedented Hack of Ukraine's Power Grid," https://www.wired.com/2016/03/inside-cunning-unprecedented-hack-ukraines-power-grid/

57. "Iranians Hacked into New York Dam," www.cnn.com/2015/12/21/politics/iranian-hackers-new-york-dam/

58. "Hackers Breach a Billion Yahoo Accounts," https://www.wired.com/2016/12/yahoo-hack-billion-users/

59. "Why Verizon Decided to Still Buy Yahoo after Big Data Breaches," https://www.wsj.com/articles/why-verizon-decided-to-still-buy-yahoo-after-big-data-breaches-1487679768

60. "Yahoo Boss Marissa Mayer Loses Millions in Bonuses over Security Lapses," https://www.theguardian.com/technology/2017/mar/02/yahoo-boss-marissa-meyer-loses-millions-in-bonuses-over-security-lapses?utm_source=esp&utm_medium=Email&utm_campaign=GU+Today+USA+-+Collections+2017&utm_term=215703&subid=20245400&CMP=GT_US_collection

61. "Tesla Driver Dies in First Fatal Crash While Using Autopilot Mode," theguardian.com/technology/2016/jun/30/tesla-autopilot-death-self-driving-car-elon-musk

62. "US Regulator Closes Tesla Death Probe; No Evidence of Defects," www.autonews.com/article/20170119/oem11/170119744/u-s-regulator-closes-tesla-death-probe-no-evidence-of-defects?utm_source=feedburner&utm_medium=feed&utm_campaign=feed%3a+autonews%2fbreakingnews+(automotive+news+breaking+news+feed)

63. "In a Huge Breakthrough, Google's AI Beats a Top Player at the Game of Go," https://www.wired.com/2016/01/in-a-huge-breakthrough-googles-ai-beats-a-top-player-at-the-game-of-go/

64. "Google's Self-Driving Car Program Odometer Reaches 2 Million Miles," https://www.wsj.com/articles/googles-self-driving-car-program-odometer-reaches-2-million-miles-1475683321

65. "Tesla Customers Have Driven 100 Million Miles with Autopilot Active," www.theverge.com/2016/5/24/11761098/tesla-autopilot-self-driving-cars-100-million-miles

66. "What's Next? V2V Communication with Connected Cars," https://www.wired.com/insights/2014/09/connected-cars/

67. "States Wire Up Roads as Cars Get Smarter," www.wsj.com/articles/states-wire-up-roads-as-cars-get-smarter-1483390782

68. "White House Sets New Fuel-Efficiency Standards for Heavy-Duty Trucks, Vans and Buses," https://www.washingtonpost.com/news/energy-environment/wp/2016/08/16/white-house-sets-new-fuel-efficiency-standards-for-heavy-duty-trucks-vans-and-buses/

69. "Vehicles Supported," derivesystems.com/efficiency/vehicles-supported/

70. "Automakers Just Lost the Battle to Stop You from Hacking Your Car," http://www.theverge.com/2015/10/27/9622150/dmca-exemption-accessing-car-software

71. Goel, A., "Fleet Telematics: Real-Time Management and Planning of Commercial Vehicle Operations," 2007.

72. "A Computer Program That Learns How to Save Fuel," www.economist.com/news/science-and-technology/21715640-piece-artificial-intelligence-can-draw-past-experience-computer

73. "Truck Platooning, Past, Present, and Future," www.truckinginfo.com/article/story/2016/04/platooning-is-on-the-way.aspx

74. "Researchers—Big Data Could Move Trucking Better," www.truckinginfo.com/channel/fleet-management/news/story/2017/03/researchers-want-to-use-big-data-to-increase-trucking-productivity.aspx?utm_campaign=topnews-20170304&utm_source=Email&utm_medium=Enewsletter

75. "Waze Digs into Your Car's Dashboard," https://www.wired.com/2017/02/waze-cars/

76. "Getting the Prices Right: An Evaluation of Pricing Parking by Demand in San Francisco," April 2013, http://shoup.bol.ucla.edu/PricingParkingByDemand.pdf

77. Milner, G., "Pinpoint: How GPS Is Changing Technology, Culture and Our Minds," 2016.

78. "Should You Run Your Business Like a Game?" http://venturebeat.com/2010/10/05/gamification-business/

79. "Sleepy behind the Wheel? Some Cars Can Tell," www.nytimes.com/2017/03/16/automobiles/wheels/drowsy-driving-technology.html

80. "The Cadillac CT6 Has a Clever Surveillance Feature That Will Give You Complete Peace of Mind," www.businessinsider.com/cadillac-ct6-surveillance-cameras-2017-3?utm_content=bufferfea18&utm_medium=social&utm_source=facebook.com&utm_campaign=buffer-ti

81. "Your Car May Be Invading Your Privacy," www.usatoday.com/story/money/cars/2013/03/24/car-spying-edr-data-privacy/1991751/#

82. "The Customer Is Always Wrong: Tesla Lets Out Self-Driving Car Data—When It Suits," www.theguardian.com/technology/2017/apr/03/the-customer-is-always-wrong-tesla-lets-out-self-driving-car-data-when-it-suits

83. "Today's Cars Are Parked 95% of the Time," fortune.com/2016/03/13/cars-parked-95-percent-of-time/

84. "Shared Autonomous Vehicles Could Increase Urban Space by 15%," www.driverless-future.com/?cat=19

85. "Amazon Is Secretly Building an 'Uber for Trucking' App, Setting Its Sights on a Massive $800 Billion Market," www.businessinsider.com/amazon-building-uber-for-trucking-app-2016-12

86. "Convoy Raises $2.5M from Jeff Bezos, Marc Benioff, Others for New On-Demand Trucking Startup," www.geekwire.com/2015/convoy-raises-2-5m-from-jeff-bezos-marc-benioff-others-for-new-on-demand-trucking-startup/

87. "The Tesla Model S Is Basically a Good-Looking IT Department on Wheels," http://jalopnik.com/the-tesla-model-s-is-basically-a-good-looking-it-depart-1558372928

88. "With Hacking, Music Can Take Control of Your Car," www.itworld.com/article/2748225/security/with-hacking--music-can-take-control-of-your-car.html

89. "Hackers Could Use Electric Vehicle Charging Stations to Cripple Cars and Grid," www.theblaze.com/news/2013/04/12/hackers-could-use-electric-vehicle-charging-stations-to-cripple-cars-and-grid/

90. "Jaguar and Shell Launch In-Car App for Cashless Gas Station Payments," http://venturebeat.com/2017/02/15/jaguar-launches-in-car-app-for-cashless-gas-station-payments/

91. "NXP's NFC Tech Lets You Unlock Cars with a Phone," www.cnet.com/roadshow/news/nxps-nfc-tech-lets-you-unlock-cars-with-a-phone/

92. "Lockheed F-35's Cybersecurity Flaws Cited by Pentagon Tester," https://www.bloomberg.com/politics/articles/2016-03-23/lockheed-f-35-s-cybersecurity-flaws-cited-by-pentagon-s-tester-im4y5nwk

93. "Comprehensive Experimental Analyses of Automotive Attack Surfaces," http://autosec.org/publications.html

94. Open Garages, *Car Hacker's Handbook*, 2014, opengarages.org/handbook/

95. "The Car Hacker's Handbook Digs into Automotive Data Security," https://techcrunch.com/2016/06/14/the-car-hackers-handbook-digs-into-automotive-data-security/

96. Smith, C., *The Car Hacker's Handbook: A Guide for the Penetration Tester*, 2016.

97. "How to Hack Your Car," https://www.forbes.com/forbes/2002/0708/148.html

98. Cornwall, H., "*Hacker's Handbook*," 1985.

99. "How a Grad Student Trying to Build the First Botnet Brought the Internet to Its Knees," https://www.washingtonpost.com/news/the-switch/wp/2013/11/01/how-a-grad-student-trying-to-build-the-first-botnet-brought-the-internet-to-its-knees/

100. U.S.C.A. § 1030 (West).

101. "Macchina: The Ultimate Tool for Taking Control of Your Car!," https://www.kickstarter.com/projects/1029808658/macchina-the-ultimate-tool-for-taking-control-of-y/faqs

102. "After Jeep Hack, Chrysler Recalls 1.4m Vehicles for Bug Fix," wired.com/2015/07/jeep-hack-chrysler-recalls-1-4m-vehicles-bug-fix/

103. "Miller and Valasek Unveil New Jeep Hack at Black Hat, Retire from Car Hacking," https://www.scmagazine.com/miller-and-valasek-unveil-new-jeep-hack-at-black-hat-retire-from-car-hacking/article/528074/

104. "Mile Run World Record Progression," www.maa.org/sites/default/files/images/upload_library/3/osslets/100multiParameterAnimation/mile_record_scatter.html

105. "Researchers Remotely Hack Tesla Model S," https://www.washingtonpost.com/news/the-switch/wp/2016/09/20/researchers-remotely-hack-tesla-model-s/

106. "GM, Hacker Say OnStar App Issue Not Completely Fixed," www.detroitnews.com/story/business/autos/general-motors/2015/07/30/gm-says-fixed-onstar-remotelink-security-issue/30877307/

107. "1,000s of Trucks, Buses and Ambulances May Be Open to Hackers," wired.com/2016/03/thousands-trucks-buses-ambulances-may-open-hackers/

108. "Houston Car Hackers Suspected of Theft of More Than 100 SUVs and Trucks," fortune.com/2016/08/06/houston-car-hackers/

109. "Android Phone Hacks Could Unlock Millions of Cars," https://www.wired.com/2017/02/hacked-android-phones-unlock-millions-cars/

110. "Script Kiddies: The Net's Cybergangs," www.zdnet.com/article/script-kiddies-the-nets-cybergangs-5000096163/

111. "How to Access the Dark Web," https://darkwebnews.com/help-advice/access-dark-web/

112. "What Is a Gantt Chart?" www.gantt.com

113. "What We Know about Friday's Massive East Coast Internet Outage," https://www.wired.com/2016/10/internet-outage-ddos-dns-dyn/

114. "Gartner Says 8.4 Billion Connected 'Things' Will be in Use in 2017, up 31% from 2016," www.gartner.com/newsroom/id/3598917

115. "Source Code for IoT Botnet 'Mirai' Released," https://krebsonsecurity.com/2016/10/source-code-for-iot-botnet-mirai-released/

116. "Cybersecurity Alliance Formed Specifically to Address IoT," www.overdriveonline.com/cybersecurity-alliance-formed-specifically-to-address-iot-the-internet-of-tru-um-things/

117. "The Car Hackers of Fremont," www.sfchronicle.com/business/article/the-car-hackers-of-fremont-10907437.php

118. "IHS: Both Fleet Age and Truck Demand on the Rise," http://fleetowner.com/truck-stats/ihs-both-fleet-age-and-truck-demand-rise

119. "Your Car's New Software Is Ready. Update Now?" https://www.nytimes.com/2016/09/09/automobiles/your-cars-new-software-is-ready-update-now.html?_r=0

120. "Software, Autonomous Tech to Raise Vehicle-Recall Stakes," wardsauto.com/technology/software-autonomous-tech-raise-vehicle-recall-stakes

121. "One in 4 Vehicles on US Roads Has Unfixed Safety Recall," www.mlive.com/auto/index.ssf/2017/03/one_in_4_vehicles_on_us_roads.html#incart_river_index

122. "Who Pays the Bill? Usually the Supplier," www.autonews.com/article/20131027/OEM11/131029913/who-pays-the-bill%3F-usually-the-supplier

123. "87% of Android Devices Insecure Because Manufacturers Fail to Provide Security Updates," www.lightbluetouchpaper.org/2015/10/08/87-of-android-devices-insecure-because-manufacturers-fail-to-provide-security-updates/

124. "Last Mile: Converging on the Future," http://automotivelogistics.media/opinion/last-mile-converging-on-the-future

125. "From Your Dorm Room to Your Car: Ethernet Is Back," https://techcrunch.com/2017/03/15/from-your-dorm-room-to-your-car-ethernet-is-back/

126. "Remote Exploitation of an Unaltered Passenger Vehicle," www.ioactive.com/pdfs/IOActive_Remote_Car_Hacking.pdf

127. "Young Millennials Top List of Worst Behaved Drivers," newsroom.aaa.com/2017/02/young-millennials-top-list-worst-behaved-drivers/

128. "Smartphone Addicts behind the Wheel Drive Car Insurance Rates Higher," https://www.wsj.com/articles/smartphone-addicts-behind-the-wheel-drive-car-insurance-rates-higher-1487592007

129. "How DDoS Attacks Work, and Why They're So Hard to Stop," www.kotaku.com/how-ddos-attacks-work-and-why-theyre-so-hard-to-stop-1676445620

130. "2016 Security Roundup—A Record Year for Enterprise Threats," https://www.trendmicro.com/vinfo/us/security/research-and-analysis/threat-reports/roundup

131. "Amazon Finds the Cause of Its AWS Outage: A Typo," https://www.wsj.com/articles/amazon-finds-the-cause-of-its-aws-outage-a-typo-1488490506

132. "The Night Zombie Smartphones Took Down 911," https://www.wsj.com/articles/how-a-cyberattack-overwhelmed-the-911-system-1488554972

133. "Incentives Need to Change for Firms to Take Cyber-Security More Seriously," www.economist.com/news/leaders/21712138-software-developers-and-computer-makers-do-not-necessarily-suffer-when-their-products-go

134. "Why Data Breaches Don't Hurt Stock Prices," https://hbr.org/2015/03/why-data-breaches-dont-hurt-stock-prices

135. "What's Happening with the Takata Airbag Recall?" www.consumerreports.org/airbags/what-is-happening-with-takata-airbag-recall/

136. Koppel, T., *Lights Out: A Cyberattack, a Nation Unprepared, Surviving the Aftermath*, 2016.

137. "Timeline: Ten Years of Russian Cyber-Attacks on Other Nations," www.nbcnews.com/news/us-news/timeline-ten-years-russian-cyber-attacks-other-nations-n697111

138. "An Unprecedented Look at Stuxnet, the World's First Digital Weapon," http://wired.com/2014/11/countdown-to-zero-day-stuxnet/

139. Zetter, K., *Countdown to Zero Day: Stuxnet and the Launch of the World's First Digital Weapon*, 2015.

140. "In Cyberattack on Saudi Firm, US Sees Iran Firing Back," www.nytimes.com/2012/10/24/business/global/cyberattack-on-saudi-oil-firm-disquiets-us.html

141. "Too Big to Value: Why Saudi Aramco Is in a League of Its Own," https://www.bloomberg.com/news/articles/2016-01-07/too-big-to-value-why-saudi-aramco-is-in-a-league-of-its-own

142. "Intel Joins Silicon Valley's Race to Make Best 'Server on Wheels' with Mobileye Deal," https://www.wsj.com/articles/intel-to-buy-mobileye-for-15-3-billion-1489404970

143. "'Porch Pirates' Stealing People's Holiday Deliveries," www.cnbc.com/2015/12/11/porch-pirates-stealing-peoples-holiday-deliveries.html

144. "Materials Transportation," https://www.nrc.gov/materials/transportation.html

145. "How Self-Driving Cars Will Threaten Privacy," www.theatlantic.com/technology/archive/2016/03/self-driving-cars-and-the-looming-privacy-apocalypse/474600/

146. *United States v. Jones*, 132 s. ct. 945, 955, 181 l. ed. *2d 911*, 2012.

147. "Cell Phone as Smoking Gun: In Court, Few Messages Are Gone for Good," https://gigaom.com/2014/02/24/cell-phone-as-smoking-gun-in-court-few-messages-are-gone-for-good/

148. Citron, D.K., *Hate Crimes in Cyberspace Paperback*, 2016.

149. "Waze Hack Lets Creeps Track You're Driving," https://www.engadget.com/2016/04/26/waze-tracking-exploit/

150. NIST 800-122, "Guide to Protecting the Confidentiality of Personally Identifiable Information."

151. "What Your Car Knows About You," www.detroitnews.com/story/business/autos/2017/02/22/auto-privacy/98281208/

152. "The Future of Biometrics Could Be in What You Type," www.edtechmagazine.com/higher/article/2016/04/keying-another-type-biometrics

153. "Cartapping: How Feds Have Spied on Connected Cars for 15 Years," https://www.forbes.com/sites/thomasbrewster/2017/01/15/police-spying-on-car-conversations-location-siriusxm-gm-chevrolet-toyota-privacy/#24a7b2032ef8

154. *Smith v. Maryland*, 442 U.S. 735, 99 S. Ct. 2577, 61 L. Ed. *2d 220*, 1979.

155. "This Doll May Be Recording What Children Say, Privacy Groups Charge," www.npr.org/sections/alltechconsidered/2016/12/20/506208146/this-doll-may-be-recording-what-children-say-privacy-groups-charge

156. "How to Stop Your Smart TV from Spying on You," https://www.wired.com/2017/02/smart-tv-spying-vizio-settlement/

157. Orwell, G., "The Telescreen Received and Transmitted Simultaneously. Any Sound That Winston Made, above the Level of a Very Low Whisper, Would Be Picked Up by It" from "1984," 1949.

158. "With Washington's Blessing, Telecom Giants can Mine your Web History," https://www.wsj.com/articles/with-washingtons-blessing-telecom-giants-can-mine-your-web-history-1490869801?mod=trending_now_1

159. "Four DARPA Projects that Could be Bigger than the Internet," www.defenseone.com/technology/2014/05/four-darpa-projects-could-be-bigger-internet/84856/

160. "Car-Hacking Research Initiative in Virginia Shows How Even Older Vehicles Could be Targeted in Cyberattacks," www.darkreading.com/attacks-breaches/state-trooper-vehicles-hacked-/d/d-id/1322415

161. "Cybersecurity Experiment to Protect Cruisers," http://dcinno.streetwise.co/2015/10/02/va-gov-mcauliffe-police-car-cybersecurity-findings-tech

162. "Senator: Your Futuristic Car is Putting Your Privacy and Security at Risk," https://www.washingtonpost.com/news/the-switch/wp/2015/02/09/senator-your-futuristic-car-is-putting-your-privacy-and-security-at-risk/

163. "SPY Car Act Hopes to Save American Cars from Digital Disaster," https://www.forbes.com/sites/thomasbrewster/2015/07/21/senators-launch-spy-car-act/#178e73409ece

164. "Penetration Test Guidance Special Interest Group PCI Security Standards Council," March 2015

165. "An Industry-Operated Environment Created to Enhance Cyber Security Awareness and Coordination across the Global Automotive Industry," https://www.automotiveisac.com

166. "Presidential Decision Directive 63," https://fas.org/irp/offdocs/pdd/pdd-63.htm

167. "January 2016 Saw the Publication of the Highly-Anticipated SAE Recommended Practice J3061," http://sae-europe.org/news-posts/new-sae-training-course-introducing-j3061/

168. "SAE Automotive Engineering – Standards News," February 2017.

169. "Joint Public Service Announcement," ic3.gov/media/2016/160317.aspx

170. "Auto Industry Turns to 'Bug Bounties' to Find Security Holes," www.autonews.com/article/20160321/OEM06/160329987/auto-industry-turns-to-bug-bounties-to-find-security-holes

171. "History of Bug Bounties," https://bugcrowd.com/resources/history-of-bug-bounties

172. "Government Accountability Office Report on Vehicle Cybersecurity," March 2016, www.gao.gov/products/GAO-16-350

173. General Services Administration (GSA), "Deploying Telematics across the Federal Fleet Request for Information (RFI)," 3/9/2017

174. "Are You Afraid Your Car Will be Taken Over?" www.csoonline.com/article/3171156/mobile-security/are-you-afraid-your-car-will-be-taken-over.html

175. "Automotive Cybersecurity Best Practices," https://www.automotiveisac.com/best-practices/

176. "MMFTA Heavy Vehicle Cybersecurity Bulletin," September 9, 2016.

177. "Federal Automated Vehicles Policy," http://www.nhtsa.gov/nhtsa/av/pdf/federal_automated_vehicles_policy.pdf

178. "US DoT Proposes Guidelines to Address Driver Distraction Caused by Mobile Devices in Vehicles," https://www.nhtsa.gov/press-releases/us-dot-proposes-guidelines-address-driver-distraction-caused-mobile-devices-vehicles

179. "Watch Out That Your Rental Car Doesn't Steal Your Phone Data," fortune. com/2016/09/01/rental-cars-data-theft/

180. "FTC v. Wyndham," http://harvardlawreview.org/2016/02/ftc-v-wyndham-worldwide-corp/

181. "ADT Agrees to End Alarm Hackability Suits with Settlement," https://www.law360.com/articles/884393/adt-agrees-to-end-alarm-hackability-suits-with-settlement

182. "NHTSA Automotive Cybersecurity," https://www.nhtsa.gov/press-releases/us-dot-issues-federal-guidance-automotive-industry-improving-motor-vehicle

183. "Regulate Cybersecurity or Expect a Disaster, Experts Warn Congress," http://money.cnn.com/2016/11/16/technology/cybersecurity-regulation-congress/

184. "Comments on Privacy Issues in Federal Automated Vehicles Policy," https://www.drivetrust.com/category/news/

185. "Proposed Rule Would Mandate Vehicle-to-Vehicle (V2V) Communication on Light Vehicles, Allowing Cars to 'talk' to Each Other to Avoid Crashes," https://www.nhtsa.gov/press-releases/us-dot-advances-deployment-connected-vehicle-technology-prevent-hundreds-thousands

186. "FCC Looks to Tackle IoT Cybersecurity through 5G Regulation," https://www.cyberscoop.com/fcc-iot-5g-cybersecurity/

187. "Snyder Signs New Michigan Self-Driving Vehicles Law," www.detroitnews.com/story/business/autos/2016/12/09/autonomous-car-law/95199544/

188. "Autonomous Vehicles - Self-Driving Vehicles Enacted Legislation," http://www.ncsl.org/research/transportation/autonomous-vehicles-self-driving-vehicles-enacted-legislation.aspx#Enacted%20Autonomous%20Vehicle%20Legislation

189. "Worried about Cybersecurity and the Connected Car? There's a Bill for That," https://arstechnica.com/cars/2017/01/worried-about-cybersecurity-and-the-connected-car-theres-a-bill-for-that/

190. "Senators Reintroduce Bill to Remove Barriers to Connected Device Industry Act," www.nextgov.com/emerging-tech/2017/01/senators-reintroduce-bill-remove-barriers-connected-device-industry/134512/

191. "Outgoing NHTSA Exec Fears Auto Safety Slide," www.detroitnews.com/story/business/autos/2017/01/20/rosekind-nhtsa/96852812/

192. "Congress Looking to Clear Hurdles to Self-Driving Cars," www.freep.com/story/news/local/michigan/2017/02/13/congress-looking-clear-hurdles-self-driving-cars/97855878/

193. "Carmakers: Don't Overregulate Self-Driving Car Testing," www.detroitnews.com/story/business/autos/2017/02/14/self-driving-hearing/97906916/

194. "Our Manifesto - Toward Tomorrow's Organically Secure Vehicle," https://fastr.org/about-us/what-is-fastr-a-manifesto/

195. Spy Car Act (2017), https://www.markey.senate.gov/imo/media/doc/2017-03-20-SPYCAR-Act-BillText-.pdf

196. "Commercial Vehicles to Join Auto-ISAC," www.prnewswire.com/news-releases/commercial-vehicles-to-join-auto-isac-300396844.html

197. "Tracking the Evolution of Automotive Cybersecurity," TU-auto web panel, Mar. 31, 2017.

198. "The Unicorn List - Current Private Companies Valued at $1B and Above," https://www.cbinsights.com/research-unicorn-companies

199. "How I Built this - Joe Gebbia," Oct. 17, 2016, www.npr.org/podcasts/510313/how-i-built-this

200. "Uber Flouts Rules with San Francisco Self-Driving Test," www.autonews.com/article/20161219/oem11/312199916/uber-flouts-rules-with-san-francisco-self-driving-test

201. "Uber's Self-Driving Cars Returning to California Roads," https://www.wsj.com/articles/ubers-self-driving-cars-returning-to-california-roads-1489011778

202. "After Mothballing Comma One, George Hotz Releases Free Autonomous Car Software," https://arstechnica.com/cars/2016/11/after-mothballing-comma-one-george-hotz-releases-free-autonomous-car-software/

203. "Uberworld," www.economist.com/news/leaders/21706258-worlds-most-valuable-startup-leading-race-transform-future

204. "Homeland Security Presidential Directive-12 (HSPD-12)," https://cio.gov/protect/identity-management-hspd-12/

205. "FIPS PUB 140-2," http://csrc.nist.gov/groups/STM/cmvp/standards.html

206. "Federal Information Security Modernization Act (FISMA)," http://dhs.gov/fisma

207. "Federal Risk and Authorization Management Program (FedRAMP)," http://fedramp.gov

208. "NIST 800-53 Revision 4," http://dx.doi.org/10.6028/NIST.SP.800-53r4

209. "U.S. Missiles Infected with Chinese Fakes," www.wnd.com/2012/06/u-s-missiles-infected-with-chinese-fakes/

210. "Secret Back Door in Some U.S. Phones Sent Data to China, Analysts Say," https://www.nytimes.com/2016/11/16/us/politics/china-phones-software-security.html?_r=0

211. "New Discovery around Juniper Backdoor Raises More Questions about the Company," https://www.wired.com/2016/01/new-discovery-around-juniper-backdoor-raises-more-questions-about-the-company/

212. "Health Information Technology," https://www.hhs.gov/hipaa/for-professionals/special-topics/health-information-technology/index.html

213. "Security Breach Notification Laws," www.ncsl.org/research/telecommunications-and-information-technology/security-breach-notification-laws.aspx

214. "NIST Releases Update to Cybersecurity Framework (CSF)," https://www.nist.gov/news-events/news/2017/01/nist-releases-update-cybersecurity-framework

215. "Framework for Improving Critical Infrastructure Cybersecurity," https://www.nist.gov/cyberframework/draft-version-11

216. "'Reasonable Security' Becomes Reasonably Clear to the California Attorney General," www.hldataprotection.com/2016/03/articles/cybersecurity-data-breaches/reasonable-security-becomes-reasonably-clear/

217. "The CIS Critical Security Controls for Effective Cyber Defense," http://sans.org/critical-security-controls/

218. "NYDFS Proposes Revised Cybersecurity Requirements for Financial Services Companies," www.drinkerbiddle.com/insights/publications/2016/12/nydfs-proposes-revised-cybersecurity-requirements

219. "Autonomous Car Technology Could Become Mandatory in Europe," www.carbuzz.com/news/2016/12/26/autonomous-car-technology-could-become-mandatory-in-europe-7737022/

220. "UK Government Finally Draws Up Laws for Autonomous Cars," https://www.engadget.com/2017/02/23/uk-government-vehicle-technology-aviation-bill/

221. "First Railway Accident," www.old-merseytimes.co.uk/huskisson.html

222. "First Military Air Casualty," www.arlingtoncemetery.net/thomaset.htm

223. "Trump Signs Executive Order to Cut, Restrict Regulations," https://www.wsj.com/articles/trump-signs-executive-order-to-cut-restrict-regulations-1485790245

224. "Trump Orders Federal Agencies to Identify Unnecessary Regulations," https://www.wsj.com/articles/trump-orders-federal-agencies-to-identify-unnecessary-regulations-1487959551

225. "Rogue Warrior" (1993) by Richard Marcinko

226. "Inside the Cyberattack that Shocked the US Government," https://www.wired.com/2016/10/inside-cyberattack-shocked-us-government/

227. "New Cybersecurity Report Gets the Hacker Perspective," http://searchsecurity.techtarget.com/news/450414361/New-cybersecurity-report-gets-the-hacker-perspective

228. "iOS Version History," https://en.wikipedia.org/wiki/IOS_version_history#iPhone_OS_1

229. "Android Version History," https://en.wikipedia.org/wiki/Android_version_history

230. "Unclassified DISA STIG List," https://www.stigviewer.com/stigs

231. The Defense Information Systems Agency (DISA) plays a critical role in enhancing the security posture of the DoD's security systems by providing the Security Technical Implementation Guides (STIGs). They contain guidance on locking down information systems.

232. "Code Bases Millions of Lines of Code," www.informationisbeautiful.net/visualizations/million-lines-of-code/

233. "When Driverless Cars Crash, Who Gets the Blame and Pays the Damages?" https://www.washingtonpost.com/local/trafficandcommuting/when-driverless-cars-crash-who-gets-the-blame-and-pays-the-damages/2017/02/25/3909d946-f97a-11e6-9845-576c69081518_story.html

234. "Mercedes, Google, Volvo to Accept Liability When Their Autonomous Cars Screw Up," http://jalopnik.com/mercedes-google-volvo-to-accept-liability-when-their-1735170893.

About the Author

Simon is co-founder of cybersecurity startup RunSafe Security. RunSafe was developed as part of DARPA's program of cybersecurity for DoD vehicles, drones, and medical devices. Simon also worked with Apple and Samsung in hardening their mobile devices for government use. He is a member of SAE's IoT Cybersecurity Committee.

Previously, he was VP of Sales at Kaprica Security (exited to Samsung), Mobile Program Director, DMI, managed mobility market leader, and Worldwide Sales and Marketing Director at Thursby Software, Apple strong security market leader. Prior executive roles include Red Hat, HP, Capgemini, a $9B hedge fund, a $50MM dot com, and a background in nuclear software engineering.

He holds a BSc (Hons.) in Physics from the University of Manchester, England, a MS in Law and Cybersecurity from the University of Maryland Carey Law, and an Executive MBA (2019) from the University of Maryland Smith Business, with CEH, CISSP, and CIPP cybersecurity and privacy certifications.

Do You Care What Time It Really Is?
A Cybersecurity Look into Our Dependency on GPS

Gerardo Trevino, Marisa Ramon, Daniel Zajac, and Cameron Mott

16.1 Background

How much do you rely on GPS daily? The U.S. Department of Homeland Security (DHS) has identified 16 critical infrastructure sectors including the energy and **transportation systems** sectors [1]. Although not apparent, the availability of position, navigation, and timing (PNT) data is increasingly vital within these critical infrastructure sectors. PNT systems have come to rely heavily on global navigation satellite systems (GNSS), including the U.S. Global Positioning System (GPS), to provide PNT data to both military and civilian systems. For example, the energy sector uses GPS-enabled time-synchronized devices to monitor and synchronize the national power grid in real time and help prevent blackouts. Malicious manipulation of GPS time data could cause the problems systems were initially trying to prevent [2]. Furthermore, the ongoing improvement of the existing systems such as the GPS and the Russian Global Navigation Satellite System (GLONASS) will provide new signals to increase performance and enable new applications. Finally, the rapid development of new satellite navigation systems including the European Galileo and the Chinese Beidou system will provide new sources of PNT data to users worldwide. In other words, GPS and other GNSS are used extensively today in critical infrastructure because they provide reliable PNT data at low cost, and these systems are not going anywhere.

From a cybersecurity perspective, the civilian GPS signals are relatively easy to jam, spoof, or manipulate [3, 4]. More concisely, you cannot protect what you do not own, and in this case, critical infrastructure asset owners do not "own" the radio frequency (RF) spectrum. Therefore, it is very difficult to counter GPS threats, and thus, the critical infrastructure that depends on GPS data is also vulnerable. In 2013 DHS released a public report evaluating the "Risks to U.S. Critical Infrastructure from Global Positioning System Disruptions [5]"; in this report, four critical infrastructure sectors including the communications, emergency services, and energy and transportation systems were evaluated by a group of experts. In this report, DHS identifies three types of GPS disruptions: (1) naturally occurring, such as space weather including extreme solar flares events; (2) unintentional, such as RF signals interfering with GPS signals; and (3) intentional, such as purposeful signal jamming[1] or signal spoofing.[2] DHS has also identified signal jamming disruptions to be more likely than signal spoofing incidents. Signal spoofing, however, was considered to have a more severe consequence than jamming since some of the systems and/or asset owners would not be able to always detect these types of attacks.

16.2 How Do Commercial Fleets Use GPS Today?

Commercial vehicle fleet owners leverage GPS to manage the operations of their fleets. Several commercially available systems claim that these GPS tracking systems improve fleet operations, increase profits by increasing utilization times, and reduce fuel costs and emissions to name a few benefits. These improvements are also derived from U.S. federal regulations that look to improve the reliability in the delivery of goods and services. It is expected that new regulations will influence the increase of data collected by commercial fleets. One example would be the requirement to log temperature information for those fleets that transport food and other perishable goods. What this means is that fleet owners and managers will continue to increase the amount of data collected (enabled by GPS) and the sample frequency of this information.

Finally, developments in autonomous vehicle fleet management will also dramatically increase the level of complexity for both the technologies used to enable autonomy and the amount of information processed and collected. GPS is currently used to provide accurate position and navigation and will continue to be a key sensor in the sensor suite used by autonomous fleets (**Figure 16.1**).

---

[1] A GPS jamming attack consists in overpowering a GPS receiver so that it can no longer receive GPS signals.

[2] A GPS spoofing attack attempts to deceive a GPS receiver by broadcasting incorrect GPS signals, structured to resemble a set of normal GPS signals, or by rebroadcasting genuine signals captured elsewhere or at a different time. These spoofed signals may be modified in such a way as to cause the receiver to estimate its position to be somewhere other than where it actually is or at a different time, as determined by the attacker [8].

FIGURE 16.1 A Class 8 Freightliner truck affectionately dubbed "Big Red" that SwRI has outfitted with autonomous capabilities including multiple GPS receivers and antennae.

© SAE International

16.3 How Could GPS Vulnerabilities Affect Fleet Vehicles?

It is obvious that commercial fleets and vehicles are not all the same. Some commercial companies own and operate their own vehicles and rarely rely on third parties to operate. Other companies run logistical operations that incorporate both. The type of goods and services can also span from the food we eat every day to critical goods and services that could affect not only businesses profits but also public safety. The vehicles themselves, the GPS systems, and the fleet management products are also different, which makes the formal assessment of GPS vulnerabilities difficult to precisely characterize. Also, the attack surface for GPS systems is large (many ways to attack) which increases the attack scenarios that could be analyzed to answer the proposed question. Proper cybersecurity and sensor technology assessments are currently underway in industry. These assessments will provide the proper evaluation of GPS vulnerabilities and other sensors used by fleet vehicles. Due to this complexity, this section addresses two different "scenarios" to illustrate the potential high-level effects from GPS vulnerabilities.

FIGURE 16.2 Vehicle GPS jammer.

© SAE International

16.3.1 GPS Jamming Scenario

Fleet owners are very familiar with the fact that their vehicles' GPS can be easily jammed. **Figure 16.2** shows an image of a portable vehicle GPS jammer found in a related internet news report [6]. Several reports have indicated that fleet vehicle drivers have attempted to "disappear" from GPS-enabled fleet tracking systems by jamming the GPS receivers while operating their vehicles. By purchasing illegal

GPS jamming systems on the internet, vehicle drivers can effectively "disappear" from the fleet management system. The location data collected by these systems are not always real time, meaning that the fleet manager only looks at the location of a vehicle at rates spanning from minutes to hours depending on how the system is set up. In some systems, the fleet manager has the option to narrow down on a vehicle to get more updated location data, but depending on how the fleet is managed, this will not always be how the vehicles are tracked. The problem with these jamming systems, aside from being illegal to operate in the United States, is that they can negatively impact other critical infrastructure assets. As these systems are not tested nor licensed by the FCC, the resulting broadcast strength and interference can vary, extending to adjacent vehicles and other GPS equipment. From an operational perspective, a jamming incident can cause, at a minimum, a set of alarms and log data that describes the event. The level of complexity could easily escalate to multiple fleet vehicles, but the result will be the same: each vehicle will lose GPS and therefore alarms will be generated, and this could cause some logistical issues at a minimum.

16.3.2 GPS Spoofing Scenario

GPS spoofing (falsely reporting the position) could affect a fleet manager by providing false indications of vehicle positions. For example, if a vehicle's position was spoofed to suddenly go from Utah to South America, multiple alerts would light up a fleet manager's operational software. Geo-fence alerts would indicate that the vehicle had violated the boundary, speed limit, and estimated time of arrival, and coordination of vehicle positions would be seriously compromised. If the spoofing attack continued to show the vehicle in the remote location, the fleet manager may be unable to utilize the fleet management software to coordinate the vehicle. GPS spoofing need not be limited to faking a vehicle's location to another continent; smaller displacements are also possible and are more difficult to detect. For example, a spoofed position could show the vehicle on route and on time while, in actuality, the vehicle was stopped on the side of the road and disabled. This would prevent or delay appropriate alerts from appearing to the fleet manager, who could otherwise respond and address the issue.

While GPS spoofing has traditionally required expensive equipment and extensive technical abilities, the availability of inexpensive software-defined radios (SDRs) and scripted attacks has increased dramatically in recent years. The hobbyist community has embraced these low-cost SDRs and started applying the technology to a variety of applications, not all of them legal and altruistic. For example, in 2015 a researcher was able to successfully demonstrate GPS location spoofing using relatively low-cost SDRs (less than $500). It is reasonable to assume that this equipment, coupled with community-developed scripts or user interfaces, could eventually be used to manipulate truck positions. In 2013, a truck driver was arrested by Oregon state police for using a device to jam fuel pumps. The driver allegedly used an electronic device to disable the metering of diesel while refueling resulting in the "sale" of over 100 gal of fuel for the price of only 1.5 gal. Consider the value a location spoofer might have if it allowed a driver to violate operating hours, speed limits, or gaps in travel. Merchandise could be illegally offloaded while the spoofer reported an even speed, allowing the driver to make up the gap later.

At a talk given at the Black Hat Security Conference, a researcher presented the results of multi-year effort to identify and track real-world GPS jamming. [7] During the course of his research, he tracked numerous GPS emanations along public roadways and intersections. When sufficiently confident the source of the emanations had been identified, the operators were typically approached and questioned. Following the presentation, he indicated company personnel with GPS-tracked vehicles were the most frequent abusers of GPS jammers. The jammers were typically used to defeat tracking or autoscheduling software used by employers to allocated mobile units more efficiently. Using jammers, the employees were able to enjoy uninterrupted breaks or avoid automatic scheduling via location. A few times, these jammers were detected near critical infrastructure locations such as power substations or airports.

16.4 Solutions, Recommendations, and Best Practices

- Fleet managers should clearly define policies regarding purchasing and operating illegal GPS jamming equipment. Guidance should be developed to communicate to drivers the financial and legal repercussions of operating equipment such as jammers. A strict corporate stance should be established to discourage use of these devices. Information is readily available at DHS and other government public websites.

- When evaluating fleet management systems, pose "what if" scenarios to the system developers. Assess awareness of GPS manipulation issues that might occur and what impacts might result. Determine if they have test results relating GPS location spoofing, jamming, mileage tampering, or other location or timing anomalies. How would your fleet operations will be impacted if GPS is not available?

- Engage a security team and identify ways fleet operations could be disrupted by various attacks. Monitoring GPS receiver data is key to properly assess if GPS is behaving correctly, for example, knowing when GPS receivers lose lock to GPS could be an indicator of unwanted anomalies. It is important to designate a security POC to monitor GPS hacker is state of the art. Attend related security conferences and develop a network of similarly minded professionals at other fleet management companies.

16.5 Key Takeaways

- GPS and receiver technologies will continue to improve as they progress through the ongoing modernization process. These modernizations will result in improved precision, robustness, and additional information. The rapid development of new satellite navigation systems including the Russian GLONASS, European Galileo, and the Chinese Beidou system will continue to provide PNT data to users worldwide. In other words, GPS and other GNSS are effective and cheap, and we will continue to use them (they are not going anywhere).

- GPS is vulnerable. Look for ways to integrate common sense limits to raise notifications when suspicious behavior is encountered.

- Commercial fleet vehicles and transportation systems rely on GPS; therefore, there are vulnerabilities that apply. It is important knowing how your fleet management system collects GPS receiver data.

- Solutions can help mitigate these problems, but it is hard to validate how effectively these solutions work since it is hard to properly expose the system to all possible scenarios.

- Engage security teams early in the product selection process. Ensure they are knowledgeable regarding GPS-related vulnerabilities. Use their expertise to identify possible shortcomings in operation critical equipment and software.

References

1. Department of Homeland Security, "Critical Infrastructure Sectors," DHS, June 1, 2017. [Online]. Available at: https://www.dhs.gov/critical-infrastructure-sectors, accessed June 1, 2017.

2. Shepard, D.P., Humphreys, T.E., and Fansler, A.A., "Evaluation of the Vulnerability of Phasor," March 19, 2012. [Online]. Available at: https://radionavlab.ae.utexas.edu/images/stories/files/papers/spoofSMUCIP2012.pdf, accessed June 1, 2017.

3. Misra, P. and Enge, P., *Global Positioning System: Signals, Measurements, and Performance*, revised 2nd ed., (Lincoln, MA: Ganga-Jamuna Press, 2010), ISBN:9780970954411.

4. Dovis, F., *GNSS Interference Threats and Countermeasures*, (Norwood, MA: Artech House, 2015), ISBN:978-1608078103.

5. Department of Homeland Security, "National Risk Estimate: Risks to U.S. Critical Infrastructure from GPS Disruptions," DHS, November 29, 2013. [Online]. Available at: http://www.gps.gov/news/2013/06/2013-06-NRE-public-summary.pdf, accessed June, 1 2017.

6. Fox News, "GPS Jammers Illegal, Dangerous, and Very Easy to Buy," *Fox News*, March 17, 2010. [Online]. Available at: http://www.foxnews.com/tech/2010/03/17/gps-jammers-easily-accessible-potentially-dangerous-risk.html, accessed June 1, 2017.

7. Gostomelsky, V., "Hunting GPS Jammers," Black Hat USA 2017, July 27, 2017. [Online]. Available at: https://www.blackhat.com/us-17/briefings/schedule/#hunting-gps-jammers-7015, accessed July 27, 2017.

8. Wikipedia, "GPS Spoofing Attack," June 1, 2017. [Online]. Available at: https://en.wikipedia.org/wiki/Spoofing_attack, accessed June 1, 2017.

About the Authors

Mr. Trevino is an experienced engineer who has worked and managed projects ranging from architecting sensor networks, data acquisition, and control systems to the development of software algorithms for embedded devices. Currently, he is using his embedded systems expertise to develop and improve critical infrastructure technologies, assess system security for commercial and government clients, and architect test equipment for different space missions under development at Southwest Research Institute (SwRI).

Mr. Trevino conducted research into the security vulnerabilities of GPS-dependent time synchronization of electrical grid control systems, developing algorithms to detect and mitigate cybersecurity attacks. He is also working to optimize the power system infrastructure for utilities clients. Mr. Trevino has been using his embedded systems expertise to design and develop data acquisition and control systems for electrical ground support equipment for different space missions, including Spectral Imaging of the Coronal Environment (SPICE) and Cyclone Global Navigation Satellite System (CYGNSS), using an array of test equipment and software tools.

Serving as the lead engineer, Mr. Trevino developed a data acquisition and control system for an electric vehicle (EV) fleet that became the first EV fleet system to be qualified to participate in the Texas energy services market by the Energy Reliability Council of Texas.

Prior to joining SwRI, Mr. Trevino interned at the National Renewable Energy Laboratory (NREL) where he helped develop a smart charger prototype for a plug-in hybrid vehicle using National Instruments hardware and software and communication protocols, including ZigBee.

Ms. Ramon is a senior research engineer with experience in embedded system development, reverse engineering, and testing of cyber-physical systems, including connected and autonomous vehicles, transportation field networked devices, smart grid devices, and flight-test systems. Her areas of expertise focus on risk management, secure design, and penetration testing. Ms. Ramon is familiar with established standards and best practices, including the NIST Cybersecurity Framework, the ISO/IEC 27000 family, and IEEE 1609.2 WAVE, as well as developing standards such as SAE J3061 and ISO/SAE 21434.

Ms. Ramon manages and develops on multiple vulnerability assessment and risk modeling projects evaluating the security postures of original equipment manufacturers' (OEMs') and supplier's product families, Critical Energy systems, and intelligent transportation systems (ITS) traffic management systems (TMS) and field equipment. Ms. Ramon has also worked as technical lead on several security penetration testing and threat analysis projects to analyze the vulnerabilities of client equipment.

Automotive-related efforts investigated secure over-the-air (OTA) processes, electronic control units (ECUs), and related components, including the controller area network (CAN) communications, binaries and source code, and wireless features including cellular, Wi-Fi, and Bluetooth capabilities. Critical infrastructure projects analyzed the vulnerabilities and protection of Global Positioning System (GPS) systems, advanced metering infrastructure (AMI), and smart grid (SG) devices for several clients including utilities and equipment vendors. ITS-related efforts included risk modeling of the traffic management systems and investigated the field network communications and hardware such as dynamic messaging signs (DMS) and traffic signal controllers.

Mr. Zajac is a senior research engineer with 11 years of experience in embedded system development, reverse engineering, and embedded system security. He has assessed the security vulnerabilities of AV control architectures and recommended changes to improve system security. He is familiar with a variety of established and developing security including J3061 and J3016, NIST Cybersecurity Framework, and E-safety Vehicle Intrusion Protected Applications (EVITA).

He developed methods of applying cybersecurity principles to conceptual AV architectures and led a team developing high-level security requirements and test procedures for AV sensors. These efforts included creating use cases specific to AV and identifying associated threats, vulnerabilities, and countermeasures.

Mr. Cameron Mott received his BS in computer science from DePauw University in 2000 and MS in computer science from New Mexico State University in 2005. He has more than a decade of experience applying innovative research efforts toward the advancement of connected and autonomous vehicles. His areas of expertise focus on system engineering, software design, and verification testing. His experience includes software and security application development for both on- and off-road vehicles with a particular focus on machine-to-machine integration. Mr. Mott also has significant experience in connected and cooperative systems, including software development of vehicle-to-vehicle (V2V) and vehicle-to-infrastructure (V2I) applications. Through his passion of pursuing the cutting-edge in intervehicle communication, he strives to improve the capability, safety, and security of tomorrow's vehicles.

Mr. Mott has held the following positions during his career: Senior Research Analyst at Southwest Research Institute (2015-present), Research Analyst at Southwest Research Institute (2014-2015), Systems Engineer at John Deere (2010-2014), Software Engineer at John Deere (2008-2010), and Software Developer at John Deere (2005-2008).

Mr. Mott is a principal investigator for a government-funded project that is working with leading automotive and research contributors to address tomorrow's security concerns for the connected automobile. In this role, he is applying his system engineering expertise to create an open-source system architecture (called Uptane) that provides secure software updates to embedded control units (ECUs) in vehicles. Mr. Mott utilizes his project management capabilities to provide on-time and on-budget deliverables for both government and commercial clients. In addition, he contributes to the development and integration of Connected Vehicle technology (via dedicated short-range communication radios) using a system-level approach to improve the capabilities of connected and autonomous vehicles. His focus is leveraging vehicle-to-any (V2X) communication to improve the performance of both autonomous vehicles and manually operated vehicles.

Publications:

Brown, M., Lemmer, S., Mott, C., and Sturgeon, P., "Subtle Anomaly Detection in the Global Dynamics of Connected Vehicle Systems," *2016 ITS World Congress*, Melbourne, Australia, October 2016.

Awards:

SWR3779 - U.S. Patent Application No. 14/732,002 filed June 5, 2015, titled "Sensor Data Confidence Estimation Based on Statistical Analysis".

CHAPTER 17

Looking Towards the Future

Gloria D'Anna

> **Rachael:** *Like our owl?*
>
> **Rick Deckard:** *It's artificial.*
>
> **Rachael:** *Of course it is.*
>
> **Rick Deckard:** *Must be expensive.*
>
> **Rachael:** *Very.*
>
> —**The movie** *Blade Runner* [1]

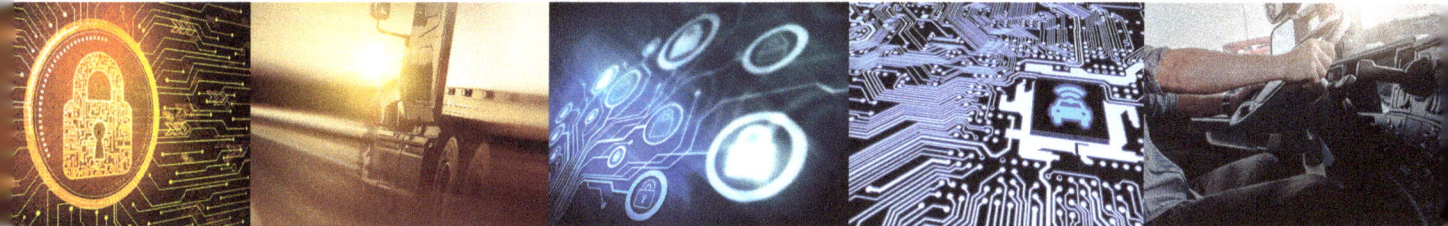

17.1 I'm a Blade Runner Fan

Whenever I think about the future, I think about the movie *Blade Runner*. I'm a big *Blade Runner* fan. I've also found that others in the cybersecurity world are *Blade Runner* fans too. Except, they have the movie poster in their office. I don't.

Blade Runner was released in June 25, 1982. The film depicted the future as dark. Los Angeles was depicted as continually besieged by rain. But, if you look at the technology used in the film, there were LED TVs on buildings, which at the time, did not exist. But, now they do. My favorite scene is where Rachel shows off her owl. It looks very real. But, it's not.

So, why do I bring up the movie, *Blade Runner*? Well, it's considered by many critics as one of the best science fiction films of all time, according to Wikipedia. I find that science fiction films and TV/Netflix plots tend to give us a visual picture of what the future holds.

If you happened to see the latest movie *Logan*, the final portrayal of Hugh Jackman as the Wolverine, there were autonomous trucks in the movie. Well, they were more like autonomous containers moving down the highway. It bothered me that there may not actually have been a powertrain in them, unless there were wheel motors. But, nonetheless, it gives you an idea that autonomous trucks are coming in high volumes.

When it comes to cybersecurity in movies, I think the most famous movie is *The Matrix* [2]. It was released March 31, 1999. The female character, Trinity, is an infamous hacker. She tends to answer ringing public telephones and vanish. You will find her famously in movie posters dressed in shiny leather pants and a shiny leather jacket.

So, what does this have to do with the future of cybersecurity of commercial vehicles? Movies portray technology as being exciting and sexy. Whereas in reality, the implementation of technology, and putting together secure systems in the commercial vehicle world, ends up being a bit of a grind. You need to put your best engineering system hat on and grind out the details. And, this is why this book is important.

17.2 **Setting Standards**

Grinding out the details consists of setting engineering statements of work (ESOWs, per industry lingo). And, engineering statements of work consist of standards. SAE International is here to help set standards.

The reader may want to consider joining the Automotive Information Sharing and Analysis Center (ISAC). Automakers proactively joined together in August of 2015 and formed this global community, the Automotive-ISAC, to share and analyze intelligence about emerging cybersecurity risks to the vehicle. The ISAC is a nonprofit organization that provides a central resource for gathering information on cyber threats and providing two-way sharing of information between the private and public sector. One company's detection of a potential attack may mean another company's prevention of a security breach.

As I write this, the United States Senate held a bipartisan hearing on connected cars. Cybersecurity was highlighted as "***Strengthen Cybersecurity***: Cybersecurity must be a top priority for suppliers and manufacturers. Legislation must address the connectivity of self-driving vehicles and potential cybersecurity threat vectors before problems arise." Education and participating in standard committees was also highlighted. As law firm Butzel Long noted in their client alert, "The hearing made clear the importance of voluntary standards development being undertaken by the Society of Automotive Engineers (SAE International), the Institute of Electrical and Electronics Engineers (IEEE), and other industry organizations. It is important for companies in the industry to encourage engineers and other technical personnel to participate in these standards organizations."

17.3 **Automotive ISAC**

Now, let's talk about the Automotive ISAC, abbreviated Auto-ISAC. The Auto-ISAC according to their website [3] is "an industry-operated environment created to enhance cybersecurity awareness and collaboration across the global automotive industry and light-and heavy-duty vehicle OEMs, suppliers, and the commercial vehicle sector." The focus of Auto-ISAC is to foster collaboration that creates a safe, efficient, secure, and resilient global connected vehicle ecosystem. In 2016, Auto-ISAC expanded to include suppliers and heavy-duty trucking. In 2017, the ISAC expanded to include the commercial vehicle sector (i.e., carriers, fleets).

The Auto-ISAC Executive Director is Faye Francy, a friend of mine. As you can see from the history of my work in commercial vehicle cybersecurity, we've been at this a long time and yet we have so much more to do. In fact, Faye was in Montreal in 2013 with me; however, at that time, she was working for Boeing and establishing an aviation ISAC. Faye has been on many of the cross-industry cybersecurity panels that we have held at SAE.

A brief history of the Auto-ISAC is as noted in the following slide:

FIGURE 17.1

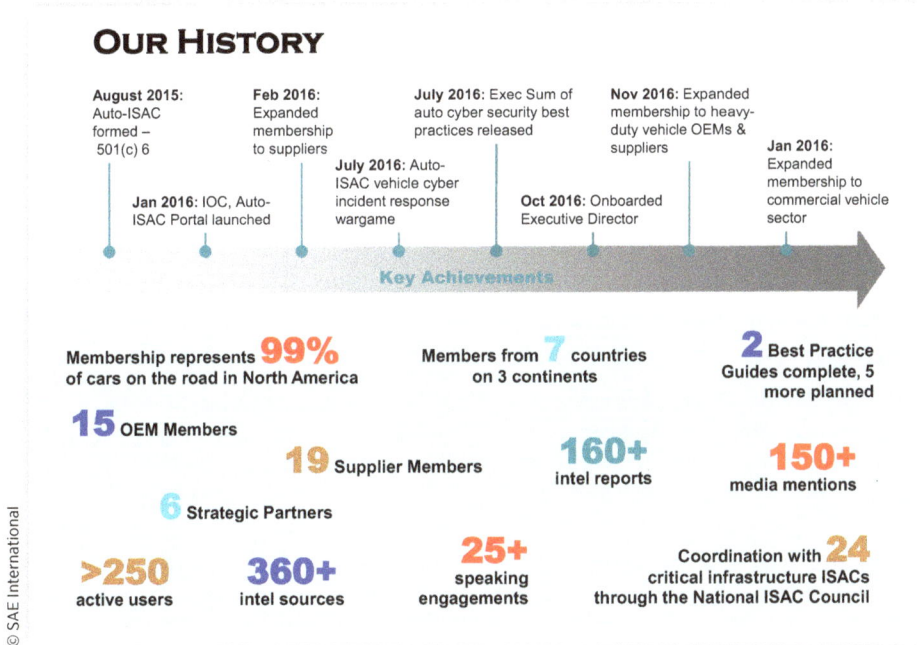

In January 2017, rather recently, the Auto-ISAC expanded membership to the commercial vehicle sector. I see this as a positive, considering that many of the technologies utilized in automotive and the commercial vehicle sector are similar.

I've also added another slide that Faye Francy had provided in an Auto-ISAC overview, to show how both the automotive and commercial vehicle sectors are being transformed due to connectivity. I like this slide, as it transitions to the next topic of commercial vehicle systems continuing to get more complicated.

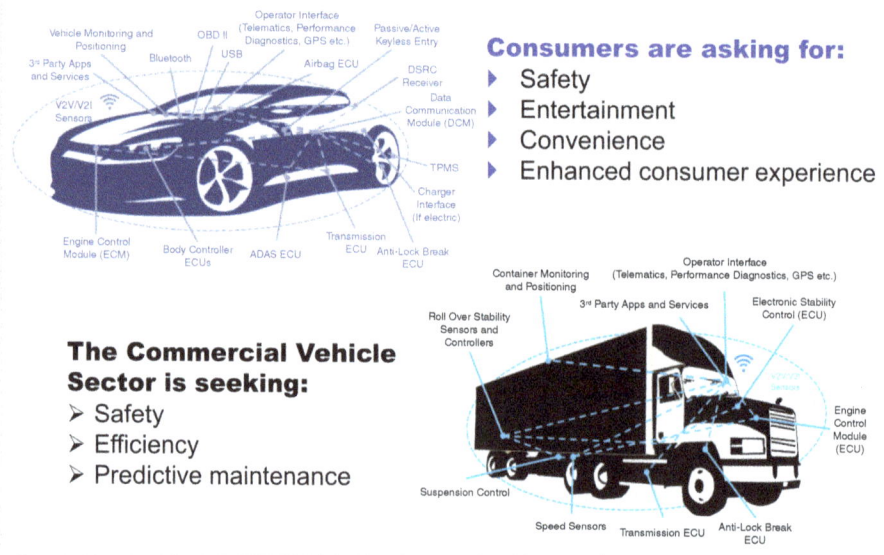

TRANSFORMING THROUGH CONNECTIVITY

Consumers are asking for:
▶ Safety
▶ Entertainment
▶ Convenience
▶ Enhanced consumer experience

The Commercial Vehicle Sector is seeking:
➤ Safety
➤ Efficiency
➤ Predictive maintenance

The Auto-ISAC is also focused on building a community of practice, and leading the industry toward better vehicle cybersecurity practices. To accomplish this goal, they conduct frequent capability development events, are building *Best Practices*, and are formalizing a *Strategic Partnership Program*. Here are some more details:

- **Capability development events**: Auto-ISAC conducts an annual tabletop exercise, quarterly workshops, and monthly analyst calls with members. It leads monthly community calls which even if your company is not a member, you can join these calls. It will run its inaugural annual *Vehicle Cyber Summit* in December 2017.

- **Best Practices**: The best practices initiative is focused on developing aspirational, living, non-required guidelines that organizations may use to mature their vehicle cyber programs. Auto-ISAC released an Executive Summary in July 2016, and is currently working on seven Best Practice Guides that cover organizational and technical aspects of vehicle cybersecurity: incident response, collaboration and engagement with third parties, governance, risk management, security by design, threat detection and protection, and training and awareness. These Guides will be released beyond our membership over time to help other industry stakeholders mature.

- **Strategic Partnership Program**: Auto-ISAC partners with vendors, associations, researchers, government, and academia through three partnership programs to cultivate relationships beyond our membership with the common goal to enhance vehicle cybersecurity.

17.4 **The Systems of a Commercial Vehicle Continue to Get More Complicated**

As noted in Chapter 4, we started out basically with a powertrain in commercial vehicles. Due to severe Environmental Protection Agency (EPA) regulations, commercial vehicles needed to add more and more electronics to optimize emissions. This work continued to grind on the commercial vehicles world with the additions of lines of code. As commercial vehicles continued to optimize their routes and customer service, telematics were added the vehicle. Currently, the telematics were noted as a "weak link" for cybersecurity. The good news is that this weak link is being fixed. These telematics systems are undergoing penetration testing by the best hackers in the business.

Then, we ran into the sexiness of hacking a car—with Charlie Miller and Chris Valasek. I say "sexy" as it got everyone excited. It also caused a large recall for Fiat Chrysler. The University of Michigan did something similar, but not as "sexy" in my opinion. As seen in Chapter 4, "Additionally, in 2016, researchers from the University of Michigan performed an experimental security analysis of a real heavy-duty truck and passenger bus vehicle network. In their work, the authors assumed physical connection to the vehicle bus through the on-board diagnostics connector, and they discovered that similar attacks against automobiles in 2010 were directly applicable to commercial vehicles." This University of Michigan paper was presented at SAE Commercial Vehicle Congress in 2016. A few in the audience groaned that all the paper showed was that J1939 CAN worked, and how it worked. Luckily the University of Michigan Researchers, in my opinion, did not do a disservice to the industry. The vehicles picked were not the latest and greatest of those issued by the commercial vehicle OEMs—Navistar, Volvo, Freightliner, or PACCAR—but were rather innocuous older vehicles. The research, however, helped ignite an awareness that commercial vehicles are also able to be hacked.

As we noted in Chapter 6, the American Trucking Associations (ATA) 2006 White Paper was published for this cybersecurity book. A big thank you to the ATA! Particularly important, the paper notes, "As the trucking industry goes, so does the United States Economy." If commercial trucks are compromised, the United States economy can also be compromised. But, does the "average" person notice? Probably not, in my opinion. We all expect to go to eBay or amazon.com and expect that whatever we order on the internet shows up at our door. Transportation of goods, even personal goods, is just "there," like "air." But, what if it wasn't.

And, it's that "what if it wasn't" that drove my curiosity to help create awareness of cybersecurity in the commercial vehicle world. And, hence, this book.

17.5 **The Good News**

The good news is that the commercial vehicle industry has woken up to this concern and is working on the issue. Maybe, not as fast as I would like to see, but it has gained tremendous momentum in the past year. We had a Commercial Cyber Truck Challenge in June 2017, which now includes commercial vehicles. This was under the leadership of Karl Heimer and Dr. Jeremy Daily. See Chapter 9.

We also have the University of Tulsa creating cybersecurity engineers for the commercial vehicle market. If you would like to know more, please contact Dr. Jeremy Daily at the University of Tulsa who has been spearheading this effort. Dr. Jeremy Daily wrote Chapter 7.

Also, the National Motor Freight Traffic Association (NMFTA) is "working to educate the transportation industry on potential threats to connected vehicle fleets. NMFTA hosts industry workshops exploring cyber security issues such as incident response, risk mitigation, data anomaly detection, and many others." See Chapter 14. Also, please see NMFTA's website [4]. Urban Jonson from NMFTA continually provides an update on NMFTA's activities at SAE's *Annual Commercial Vehicle Engineering Congress*.

17.6 **Telematics**

There also has been tremendous focus on shoring up cybersecurity in telematics. I give a kudo to Geotab for helping to coordinate activities. Geotab, from a good design perspective, had looked at end-to-end cybersecurity for their telematics solutions from the onset. See Chapter 8.

And, as a word of caution, please be careful what you are plugging into your OBD-II ports. Not knowing is like putting a syringe of unknown substance into your arm. Would you do that?

17.7 **Cybersecurity as an Enabler for New Technologies**

Sometimes when I mention cybersecurity for commercial vehicles, I get that "eye roll." I consider cybersecurity to be "clean data" similar to what we saw for SAE standards for "clean air," the emission regulations needed to have been put in place in 2007 and 2010 by the United States EPA.

I happened to have had a conversation with Pete Tseronis [5], the former Department of Energy's first Chief Technology Officer. In that role, Pete was responsible for providing strategic directions and vision through the department and was charged with establishing a formal and sustainable federal technology deployment program that offers secure transformative innovative solutions to tough challenges. Our conversation went something like this, "Shouldn't we look at cybersecurity as an enabler for new technologies?" "Oh, that would be an excellent topic for a speech." So, Pete came to Detroit for *SAE WCX 2017* (*SAE World Congress and Exhibition 2017*) to talk about that very subject. Pete's outline for his speech reads as follows:

Self-Driving Cars: The Result of Secure Transformative Innovation

It is widely believed that autonomous vehicles will transform how we live, work, and, of course, commute. Over the next decade, pilot deployments of safe and climate smart autonomous vehicles are expected to generate enriched, faster, and cleaner transportation networks. However, the introduction of connected cars and the "Internet of Things" creates a new set of risks, new security requirements, and new costs. As the cyber ecosystem becomes more complex, the automotive industry must balance the promise of increasingly feature-rich communications with the realization that more hackers will be motivated to develop attacks and exploitation tools.

Pete Tseronis, Dots and Bridges

© SAE International

17.8 Department of Energy Work on Cybersecurity for Vehicles

As part of the Department of Energy work on cybersecurity, I hope you read Chapter 10. This describes the potential of electric commercial vehicles (yes, they are coming to market. I'm sure you have read about Tesla's Truck) shutting down the electrical grid. Reminds me of the TV show "Evolution"—where the power grid has been down for quite some time. As we go to press, I had an opportunity to review the DOE's work in this area at Argonne National Lab. It was great to see their work in person, considering that I have reviewed their papers at DOE's Annual Merit Review (AMR) held each year in Washington, D.C.

17.9 Commercial Truck Platooning

As I was putting this book together, I had hoped to have had a chapter written by Ryan Gerdes. Professor Gerdes is an Assistant Professor at Virginia Tech's Bradley Department of Electrical and Computer Engineering. But, timelines get in the way.

Professor Gerdes' recent projects "have focused on security automated transportation systems, exploring ways to identify devices based upon their electrical signature, and finding ways to exploit and secure cyber-physical systems" [6].

In one of his papers in 2013, _CPS: An Efficiency-Motivated Attack Against Autonomous Vehicular Transportation_, Professor Gerdes explores, "Next-generation transportation technologies will leverage increasing use of vehicle automation. Proposed vehicular automation systems include cooperative adaptive cruise control and vehicle platooning strategies which require cooperation and coordination among vehicles. These strategies are intended to optimize throughput and energy usage in future highway systems, but as we demonstrate, they also introduce new vulnerabilities."

One of his 2015 papers, ASIACCS2015 paper _Vehicular Platooning in an Adversarial Environment_ by S. Dedras, R. M. Gerdes, and R. Sharma in the conference _ACM Symposium on Information, Computer and Communications Security_, discusses safety concerns of operating platoons in adversarial environments. "In this paper, we show that a single, maliciously controlled vehicle can destabilize a vehicular platoon, to catastrophic effect, through local modifications to the prevailing control law. Specifically, by combining changes to the gains of the associated law with the appropriate vehicle movements, the attacker can cause the platoon to oscillate at a resonate frequency, causing accidents that could result in serious injury or death."

So, keep an eye out for Professor Gerdes' research.

The good news is that the technology that is being utilized on automobiles for vehicle to vehicle communication and adaptive cruise control is similar enough to what is being utilized for commercial vehicles. Hence, there is a volume learning curve for cybersecurity as well as engineering of components that can be shared across automotive and commercial vehicles. The big difference is that commercial vehicles are just that much bigger and that cars and commercial trucks are in traffic together.

Have you ever driven a car in Chicago traffic with commercial trucks surrounding you? Hence, my worry about commercial vehicles and cars sharing the road together in some sort of automation.

17.10 So, Why Is Platooning Such a Big Deal?

When I worked at Ricardo Engineering, we worked on many advanced technology projects for commercial trucks. Some of these projects were for Michael Roeth and David Schaller. These two favorite people of mine now work at the North American Council for Freight Efficiency (NACFE) [7].

> _"The Mission of the NACFE is as follows: The NACFE is dedicated to doubling the efficiency of North American good movement. We pursue this goal in two ways: By improving the quality of the information flow and by highlighting successful adoption of technologies."_
>
> _www.nacfe.org_

In the summer of 2016, I was looking for data on the efficiency of truck platooning. I contacted Mike Roeth to see what he was up to. He said that the report was coming. Officially, September 28, 2016, NACFE published their report.

"On September 28, 2016, NACFE completed (their) first Confidence Report on an emerging technology, Two-Truck Platooning. It looks at the benefits, consequences and payback expected for linking two trucks together at about a 50 feet following distance. The report concludes that Two-Truck Platooning is a profitable first step on the pathway to more autonomous trucking."

The report itself is downloadable free from the NACEFE website [7]. I've highlighted a portion of the Executive Summary here:

"FUEL SAVINGS OF PLATOONING Without question, truck platooning is a valid method of reducing fuel consumption for tractor-trailers engaged in long-haul applications. Once the trucks have moved into close following distances, all of the engaged vehicles receive a significant fuel economy boost thanks to increased aerodynamic efficiencies. The lead vehicle, which bears the brunt of the aerodynamic load, typically sees only a modest fuel economy boost. But the trailing truck in a platoon, which is now operating in a low air pressure aerodynamic "sweet spot," can see significant increases in fuel economy performance at highway speeds. Moreover, overall fleet operations remain largely intact in terms of vehicle routing and operations. The potential fuel consumption savings versus an isolated single vehicle varies depending on the separation distance of the trucks (as shown for the lead vehicle in Figure ES1). Multiple fuel consumption tests have been conducted over the past few decades to better understand this efficiency improvement. A separation distance of 40 to 50 feet could lead to average savings of about 10% for the following vehicle and 4% for the lead vehicle. However, real-world factors such as congestion, terrain, weather, and road construction will reduce these savings, so fleets will have to estimate this reduction depending upon the routes on which they plan to operate the trucks; a reasonable estimate would be a reduction of about a quarter of the savings, but very little data exists for this prediction. A fleet must also apply the percentage of operating time that the truck equipped with platooning will actually be involved in a platoon, which NACFE research suggests will be less than 100%. If that were 75%, then the real-world, expected savings would be on the order of 4% average for both trucks. Even if a truck is platooning 50% or less, it still represents significant potential improvement in fuel use for the following vehicle in a platoon."

Simply stated, Two Truck Platooning is a viable option for FUEL SAVINGS for fleets. Four percent for the lead vehicle, and 10% for the following vehicle. This is a significant potential improvement for fuel use for commercial vehicles. However, to implement this, we must consider the cybersecurity aspects of the technology from an end-to-end perspective.

"We must look at Cyber Security as an Enabler for New Technologies."

—*Gloria D'Anna*

17.11 So What Have We Learned from This Book?

Hopefully, you have found this book informative as well as interesting. My goal was to have this read as a nonfiction novel, and not as a boring engineering overview of cybersecurity for commercial vehicles. If you utilize commercial vehicles for business, you realize the importance of the asset, and now, you can visualize that the use of your asset has cybersecurity risks.

And, mitigating cybersecurity risks is not such an easy thing, but a continuous grind to keep improving cybersecurity, or patching risks as they occur. Is it sexy? No. Does it have severe risks? Yes.

We also know that technology is moving at a fast pace. We have the potential of utilizing vehicle platooning, to provide a rather large cost savings in fuel efficiency for fleets. But we also need to secure the sensors, CAN, ECUs, etc., that are utilized for this system. And, secure the supply chain from where these products stem.

We also know that not everyone wants to talk about their cybersecurity issues. This is why this book is full of SAE standards, and chapters written by academics, and even by the state government and the U.S. government.

My goal was not to go "open kimono" with this book. (Why let your potential threats (hackers) have a leg up on you?) The goal was to put out in public a book that shows that cybersecurity for commercial vehicles is being taken seriously by everyone in the industry and that the industry is continuing to work on the issues. And, my belief is that we will continually grind out the cyber issues that affect us. We will be putting in mitigating technologies and policies as it affects commercial vehicles.

I am also personally rather happy with how cybersecurity is progressing for commercial vehicles. We started this topic of cybersecurity for commercial vehicles at SAE in 2013 in Montreal, Quebec. We are now in 2018, the sixth year of this topic at SAE. For the first few years, I utilized my various personal contacts to start to bring this topic to the public through SAE. For several years, I felt as if I was "pushing a rope."

There was some interest, but not much. Academic papers were few and far between. I'd like to thank my business partner Billy Jones for helping me get the FBI to join our Technical Expert Panel Discussion: With connectivity, comes risks - cybersecurity and safety" 2015 SAE World Congress.

17.12 And Then, Something Happened

Then, in the September of 2015, something happened. The Technical Expert Panel for SAE Commercial Vehicles Cybersecurity went to the bar after SAE ended. We looked at each other, and said, "How do we really solve this problem?" And, that is how the dam broke.

I'd like to personally thank Dr. André Weimerskirch, who has been on this journey with me since Montreal in 2013. I'd also like to thank Dr. Jeremy Daily; Dr. Dan Massey, formerly at Department of Homeland Security, and Kevin Harnett from Department of Transportation—VOLPE Center who went out to the bar that night.

17.13 *SAE World Congress 2017*

In April 2017 at *SAE World Congress* in Detroit, we had a rather large technical panel, consisting of the U.S. government—Department of Homeland Security and the Department of Transportation—VOLPE Center, academics, and industry. This panel could discuss both cybersecurity in automobiles as well as commercial vehicles.

This technical panel was part of the electronics track for SAE. The room, which was quite large, consisted primarily of electrical engineers. How do I know? I asked. The room was packed, with about 200 people. Interest was high.

I always start out with the question, "What keeps you up at night" (in regard to cybersecurity).

As per the SAE publication, Dr. Dan Massey, answered, "I don't think we'll have a gradual change." "I'd love to see a slow progression starting with one or two one-off attacks. I fear we won't have the opportunity-that it will go from seeing nothing but a few cyber-attack demonstrations to tens or hundreds of thousands of vehicles to be concerned about."

FIGURE 17.4

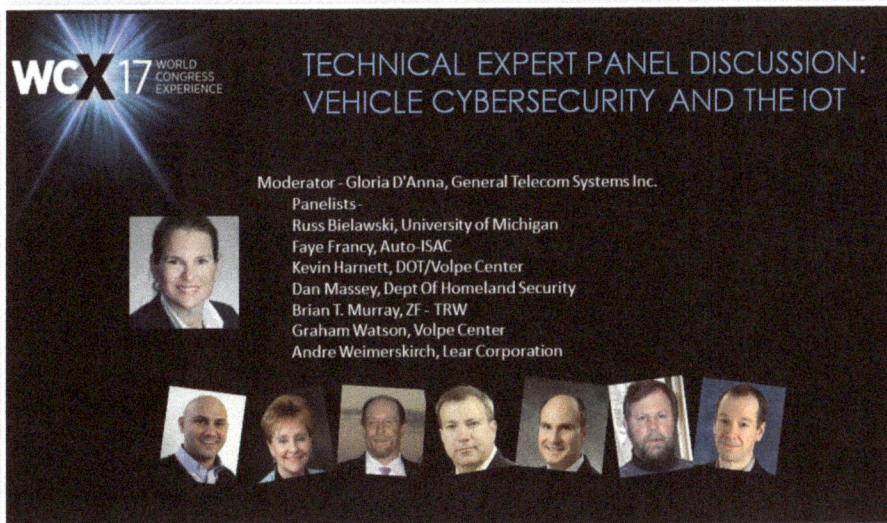

With that sobering thought, go back to Chapter 2, and read "Should We Be Paranoid?" And come join us at SAE to work on this cybersecurity challenge.

17.14 **As We Go To Press**

As we go to press, we have just held the 2017 SAE Commercial Vehicle Congress in Chicago, Illinois.

FIGURE 17.5

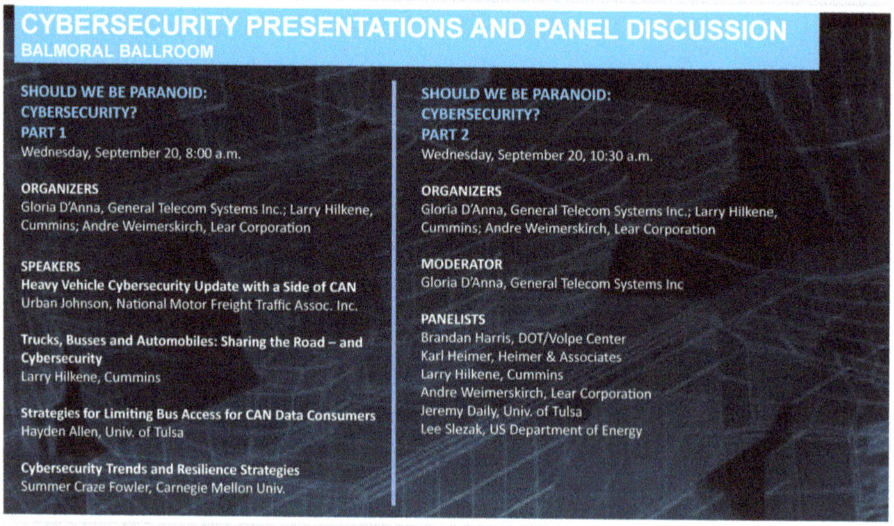

CYBERSECURITY PRESENTATIONS AND PANEL DISCUSSION
BALMORAL BALLROOM

SHOULD WE BE PARANOID: CYBERSECURITY? PART 1
Wednesday, September 20, 8:00 a.m.

ORGANIZERS
Gloria D'Anna, General Telecom Systems Inc.; Larry Hilkene, Cummins; Andre Weimerskirch, Lear Corporation

SPEAKERS
Heavy Vehicle Cybersecurity Update with a Side of CAN
Urban Johnson, National Motor Freight Traffic Assoc. Inc.

Trucks, Busses and Automobiles: Sharing the Road – and Cybersecurity
Larry Hilkene, Cummins

Strategies for Limiting Bus Access for CAN Data Consumers
Hayden Allen, Univ. of Tulsa

Cybersecurity Trends and Resilience Strategies
Summer Craze Fowler, Carnegie Mellon Univ.

SHOULD WE BE PARANOID: CYBERSECURITY? PART 2
Wednesday, September 20, 10:30 a.m.

ORGANIZERS
Gloria D'Anna, General Telecom Systems Inc.; Larry Hilkene, Cummins; Andre Weimerskirch, Lear Corporation

MODERATOR
Gloria D'Anna, General Telecom Systems Inc

PANELISTS
Brendan Harris, DOT/Volpe Center
Karl Heimer, Heimer & Associates
Larry Hilkene, Cummins
Andre Weimerskirch, Lear Corporation
Jeremy Daily, Univ. of Tulsa
Lee Slezak, US Department of Energy

© SAE International

We started with Urban Jonson of NMFTA providing an update on their progress. Then, we were followed by Larry Hilkene of Cummins and Hayden Allen, a student of Dr. Jeremy Daily's at the University of Tulsa. Hayden Allen is a wonderful example of bringing up cybersecurity talent from the University. We also added an update by Karl Heimer and Dr. Jeremy Daily of the 2017 Commercial Cyber Truck Challenge—and encouraged the audience to participate in 2018.

The Technical Expert Panel consisted of Brendan Harris from DOT/Volpe, Karl Heimer from Heimer & Associates, Larry Hilkene of Cummins, Dr. André Weimerskirch of Lear Corporation, Dr. Jeremy Daily of the University of Tulsa, and Dr. Lee Slezak of the Department of Energy. And, as always, I am moderating, to create an "interesting" discussion.

FIGURE 17.6

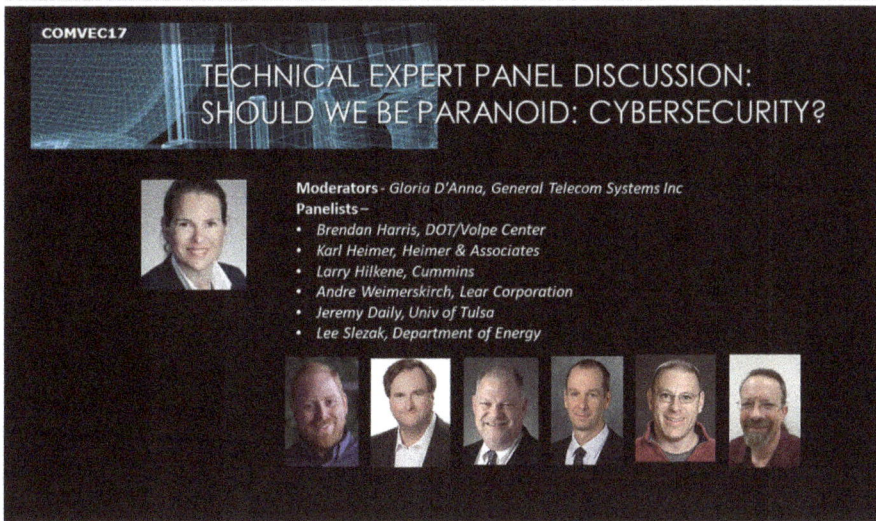

One of my final questions to the Expert Panel was, "Are we moving fast enough in cybersecurity for commercial vehicles?" The consensus was, "No, we are not moving fast enough. Yet, we've made significant progress in 18 months."

And as Karl Heimer will always remind us, "cybersecurity is a verb." It continues to move and morph.

References

1. *Bladerunner.* Directed by Ridley Scott. Produced by Ridley Scott.

2. *The Matrix.* Produced by Lilly Wachowski and Lana Wachowski. Directed by Lilly Wachowski and Lana Wachowski. By Lilly Wachowski and Lana Wachowski.

3. www.automotiveisac.com.

4. http://www.nmfta.org/pages/HVCS.

5. https://energy.gov/contributors/peter-j-tseronis.

6. https://www.ece.vt.edu/news/article/gerdes-joins-ece.

7. www.nacefe.org.

Gloria D'Anna is an engineer, entrepreneur, and multiple patent award holder—an expert in vehicle engineering and cybersecurity. She likes to solve problems, from future light commercial vehicles, to beefing up the cybersecurity of vehicles, always rolling out new tech, reducing inefficiencies, and driving business.

She began her career at GM, winning an MBA Fellowship to the University of Chicago, moving on to Ford, Navistar, Textron, Eaton, and Ricardo. Later, she led sales at three successful startups, addressing challenges from school safety to building connected devices for law enforcement.

Gloria has been working with SAE for the last 6 years, creating and moderating popular and educational cybersecurity technical sessions from Commercial Vehicle markets to the Internet of Things. Her 2018 book, SAE's *Cybersecurity for Commercial Vehicles* is her first book.

She is the recipient of SAE International's 2018 Forest R. McFarland Award for Automobile Electronics Activity.

She is currently the CEO of General Telecom Systems, a private telecommunications company, on the Board of Atmos XR, an advisor to 202 Partners, a lead mentor for Techstars Mobility and a member of the Uptane Advisory Board.

She likes trucks.